Cars and Carbon

Theodoros I. Zachariadis
Editor

Cars and Carbon

Automobiles and European Climate Policy
in a Global Context

Editor
Dr. Theodoros I. Zachariadis
Department of Environmental
　Science & Technology
Cyprus University of Technology
P.O. Box 50329
3603 Limassol
Cyprus
t.zachariadis@cut.ac.cy

ISBN 978-94-007-2122-7　　　　e-ISBN 978-94-007-2123-4
DOI 10.1007/978-94-007-2123-4
Springer Dordrecht Heidelberg London New York

Library of Congress Control Number: 2011941864

© Springer Science+Business Media B.V. 2012
Chapter 3 is published with kind permission of © OECD/IEA, 2012. All rights reserved
No part of this work may be reproduced, stored in a retrieval system, or transmitted in any form or by any means, electronic, mechanical, photocopying, microfilming, recording or otherwise, without written permission from the Publisher, with the exception of any material supplied specifically for the purpose of being entered and executed on a computer system, for exclusive use by the purchaser of the work.

Printed on acid-free paper

Springer is part of Springer Science+Business Media (www.springer.com)

Contents

1   Introduction   1
Theodoros I. Zachariadis

Part I   Background – Automobiles and Climate Change

2   The Attractiveness of Car Use   19
Arie Bleijenberg

3   The Importance of Passenger Cars for Global Greenhouse
Gas Emissions – Today and Tomorrow   43
Lew Fulton

Part II   European Union Policies

4   The Past and the Future of EU Regulatory Policies
to Reduce Road Transport Carbon Emissions   73
Karl-Heinz Zierock

5   Fuel Policies in the EU: Lessons Learned from the Past
and Outlook for the Future   97
Sandrine Dixson-Declève

6   Fuel Taxation, Regulations and Selective Incentives:
Striking the Balance   127
Per Kågeson

7   The Right EU Policy Framework for Reducing Car
CO_2 Emissions   153
Jos Dings

Part III   National Policies

8   CO_2-Based Taxation of Motor Vehicles   181
Nils Axel Braathen

9   Fuel Taxation in Europe   201
Jessica Coria

10	**Passenger Road Transport During Transition and Post-transition Period: Residential Fuel Consumption and Fuel Taxation in the Czech Republic** Milan Ščasný	221
11	**Accelerated Introduction of 'Clean' Cars in Sweden** Muriel Beser Hugosson and Staffan Algers	247
12	**Making People Independent from the Car – Multimodality as a Strategic Concept to Reduce CO_2-Emissions** Bastian Chlond	269
13	**National Road User Charging: Theory and Implementation** Bryan Matthews and John Nellthorp	295

Part IV   The International Context

14	**Mobility Management Solutions to Transport Problems Around the World** Todd Litman	327
15	**Automobiles and Climate Policy in the Rest of the OECD** Michael P. Walsh	355
16	**Transport and Climate Policy in the Developing World – The Region that Matters Most** Cornie Huizenga and James Leather	371
17	**Epilogue – The Future of the Automobile: CO_2 May Not Be the Great Decider** Lee Schipper	393
Index		417

Contributors

Staffan Algers Department of Transport Sciences, Centre for Transport Studies, Royal Institute of Technology, 100 44 Stockholm, Sweden, staffan.algers@abe.kth.se

Arie Bleijenberg TNO, Schoemakerstraat 97, 2628VK Delft, The Netherlands, arie.bleijenberg@tno.nl

Nils Axel Braathen OECD, Environment Directorate, F-75775 Paris, France, Nils-Axel.Braathen@oecd.org

Bastian Chlond Institute for Transport Studies, Karlsruhe Institute for Technology, 76128 Karlsruhe, Germany, Chlond@kit.edu

Jessica Coria Department of Economics, University of Gothenburg, SE 405 30 Gothenburg, Sweden, Jessica.Coria@economics.gu.se

Jos Dings Transport & Environment (T&E), 1050 Brussels, Belgium, jos.dings@transportenvironment.org

Sandrine Dixson-Declève Cambridge Programme for Sustainability Leadership, B-1050 Brussels, Belgium, sandrine.dixson@cpsl.cam.ac.uk

Lew Fulton International Energy Agency, Paris 75015, France, lew.fulton@iea.org

Muriel Beser Hugosson Department of Transport Sciences, Centre for Transport Studies, Royal Institute of Technology, 100 44 Stockholm, Sweden, muriel@kth.se

Cornie Huizenga Partnership for Sustainable, Low Carbon Transport and Asian Development Bank, 200051 Shanghai, China, cornie.huizenga@slocatpartnership.org

Per Kågeson Centre for Transport Studies (CTS), Royal Institute of Technology, SE-100 44 Stockholm, Sweden, kageson@kth.se

James Leather Partnership for Sustainable, Low Carbon Transport and Asian Development Bank, 200051 Shanghai, China, jleather@adb.org

Todd Litman Victoria Transport Policy Institute, Victoria, BC, Canada V8V 3R7, litman@vtpi.org

Bryan Matthews Institute for Transport Studies, University of Leeds, Leeds LS2 9JT, UK, b.matthews@its.leeds.ac.uk

John Nellthorp Institute for Transport Studies, University of Leeds, Leeds LS2 9JT, UK, j.nellthorp@its.leeds.ac.uk

Milan Ščasný Environment Center, Charles University Prague, 162 00 Prague 6, Czech Republic, milan.scasny@czp.cuni.cz

Lee Schipper (Deceased) Precourt Energy Efficiency Centre, Stanford University, Stanford, CA, USA; Global Metropolitan Studies, University of California, Berkeley, CA, USA

Michael P. Walsh International Council on Clean Transportation, Washington, DC, USA, mpwalsh@igc.org

Theodoros I. Zachariadis Department of Environmental Science & Technology, Cyprus University of Technology, 3603 Limassol, Cyprus, t.zachariadis@cut.ac.cy

Karl-Heinz Zierock EnviCon – Environmental Consultancy, Berlin 41, Germany, Dr.Karl-Heinz_Zierock@t-online.de

List of Figures

Fig. 2.1	Average mobility (km/day) per person 1800–1990 in France (Grübler 1999; reproduced with permission from the International Institute for Applied System Analysis, IIASA)	22
Fig. 2.2	Average travel time (hour/day) per person related to income level (Schäfer and Victor 2000)	23
Fig. 2.3	Share of public transport (*dashed*) and percentage of trips (*solid*) related to the travel time ratio public transport/car (van den Heuvel 1997)	25
Fig. 2.4	Mobility forecast in kilometers per person per day for Western Europe. HST: high-speed train	28
Fig. 2.5	Average door-to-door speed of passenger travel, with and without aviation (Dutch population; Verkeer en Waterstaat 2002)	29
Fig. 2.6	Share of urban population 1950–2050, World and Europe (UNPD 2008)	32
Fig. 2.7	Sequence in new infrastructures and urban development. Railway stations were built just outside the old towns followed by city growth around the stations. A century later ring roads were built just outside the cities and generated urban growth along the ring roads (Kwantes and Govers 2007)	33
Fig. 3.1	World transport energy use by mode, 1971–2006	45
Fig. 3.2	Motorized passenger travel split by mode, 2005	46
Fig. 3.3	GHG efficiency of different modes, freight and passenger, 2005	47
Fig. 3.4	Passenger mobility (trillion passenger kilometers) by mode, year and scenario	50
Fig. 3.5	GHG intensity of passenger transport in 2005 and 2050, Baseline and BLUE Map scenarios	51
Fig. 3.6	Evolution of global transport energy use by fuel type, worldwide	52

Fig. 3.7	LDV annual sales by technology and scenario	53
Fig. 3.8	Evolution of LDV sales by technology type in the BLUE Map scenario	53
Fig. 3.9	Vehicle stocks by technology and scenario	54
Fig. 3.10	New LDV tested fuel economy for selected regions	55
Fig. 3.11	LDV energy use by scenario	56
Fig. 3.12	Transport energy use in the baseline and BLUE scenarios by fuel type in OECD Europe	57
Fig. 3.13	OECD Europe's GHG emissions evolution by transport mode	58
Fig. 3.14	Passenger light duty vehicles sales by technology type in OECD Europe in the baseline and BLUE Map scenarios	59
Fig. 3.15	Average CO_2 emissions trends through 2008 with targets enacted or proposed thereafter by region	61
Fig. 3.16	Country targets for EV/PHEV annual sales as of December 2010	65
Fig. 5.1	Average GHG emissions factor by fuel	100
Fig. 5.2	The 'systems approach'	102
Fig. 5.3	Predicted air pollution reductions from auto oil I and II	103
Fig. 5.4	Life cycle carbon intensity of average EU diesel	116
Fig. 6.1	The maximum average highest permissible CO_2 emission from new cars in the EU in 2015 for vehicles of varying weight, g/km	134
Fig. 6.2	Average CO_2 emission (g/km) from cars registered in France before and after the bonus-malus reform	145
Fig. 7.1	Trend since 1980 in weighted, inflation-corrected tax on petrol and diesel in eight EU member states (representing two thirds of the EU market)	159
Fig. 7.2	Differences in taxes on petrol and diesel vs. the share of diesel in new car sales in 2009	160
Fig. 7.3	Progress in sales-average CO_2 emission in g/km per carmaker in 2009 compared with 2008, and split between demand-side changes and technology changes. Carmakers are sorted on the basis of their technology-only performance	167
Fig. 7.4	What carmakers need to do between 2008 and 2015, and what they did in 2009	167
Fig. 7.5	Fleet-average weight and fleet-average CO_2 emissions by carmaker in 2009, compared with EU target curve for 2015	168
Fig. 7.6	CO_2 reduction cost curves	170
Fig. 7.7	CO_2 reduction cost curves	171
Fig. 7.8	Official CO_2 figures (*horizontal axis*) vs. real-world CO_2 emissions in 140,000 Travelcard vehicles	172
Fig. 8.1	CO_2-related tax rates in one-off taxes on motor vehicles. Tax rates per vehicle, petrol-driven vehicles	183

List of Figures   xi

Fig. 8.2	CO_2-related tax rates in one-off taxes on motor vehicles. Tax rates per vehicle, diesel-driven vehicles	183
Fig. 8.3	CO_2-related tax rates in one-off taxes on motor vehicles. Tax rates per vehicle, petrol-driven vehicles, selected tax rate range	184
Fig. 8.4	CO_2-related tax rates in recurrent taxes on motor vehicles. Tax rates per year, petrol-driven vehicles	185
Fig. 8.5	CO_2-related tax rates in recurrent taxes on motor vehicles. Tax rates per year, diesel-driven vehicles	186
Fig. 8.6	CO_2-related tax rates in one-off taxes on motor vehicles, per tonne CO_2 emitted over the lifetime of a vehicle, petrol-driven vehicles	188
Fig. 8.7	CO_2-related tax rates in one-off taxes on motor vehicles, per tonne CO_2 emitted over the lifetime of a vehicle, diesel-driven vehicles	188
Fig. 8.8	CO_2-related tax rates in recurrent taxes on motor vehicles, per tonne CO_2 emitted over the lifetime of a vehicle, petrol-driven vehicles	192
Fig. 8.9	Average CO_2-related tax rates in recurrent taxes on motor vehicles, per tonne CO_2 emitted over the lifetime of a vehicle, diesel-driven vehicles	192
Fig. 8.10	CO_2-related tax rates in recurrent taxes on motor vehicles, per tonne CO_2 emitted over the lifetime of a vehicle, petrol-driven vehicles; 7% discounting	193
Fig. 8.11	Total CO_2-related tax rates in taxes on motor vehicles, per tonne CO_2 emitted over the lifetime of a vehicle, petrol-driven vehicles	194
Fig. 8.12	Marginal CO_2-related tax rates in one-off taxes on motor vehicles, per tonne CO_2 emitted over the lifetime of a vehicle, selected countries	195
Fig. 10.1	Emission of particulate matter, CZE 1995–2008	224
Fig. 10.2	Fuel consumption and its drivers (1993 levels = 100)	227
Fig. 10.3	Stock of registered passenger cars, age and engine size structure	228
Fig. 10.4	Ex post measurement of progressivity of fuel taxes using the Suits and Jinonice index	236
Fig. 11.1	Clean car model supply 2006–2009	252
Fig. 11.2	Fuel type market shares for new cars bought in Sweden from 2004 to 2009	253
Fig. 11.3	Fuel price development 2005–Sept 2010	254
Fig. 11.4	The development of exempt passages over the congestion charges cordon in Stockholm (City of Stockholm Traffic Administration 2009)	255
Fig. 11.5	The development of clean car sales in Stockholm (*bars*) and Sweden as a whole (*line*) (City of Stockholm Environment and Health Administration 2009)	256

Fig. 11.6	E85 price advantage and monthly consumption, Jan 2006–July 2010 (season adjusted)	258
Fig. 11.7	Example of the Swedish car fleet composition in year 2015	261
Fig. 11.8	Total CO_2 emission 2006–2020 (million tons per year)	265
Fig. 12.1	Energy consumption of transport in Germany 1960–2008 in Petajoule per year (Knoerr et al. 2010)	270
Fig. 12.2	Different developments in modal behavior for young and old people (Chlond et al. 2009)	274
Fig. 12.3	Competing with the car by a kit of complementing modes	280
Fig. 12.4	Development of car-sharing participants in Karlsruhe (Stadtmobil 2010)	289
Fig. 12.5	Development of car-density figures in different parts of Karlsruhe 1980–2005 (based on data of the Office of statistics of the City of Karlsruhe)	291
Fig. 13.1	Range of climate change damage functions considered by Stern (2006)	300
Fig. 13.2	Efficient greenhouse gas and congestion charges	301
Fig. 13.3	Marginal external cost of GHG emissions, per kilometre and per minute, by traffic speed	303
Fig. 14.1	Cycle of automobile dependency and sprawl	330
Fig. 14.2	U.S. state per capita GDP and VMT (VTPI 2009)	350
Fig. 14.3	Per capita GDP and transit ridership (VTPI 2009)	350
Fig. 14.4	GDP versus fuel prices, countries (Metschies 2005)	351
Fig. 15.1	New (2015) Japanese standards compared to previous (2010) standards	357
Fig. 15.2	New versus old Japanese vehicle emission test cycles	358
Fig. 15.3	Actual fuel economy performance versus CAFE standards in the US	360
Fig. 15.4	Passenger vehicle GHG emissions fleet average performance and standards by region	367
Fig. 15.5	Passenger vehicle fuel economy fleet average performance and standards by region	368
Fig. 16.1	Development of traffic activity (passenger and freight), 2005–2050 for OECD and Non-OECD countries	376
Fig. 16.2	Total vehicles and motorization index 2005–2035	376
Fig. 16.3	Greenhouse gas emission estimates for India from different reports	377
Fig. 17.1	On-road emissions per km from automobiles for eight European countries	397
Fig. 17.2	Fuel use per capita from automobiles vs. GDP per capita (the latter expressed in constant purchasing power-adjusted US dollars of year 2000)	398

Fig. 17.3	The real cost of fuel for driving 1 km in Europe, 1970 to 2008. Cost is expressed in constant purchasing power-adjusted US dollars of year 2000, per 100 km	399
Fig. 17.4	Symbolic diagram of evaluation of policy or technology impact. Adapted from Schipper et al. (2009a)	402
Fig. 17.5	Motorization and per capita GDP	403
Fig. 17.6	Traffic Jam in Beijing December 2005; there are more than twice as many cars in 2011	404
Fig. 17.7	Pune, India, 2004. Loans for two wheelers, but no sidewalks in front of the bank	405
Fig. 17.8	Two wheelers in mixed traffic in Hanoi	405
Fig. 17.9	'Air cooled transport'? Colectivo on the Pan American Highway outside of San Salvador, El Salvador, 2000	406
Fig. 17.10	Mexico City. Cars in the counterflow bus lane darting out of the way as the bus plows forward	407
Fig. 17.11	CO_2 emissions from all traffic in the insurgentes corridor of Mexico City in 2005, before and after Metrobus was established	408
Fig. 17.12	Sorting out life cycle emissions: Which registration plate is closer to the truth?	411
Fig. 17.13	Automobile choices	412

List of Tables

Table 3.1	Scenario descriptions and main assumptions	48
Table 3.2	Comparison of EV/PHEV-related policies in several countries as of October 2010	67
Table 4.1	Elements of European commission's 2007 strategy on CO_2 reduction from passenger cars	83
Table 5.1	Proposed default values (European Commission 2011a)	118
Table 6.1	Malus (€) on first registration of passenger cars in France	136
Table 6.2	Bonus (€) to first registration of passenger cars in France	136
Table 7.1	Improvement of 'best practice' diesel cars between 2007 and 2011. The basis for the data has been the 2007 and 2011 editions of the Dutch 'Brandstofverbruiksboekje' ('fuel consumption booklet')	170
Table 8.1	Subsidies in one-off vehicle taxes per tonne CO_2 'saved', calculated based on emission reductions compared to the lowest-emitting vehicles not being subsidized	189
Table 9.1	Fuel taxes and fuel prices in selected countries	204
Table 9.2	Share of diesel cars in the car stock 1995 and 2006	205
Table 10.1	Fuel consumption and expenditures (households with zero fuel expenditures excluded)	226
Table 10.2	Passenger car ownership and the age of the fleet	230
Table 10.3	Car ownership and fuel expenditures across household income deciles	232
Table 10.4	Car ownership and fuel expenditures across other household segments	233
Table 10.5	Probability of owning a car, logit model	234
Table 10.6	Estimates of price and income elasticity for motor fuel demand	238

Table 10.7	Definition of policy scenarios	240
Table 10.8	Effect of *Scenario 2a* on several household segments and public finances	241
Table 10.9	Effect of fuel taxation on household expenditures and welfare	243
Table 11.1	Fuel type market shares (%) for new clean vehicles	253
Table 11.2	Top five selling models of clean vehicles in 2008 and 2010	254
Table 11.3	Yearly circulation tax example	259
Table 11.4	Fuel shares for private buyers and companies	259
Table 11.5	Average fuel consumption development	260
Table 13.1	CO_2 differentiated car ownership tax in the UK	297
Table 13.2	Marginal external costs of car use in Great Britain, at peak/off peak times, low estimates, 1998	299
Table 13.3	Progress relative to EU targets for CO_2 emissions from newly-registered cars	301
Table 13.4	Marginal costs of road use	302
Table 13.5	Transport performance in EU25 countries for N, F and E scenarios, relative to 2000 (=100)	305
Table 13.6	Revenues, welfare change and transport use (EU27+4) from GRACE project	306
Table 13.7	Key findings from the UK national road user charging feasibility study	307
Table 13.8	Results of a snapshot survey	310
Table 14.1	Mobility management strategies (VTPI 2010)	328
Table 14.2	Parking management strategies (VTPI 2010)	336
Table 14.3	Mobility management strategies (Litman 2007)	340
Table 14.4	Comparing strategies (Litman 2007)	342
Table 14.5	Indian cities mode split, 2007	347
Table 15.1	U.S. new-car fuel efficiency standards (cafe) (miles per U.S. Gallon)	359
Table 15.2	Measured CAFE performance	361
Table 16.1	Examples of application avoid-shift-improve approach	382
Table 16.2	Overview of transport projects in existing climate instruments	387

List of Box

Box 3.1 France's Plan to Launch EVs 65

Chapter 1
Introduction

Theodoros I. Zachariadis

Transportation is a major contributor to global energy consumption and greenhouse gas (GHG) emissions, accounting for about one fourth of total energy-related carbon dioxide (CO_2) emissions worldwide. Together with power generation, it is the fastest growing sector in the world. But unlike power generation, whose emissions may be easier to control because they come from a few thousand power plants around the world and because low-carbon or zero-carbon energy sources are already available on a large scale, transport emissions are created by the individual tailpipes of more than one billion motor vehicles (mostly passenger cars) as well as from fuel combustion in airplanes and ships, depending almost entirely on petroleum products with still limited low-carbon alternatives. The global car population is projected to exceed two billion by the year 2050, mainly due to increased car ownership in China, India and other rapidly growing economies (IEA 2009, Sperling and Gordon 2010). And car travel is among the economic activities that are least responsive to price changes: increased mobility improves the standard of living, and automobiles are associated with freedom and comfort. Most citizens of the world wish to have the opportunity to use a car – but can this wish be made compatible with the increasingly strained carrying capacity of the earth and the associated climate challenges?

It is quite simple to calculate car carbon emissions: multiply the number of cars with the average distance travelled by each car, the amount of fuel consumed by a car per kilometer travelled and the carbon content of each fuel, which determines the amount of carbon emitted during combustion of that fuel. These four factors indicate also the options policymakers have in order to curb emissions – they have to reduce or mitigate the growth rate of one or more of these factors: car ownership, use of each car, fuel intensity (the inverse of fuel economy) and fuel carbon content respectively. The first two of these factors, which together amount to total automobile use, are most difficult to tackle because they are associated with individual preferences

---

T.I. Zachariadis (✉)
Department of Environmental Science & Technology, Cyprus University of Technology,
3603 Limassol, Cyprus
e-mail: t.zachariadis@cut.ac.cy

and living standards. The latter two factors – fuel intensity and carbon content – are more prone to technological solutions that may not compromise comfort and personal welfare. Unsurprisingly, most international carbon mitigation policies have primarily addressed these two parameters; this was also a natural continuation of earlier successful attempts to reduce emissions of conventional air pollutants such as carbon monoxide, sulfur dioxide and nitrogen oxides.

Unlike air pollutants, however, CO_2 is not an unnecessary by-product of fuel combustion that can be eliminated by using cleaner fuels and exhaust treatment technologies; it is the main product of fossil fuel combustion, which forms the basis of our economic welfare. Apart from some technological measures that can modestly reduce the growth rate of CO_2 emissions – such as technological improvements in combustion efficiency and a shift toward the use of fossil fuels with lower carbon content (e.g., from petroleum products to natural gas) – a real technological breakthrough is required if automobile CO_2 emissions are to decrease significantly in the future, in line with the stated global objective to contain average global temperature increase to two degrees Celsius (compared to pre-industrial standards) by the year 2050.

For historical and political reasons, the European Union (EU) has attempted to assume a leading role in climate change mitigation worldwide. EU climate policies that have addressed passenger cars have mainly focused on the technological aspects mentioned above – improving fuel economy and reducing the carbon content of fossil fuels used. Thus, apart from initiatives to inform citizens about the fuel consumption of cars (aiming to increase public awareness), a voluntary commitment of the auto industry to reduce CO_2 emissions of new cars was agreed in the mid-1990s. A decade later, it became apparent that this agreement would not deliver the emission reductions it was meant to, which led the EU in 2009 to implement mandatory regulations on car CO_2 emissions and demand a minimum penetration of biofuels as automotive fuel blends. At the same time, as taxation remains at the competence of each EU member country, there are attempts to partly harmonize vehicle taxation and shift it in order to be more favorable to low-CO_2 cars. High taxes on motor fuels, although not designed for this purpose and despite the low responsiveness of car travel to fuel prices mentioned above, may currently be the most effective climate mitigation policy in the continent.

Meanwhile it has become apparent that, in order to make real progress in curbing automobile carbon emissions, it is necessary to enrich policy options with non-technological interventions in the first two factors of the emissions 'equation' mentioned above, i.e., in car ownership and use. In response to this need various policies addressing personal transportation have been initiated at a regional or local level across Europe. Such measures comprise urban road charges, 'ecodriving' seminars, speed restrictions in urban areas, and environmental zones where access to high-emission vehicles is prohibited. Although most of these actions have primarily intended to tackle other, more localized negative impacts of car travel such as congestion, accidents, noise and air pollution, they usually contribute to

1 Introduction

CO_2 reductions as well.[1] According to 2009 Economics Nobel Laureate Elinor Ostrom, a 'polycentric' approach to climate change is required if we are to achieve meaningful emission reductions worldwide (Ostrom 2009); localized measures may thus prove to be a critical ingredient of such a 'polycentric' approach in the transport sector, where billions of individual drivers are involved. And whereas it is desirable to achieve global economy-wide greenhouse gas mitigation agreements in which transportation will have its 'fair share' of obligations, a portfolio of smaller scale actions such as those mentioned here may be more realistic within a complex world with different circumstances and priorities (Barrett and Toman 2010).

This book intends to shed light into the lessons that can be learned from the European experience to mitigate carbon emissions from private cars in the last two decades. Inevitably, it cannot focus on all aspects mentioned above at the same time. Its emphasis is on EU-wide and national policies, not on local measures, because they are crucial not only for the formulation of actions on a local scale, but also for the future EU position in the negotiation of international climate change mitigation actions. However, as there are particularly interesting and promising success stories of local initiatives, the book also looks into such stories – in Europe and elsewhere in the world – and attempts to derive general implications for policymakers. And while our focus is on Europe, we should not lose sight of the global picture; therefore we have attempted to frame all analyses in the context of global policies.

This collective work attempts to distinguish between EU-wide and nation-wide policy responses. This is not always straightforward because of the interaction between these two policy-making levels. However, as many initiatives remain at the discretion of national governments, it is appropriate to examine these two levels in a distinct manner. Broadly speaking, technical regulations such as technology and fuel standards are primarily determined at the EU level, but fiscal measures are largely decided by national authorities – and some of these measures may even contradict stated EU-wide carbon mitigation objectives. The book thus reports not only on types of policies but also on the challenges associated with harmonizing different policy levers toward a common target. Its aim is not merely to present the various policy options but to critically assess them in light of the experience gained during the last two decades in Europe, and keeping in mind the future of climate policies worldwide. These critical reflections address both the nature of each policy measure and the way policies are implemented in the real world.

The book is basically organized in four parts: Part I provides the background of the 'cars and climate policy' topic; Part II reports on and evaluates EU-wide policies of the past with an outlook to the future; Part III gives examples of national fiscal policies as well as other national initiatives aiming at sustainable mobility, discussing their effectiveness in tackling car carbon emissions and their distributional impacts; and Part IV describes the international scene – both the non-European

---

[1] In fact, economists have calculated that the social costs caused by automobile use due to congestion and accidents are more significant than those related to the emission of greenhouse gases from motor vehicles (Parry et al. 2007).

industrialized world and the developing world – in which transport policy makers have to act. The following paragraphs provide a more detailed account of each chapter.

Following this introductory chapter, Part I sets the stage by putting the whole discussion in context. Chapter 2 focuses on the first two parts of the emissions equation, which together constitute the main variable 'car use' expressed in total kilometers travelled, whereas Chapter 3 deals in more detail with the two latter parts of the equation, i.e. with the prospects of low-carbon vehicle technologies and fuels.

In Chapter 2, Arie Bleijenberg states clearly that we should not expect car traffic to stop growing in Europe (and the rest of the industrialized world) any time soon. Increasing travel speed is a strong driving force behind the growth in mobility and car use, and this trend is unlikely to be reversed. Although investments in public transport and high speed rail links as well as better urban planning are useful in specific circumstances, they will probably not have a strong effect on the growth rates of mobility and car use at national and European level. The author suggests that technology can provide the necessary CO_2 emission reductions in Europe through a combination of very fuel efficient cars with low-carbon fuels. To achieve this combination, however, strong political will is necessary in order to set stringent emission and fuel standards, adopt pricing for all transport services so as to reflect marginal social costs and evaluate plans for future urban infrastructure investments with the aid of a proper social cost benefit analysis. Finally, he notes that the developing world has more options to switch to low-carbon transport since mass motorization is still at an early phase and there is still time to adopt more sustainable mobility options.

Lew Fulton shows in Chapter 3 where the transport sector stands in terms of greenhouse gas emissions and where it heads to in the coming decades. Under business-as-usual assumptions, worldwide passenger kilometers and GHG emissions are expected to double by 2050, whereas they are projected to remain stable in Europe, despite improving fuel economy of new cars. However, scenarios prepared by the International Energy Agency and presented in this chapter give some signs of hope: exploiting the full potential for fuel economy improvements in current technologies and widespread adoption of new technology vehicles and low-carbon fuels could cut CO_2 emissions from cars around the world by more than half in 2050, compared to 2005 levels. To explain how this can be achieved, the author deals in more detail with the prospects of low-carbon vehicle technologies and fuels. The focus here is not on a detailed account of specific technologies and fuels – this is provided extensively in other publications such as the excellent book of Schäfer et al. (2009) – but rather on an outline of the policies required to develop the necessary infrastructure and enable the widespread adoption of low-carbon engines and fuels within a tight time schedule. Although an unprecedented automobile market transformation is required for all this to be achieved, the author is optimistic because a related market transformation is already under way: a new direction for fuel economy of today's vehicles, now clearly on a path toward much more efficient vehicles in the future – in contrast to the stagnation in fuel economy levels that prevailed in recent decades.

1 Introduction

The next two chapters provide a historical account and a future outlook of EU policies in the field of CO_2 regulations and fuel standards respectively. Their authors have actively participated in the relevant policy analyses, discussions and consultations in or around EU institutions for many years. In Chapter 4, Karl-Heinz Zierock outlines the milestones of EU climate policies aimed at automobiles, starting already in the late 1980s and reaching the more extensive decarbonization strategy that unfolded publicly in 2007, including mandatory CO_2 emissions standards, a low-carbon fuel standard and several additional provisions. The author provides background information on how decisions have been made in EU legislative and executive bodies, and concludes that the key for the long-term success of transport decarbonization lies outside the transport sector, namely in the production of renewable electricity that may be used as future transport fuel or as an energy source for the generation of low-carbon fuels. He finds, however, that the way toward realizing this vision is full of obstacles and requires re-inventing motorized transport, entailing significant changes in the automobile industry and even more drastic changes in the oil industry.

Sandrine Dixson-Declève reinforces Zierock's argument in Chapter 5, which looks at the lessons learned in the area of cleaner fuels as well as the problems confronting policy makers in trying to move toward low-carbon fuels. After an extensive account of past policies and an analysis of the elements and challenges of current EU fuel policy, she underlines that we are confronted with the need to take an entirely new approach to liquid fuels and their role in society; the feat ahead, she says, is far more complex than any challenge that the fuels industry and policy makers have faced before. The author reminds us of the 'quantum leap' industry and European governments have made over the last decade by working together to develop new more environmentally friendly vehicles and fuels as a result of the Auto Oil Programs. In view of this successful track record, she stresses the need to establish a comprehensive stakeholder program similar to the Auto Oil Program in order to agree on sustainability criteria for low-carbon fuels and to ensure the adoption of appropriate fuel legislation. She concludes that only through bold political and industry action to address real low-carbon options and innovative solutions will Europe enable transformational change in the fuels industry and meet its 2050 GHG emission reduction targets.

The discussion on the most appropriate policy instrument (fuel taxes, regulations and/or complementary measures) for curbing automobile GHG emissions continues unabated: theoretical arguments are countered by practical considerations, engineering approaches are questioned by economists and vice versa. Per Kågeson provides in Chapter 6 a thorough overview of this policy debate, drawing from findings of studies around the world. Based on the lessons learned worldwide, but also from recent research results, he offers crucial guidelines to policymakers: harmonize incentives across Europe in order to allow the industry to adapt faster; regulate energy use rather than CO_2 emissions per kilometer; offer technologically neutral incentives, with some cautious extra support to very promising technologies; apply continuous functions to calculate incentives such as vehicle taxes or subsidies, and avoid thresholds and notches; treat private and company cars in the same way; allow for different treatment according to vehicle size, not vehicle weight; and determine the magnitude of the incentives so that they correspond to the social cost of the problem they are supposed to tackle – i.e. the marginal damage costs of climate change. Finally, the author offers a proposal for a harmonized European system of taxes and incentives, which can provide efficient incentives to reduce car fuel consumption and carbon emissions. Although this chapter refers to policies that can be adopted at both EU-wide and national level, it is included in this part of the book since the author's proposals are based on the judgment that the most cost-effective solutions are associated with harmonized EU-wide policies that send clear messages to both consumers and the auto and oil industry.

Chapter 7 evaluates EU policies from the point of view of an environmental non-governmental organization that has monitored the EU policy making process in the field of transport and the environment for the last 20 years. In line with the focus of Part II of the book, Jos Dings discusses the past and the future of EU-wide regulations, i.e., carbon-related regulations for vehicles and fuels and fuel taxation. He argues that the co-existence of command-and-control policies with fuel taxes is economically justified, and that carbon prices in transportation should be higher than those applied to industrial sectors through the EU Emissions Trading System. He addresses in detail the loopholes that exist in the current automobile CO_2 legislation, which in his opinion may considerably compromise its environmental effectiveness, and provides recommendations for alleviating these problems. He also questions the *ex ante* estimates provided by auto manufacturers on the costs of compliance with stricter environmental regulations, which he finds highly exaggerated, and cautions against relying on the current test procedures to determine car CO_2 emissions. He recommends that fuels should be taxed on the basis of their well-to-wheel carbon footprint, which in turn requires a strong improvement in the carbon accounting of fuels. He further explains that future automobile regulation should not be based on a CO_2 emission standard but rather on an energy efficiency standard in order to be fair toward the automobile industry and to account for the increased penetration of hybrid and electric cars. The paper also suggests raising minimum tax rates for diesel fuel and addressing the 'diesel tourism' phenomenon that prevents EU member states from taxing diesel more aggressively.

Part III of the book presents and discusses policy options that, although implemented in many European countries, are determined at national level and hence are not characterized by harmonization across the continent. Such policies are vehicle and fuel taxes, incentives to encourage the use of low-carbon cars or to reduce the use of cars altogether, and road charging schemes. In the case of vehicle and fuel taxes, which are everywhere similar in nature, a general overview of regulations around Europe is provided, whereas other measures are much more specific to the country or even the city adopting them. Therefore, Chapters 8 and 9 offer an outline of vehicle taxes and fuel taxes respectively as existing in early 2011, while Chapters 10 through 13 present specific case studies that can provide useful conclusions to policy makers.

In Chapter 8, Nils-Axel Braathen describes vehicle taxation policies implemented in Europe, in which the tax (a one-off registration tax paid at the purchase of a new car and/or a vehicle tax paid annually by all licensed cars) is determined on the basis of a car's CO_2 emission levels. Such taxes have been increasingly adopted in European countries in the 2000s, partly replacing older taxes levied on cars according to their weight, price or engine size, in order to encourage the purchase of low-carbon vehicles. The author also calculates the implied tax rates of these policies (expressed in Euros per tonne of CO_2 emitted or abated over the lifetime of a vehicle). He notes that, while other fiscal measures such as fuel taxes and road charging would be the first-best approach, some CO_2-related tax rate differentiation of motor vehicles can be useful if political economy constraints (i.e., low public acceptance) make it difficult to put in place an 'ideal' system. However, he finds the degree of tax differentiation applied in some countries to be disproportionately high compared to the marginal abatement costs of CO_2 mitigation options in other economic sectors.

In a similar fashion, Jessica Coria provides in Chapter 9 a comprehensive account of motor fuel taxation in Europe. This measure remains at the discretion of national governments, and only a minimum tax level is determined at the EU level in order to avoid excessive differences between member countries. She summarizes the fuel tax rates applied in different EU countries by the end of year 2010 and mentions how these rates are related to per capita income and per capita government expenditure in each country. She points out that, although most fuel tax regimes have been designed in order to generate public revenues, fuel demand and CO_2 emissions would have been much higher in the absence of the existing high fuel taxes in Europe. After reviewing the literature on price elasticities of fuel demand, which are crucial for the estimation of the effect of a fuel tax on automobile energy use and carbon emissions, the author examines issues of political economy – to what extent consumers are willing to accept higher fuel taxes – as well as distributional aspects – whether a fuel tax increase affects proportionately more the rich or the poor. Obviously, concerns that a fuel tax rise puts a larger burden on lower-income households (a concern that is not always confirmed by empirical research) render fuel taxation unpopular among citizens.

These distributional aspects of fuel taxes are illustrated in a case study in Chapter 10. Milan Ščasný explores household expenditures on transport fuel by income group during the 1990s and 2000s in the Czech Republic. He uses two different indices to measure the progressivity of fuel expenditures, and finds these to be almost uniform (neither strongly progressive nor strongly regressive) across income groups. He also analyzes the effects of changes in automotive fuel taxation on household expenditures, depending on how the increased public revenues are recycled in the economy – through reductions in personal income tax rates, social security contributions of workers or tax credits. The overall impact is quite small and the burden to households varies according to social status and the size of residence of each household rather than across income deciles. These results are in line with those of other empirical analyses in industrialized countries and reinforce the view outlined in Chapter 9: fuel taxation – a strong economic instrument of climate policy – should not be abandoned on the grounds of equity concerns as the latter may not be justified – or can be alleviated through targeted interventions to those types of households that will be adversely affected.

The second case study comes from Sweden. Muriel Beser Hugosson and Staffan Algers describe in Chapter 11 national policies that have been designed in order to accelerate the introduction of clean cars, i.e., cars with low CO_2 and air pollutant emissions, including those powered by ethanol blends and gas. In a country which, for many years, possessed the heaviest and highest CO_2-emitting cars in Northern and Western Europe, such measures are important for reducing automobile carbon emissions. The article outlines first the institutional and fiscal measures taken by the government in order to prepare the market for the penetration of clean cars. Then it describes those regulatory and fiscal measures implemented to encourage the purchase of clean cars, and reports on the changes induced on the supply side as well: an increasing number of low-CO_2 models entered the Swedish market after 2005, particularly compact diesel and ethanol powered cars. The authors evaluate each one of the measures mentioned in the paper, and highlight problems associated with these initiatives: although sales of new clean cars in Sweden rose impressively between 2005 and 2010, the shares of low-carbon cars are sensitive to fuel prices, and 'flexifuel' cars, which can run on either petrol or a petrol-ethanol blend, may be run on pure petrol most of the time if petrol prices are favorable – thus diminishing a large part of the emissions benefit. Finally, the need for detailed policy simulation tools is explained in order to support policy makers in their decisions.

Perhaps contrary to what other authors describe, Chapter 12 paints a more optimistic picture on the prospects of changing travel behavior. Bastian Chlond claims that the stabilization of transport CO_2 emissions in Germany since the year 2000 is attributable mainly to the increased use of public transport modes and bicycles. He describes the gradual shift of German society, from absolute car dependence in earlier decades to a slow relative decline in the use of private cars and a corresponding rise in utilization of other passenger transport modes. The author explains this slow paradigm shift as a combination of a demographic process, whereby young generations get used to driving less because there are plenty of alternatives to the automobile; economic policies such as financing public transport infrastructure and raising fuel taxes; and urban planning choices that discourage urban sprawl and enable a 'cultural' change, which reduces the symbolic status of car ownership and gives more emphasis to environmental protection and a healthy lifestyle. Next, the article outlines the basic ingredients of a strategy to create a multimodal transport system, which comprises a number of policies that increase the attractiveness of public transport and non-motorized travel, while at the same time reducing the attractiveness of car use. Then the author focuses on the particular case of the German city of Karlsruhe and refers to the specific institutional and regulatory arrangements that have helped make this city the most prominent example of such a shift toward multimodal behavior in Germany. To a cautious reader who might consider this case study to be the exception rather than the rule, the author responds that Karlsruhe is not an exception but just a pioneer among many German cities that follow in the same direction. 'This gives hope for optimism', he concludes.

Urban road charging systems have been implemented in a number of European cities. Currently there are plans to apply such schemes on a national basis too. Although such systems attempt to tackle multiple types of externalities of car use (congestion, accidents, noise and air pollution), they are increasingly mentioned as carbon mitigation policy options as well. In Chapter 13, Bryan Matthews and John Nellthorp review the theory and practice of implementing nationwide road charging systems. They point out that national road user charging appears to offer a holistic solution for tackling transport externalities; therefore they explore the role of climate change costs in this debate, and what impact on climate change such a solution might have. They review the theory on which the case for national road user charging is based and provide a global overview of attempts at implementation in a number of countries. Their survey shows that apart from Singapore, the city state that has adopted national road user charging, at least another ten countries have adopted or are considering adopting nationwide road charges – though not always targeting passenger cars. The authors focus on areas of progress as well as on the sticking points with this policy, and discuss how public acceptability barriers can be overcome through careful design. They conclude that if national road user charging is adopted, this charge could also be used for charging CO_2 emissions instead of using the fuel tax for this purpose.

Despite the book's focus on Europe in all previous chapters, it is evident that European policy makers cannot act alone – particularly if the EU aspires to maintain its leading role in climate policy. Climate change is a truly global problem; the contribution of Europe to global anthropogenic GHG emissions is less than one fifth and expected to decline in the future due to the rise of emissions in developing nations; and the automobile and oil industries are globalized to a very large extent. Therefore, Part IV of the book is devoted to the international context of the 'cars and climate policy' topic. It comprises two chapters on major automobile-related climate policies in the rest of the industrialized world and the developing world respectively, which deal primarily with technical measures – fuel economy regulations and fuel standards; and one chapter on non-technical measures, i.e., mobility management practices around the world.

In Chapter 14, Todd Litman explores the role that mobility management can play in a sustainable and economically efficient transport system. He describes the background of such measures: Many current policy and planning practices tend to favor mobility over accessibility and automobile travel over alternative modes, which often results in economically excessive motor vehicle travel. Similarly to what has been described in the previous chapter, he notes that a paradigm shift is occurring among transport planners, from the current mobility-based to accessibility-based planning; this calls for mobility management strategies that increase (and provide incentives to use) transport options such as walking, cycling and public transport, and enable appropriate land use planning so as to improve accessibility. The chapter outlines the basic principles of proper mobility management strategies, discusses the critiques toward these concepts and provides a number of case studies from around the world where elements of such policies have been implemented. The example of the city of Karlsruhe, presented earlier in Chapter 12, seems to be compatible with the best cases presented by Litman. According to the author, if these strategies are implemented appropriately and in a cost-effective fashion, they can reduce motor vehicle travel by 30–50% compared with what results from conventional policies and planning practices, and make people better off. He concludes by underlining that transport planning reforms in line with the new accessibility-based paradigm are particularly appropriate in developing countries to support economic development as well as environmental and social equity objectives.

Chapter 15 focuses on non-European OECD countries. Michael P. Walsh explains that there has been a fundamental change in the approach to regulating fuel economy or GHG emissions from road vehicles over the past decade, which was mainly induced by concerns of human-induced climate change. The number of countries adopting some form of regulation has grown dramatically. Moreover, the form of the fuel economy or GHG standard is starting to shift away from a mass-based approach toward a footprint-based approach, which will open up additional opportunities to take advantage of lightweighting as a key element of a control strategy. The chapter describes the history and most recent developments (up to the beginning of 2011) on such standards from non-EU OECD countries around the world, namely in Canada, Japan, South Korea and the United States. Other OECD countries such as Australia and Mexico are also considering the implementation of similar standards. While command-and-control standards are expected to remain the backbone of control efforts, economic incentives or disincentives including fuel taxes are expected to play a more important role in the future than they do today.

In Chapter 16, Cornie Huizenga and James Leather assume the difficult role to describe the situation in the developing world and propose policy solutions. They highlight the importance of the developing world in terms of their growth in transportation GHG emissions, which underlines the urgent need for low-carbon solutions. The chapter describes the currently dominant planning paradigm as the 'Predict and Provide' approach, which has been financially supported by Multilateral Development Banks and has focused almost exclusively on building sufficient road infrastructure for new cars and trucks, thereby ignoring other transport modes and leading overall to unsustainable solutions. In order to move to a low-carbon, sustainable transport future the authors emphasize the need for a paradigm shift from the 'Predict and Provide' approach to an 'Avoid-Shift-Improve' approach; this can enable both controlling the growth in motorization and providing alternative transport modes to meet the rising demand for welfare-improving mobility in the developing world. They proceed with recommendations for the shaping of external assistance policies in the future, in terms of private investments, development assistance from bilateral or multilateral funding mechanisms, as well as climate-related financing instruments. The authors point out that the developing world has the possibility to opt for a leapfrog approach to transport and climate change, which will be required if the transport sector is to meet the drastic global GHG emission reductions required up to 2050.

Lee Schipper provides an eloquent epilogue to this book in Chapter 17. The future of the transport sector will greatly affect the future of carbon emissions, as it is the fastest rising source of CO_2 emissions in the world. At the same time, he notes, high CO_2 emissions are only one of the symptoms of poor urban transport, particularly in cities of the developing world; light duty vehicles are at the centre of broader urban transport problems such as congestion, accidents and air pollution.

If we are to attack these problems effectively, Schipper claims, we have to frame the issue as a transport problem and not merely as a CO_2 problem. In fact, some of the largest benefits of CO_2 reduction come as indirect benefits of other strategies to improve transportation. Although technology improvements to cars – such as greater fuel economy and use of low-carbon fuels – are important, technology per se is the smallest uncertainty; the major problem is the future growth in global vehicle kilometers travelled. Therefore, a sustainable long term approach will involve a coordinated effort encompassing efficient vehicles and low-carbon fuels, congestion pricing and other strategies to reduce externalities, provision of viable public transport options, and promoting land use policies that discourage automobile use. At any rate, he concludes, in line with many other authors of this collective volume, the future of the automobile cannot be like its past. The future will be grim if individuals, their elected officials and stakeholders in fuel and vehicle companies continue as if there are no profound problems confronting the choices automobiles give their users.

Trying to distil the analyses and viewpoints presented in the chapters of this book in order to come up with some broad conclusions is a daunting task. However, I would single out four major findings:

1. *Irrespective of the GHG emission mitigation effort in other economic sectors, global transport emissions should decrease greatly in the coming decades if the two-degree-Celsius objective is to be met.* Research shows that transportation is not the sector of top priority for reducing GHG emissions since marginal carbon abatement costs in other sectors of the economy may be lower (McKinsey 2010, Proost 2008). Nevertheless, if global climate forecasts are able to capture the relationship between GHG emissions, GHG concentrations and temperature changes with reasonable accuracy, it is not justified for policy makers to postpone transport-related climate policies until other sectors have assumed their 'fair share' of mitigation effort. Keeping in mind the potential bias in assessing costs of stringent climate policies (Tavoni and Tol 2010), and notwithstanding the lively discussions on how to discount the distant future, it is clear that if there is indeed a probability for catastrophic climate change, even if very low, action must not be delayed (Weitzman 2009).
2. *To reduce automobile GHG emissions we need mandatory regulations, which should go hand-in-hand with fiscal policies as well as local and national economic incentives.* Many economists might disagree with this finding: there is ample theoretical and empirical evidence that a fuel/carbon tax is a more efficient solution than a command-and-control regulation of GHG emissions, with considerably lower transaction and enforcement costs (Austin and Dinan 2005, Parry et al. 2005, Sallee 2010). The design of such regulations often makes things even worse since regulations include thresholds and notches, and their implementation induces short-term producer and consumer behavior that reduces their effectiveness (Sallee and Slemrod 2010). Despite these well-known limitations, and irrespective of the continuing debate as to whether consumers undervalue fuel economy savings (Greene 2010), one thing is certain: raising fuel taxes is

unpopular, hence very few governments implement tax increases. If, as explained in the previous paragraph, we must act soon to curb automobile GHG emissions, there is little point in waiting for 'enlightened' leaders to risk their political future by raising fuel taxes; if a first-best option is infeasible, second-best or even third-best policies are better than nothing.[2]

Moreover, economic analysis has its caveats too: Firstly, a regulation is often considered to reduce welfare because it may induce consumers to purchase products different than the ones they would 'ideally' prefer. As Hanemann (2008) has pointed out, however, consumer preferences are not fixed (as neoclassical economic theory assumes) but evolve; mandated constraints that were once considered to affect consumer welfare are not regarded as welfare-reducing any longer. For example, citizens who were initially disturbed by an anti-littering law or non-smoking obligations have adapted over the years so that a littering or smoking ban may not be considered adverse to their welfare any longer. In a similar fashion, if a GHG regulation makes some consumers purchase smaller, less powerful or less convenient automobiles this welfare loss may be negligible after some time.

A second caveat of the simple economic rationale that states 'if an externality exists you just have to impose a (Pigovian) tax to reduce it' is illustrated by Acemoglu et al. (2010). Instead of implementing just a carbon tax, a policy combining such a tax with economic support for research & development in low-carbon technologies may achieve an environmental objective at a lower cost than the tax-only policy. This finding seems to reinforce what was stated above: carefully designed regulations may be a reasonable way forward for climate policy in transport, particularly in Europe where fuel taxation is already high. Economic policies such as CO_2-related vehicle taxes, road charging schemes and local incentives toward sustainable mobility – as long as they do not imply unreasonably high carbon abatement costs – are necessary complements to regulations; and, as stated earlier in this chapter, a 'polycentric' approach comprising measures at international, national and local level may be more appropriate for addressing the transportation-climate problem in the real world.

3. *There are some indications that 'conventional wisdom' transportation forecasts may not apply any more.* The evolution of automobile use and GHG emissions have clearly followed a business-as-usual path up to now, in contrast to more optimistic scenarios of earlier decoupling of travel demand from economic growth, which have not materialized. However, some things seem to be different now. Firstly, market transformations are happening around the industrialized world, leading to significant improvements in fuel economy and perhaps to changes in citizens' behavior toward the private car. Secondly, space restrictions in densely populated areas of the developing world may be restraining the growth

---

[2] See also Flachsland et al. (2011) for a discussion of alternative climate policy instruments in road transportation.

in automobile use already now, at low levels of motorization. Thirdly, rapidly developing economies such as China and India are adopting (or are expected to adopt soon) fuel economy regulations that would not have been expected a few years before. On the other hand, although such optimism is justified to some extent, one should always keep in mind that tremendous effort is required in order to accelerate technological improvements and restrain growth in automotive travel demand in the long run.

4. *The industrialized world can contribute to automobile GHG abatement thanks to near-saturation motorization levels, but the great hope comes from the developing world.* The overwhelming share of the increase in transportation GHG emissions will come from developing countries in the next decades. In contrast to the industrialized world, where travelling habits have been formed throughout the years and are changing only slowly, and where it takes time to replace the vehicle fleet with low-carbon cars, citizens of developing nations are just starting to own automobiles on a large scale. Hence, if national policies nudge consumers toward buying low-carbon vehicles and encourage sustainable mobility practices, it should be possible for large parts of the developing world to leapfrog to a sustainable transport path. International financing institutions can significantly contribute to this target by directing funds toward low-carbon investments in public and non-motorized transport infrastructure. Avoiding urban sprawl through smart land use policies and directing investments toward dense urban areas is not only environmentally sustainable but also seems to promote economic growth; Chapter 2 has touched upon this topic, and Glaeser (2011) provides compelling evidence for the existence of this effect worldwide. Therefore, even if it takes a long time for the industrialized world to adjust, developing economies have the opportunity to shape their future with more sustainable, low-carbon mobility patterns.

We have strived to keep this book easily readable but also widely informative. The authors have attempted to write their chapters in a manner that is partly technical (which is inevitable due to the nature of the topics discussed herein) and partly accessible to a wider public having only basic familiarity with the transportation, energy and climate change terminology. To the extent possible, we have avoided providing equations and complex tables and charts. For readers interested in more technical aspects of a topic, a large variety of scholarly papers is available in academic journals. If this book is of some value to researchers, students and policy practitioners, it will be because of the reviews and recommendations made by its authors, encompassing – hopefully – most of the important aspects of the 'cars and climate policy' debate.

It was also impossible to avoid some technical language related to the decision-making process in the EU; we have nevertheless tried to keep this jargon to a minimum in order to maintain the interest of non-expert and non-European readers, without compromising the need to describe the policy-making process with reasonable accuracy. We hope – and readers will judge by themselves – that we have not entirely failed in this attempt.

I am indebted to all chapter authors, widely known professionals with mostly long experience in the analysis and formulation of transportation-related climate policies around the world, who have offered to contribute to this book with their knowledge and intellect. The formidable task of compiling such a collective work would have entirely failed if it were not for the willingness of these great analysts to devote some of their scarce time to the success of the book. I am also grateful to Gay Christofides for her excellent editing work and Panayiotis Gregoriou for his superior technical assistance. Finally, I would like to dedicate this volume to Lee Schipper, who passed away in August 2011, soon after he finalized the epilogue (Chapter 17) of this book. Together with hundreds of other people around the world, I have benefited enormously from his warmth, his encouragement and his incredible energy in analyzing transportation issues.

References

Acemoglu D, Aghion P, Burztyn L, Hemous D (2010) The environment and directed technical change. FEEM Working Paper (Nota di Lavoro) 93.2010, Fondazione Eni Enrico Mattei, Venice

Austin D, Dinan T (2005) Clearing the air: the costs and consequences of higher CAFE standards and increased gasoline taxes. J Environ Econ Manage 50:562–582

Barrett S, Toman M (2010) Contrasting future paths for an evolving global climate regime. Policy Research Working Paper 5164, The World Bank, Washington, DC

Flachsland C, Brunner S, Edenhofer O, Creutzig F (2011) Climate policies for road transport revisited (II): closing the policy gap with cap-and-trade. Energy Policy 39:2100–2110

Glaeser E (2011) Triumph of the city: how our greatest invention makes us richer, smarter, greener, healthier, and happier. Macmillan, London

Greene DL (2010) Why the market for new passenger cars generally undervalues fuel economy. Discussion Paper No. 2010-6, OECD/International Transport Forum, Paris

Hanemann M (2008) Climate change policy: a view from California. Presentation at the annual conference of the European association of environmental and resource economists, Gothenburg, Sweden

IEA (2009) Transport, energy and CO_2: moving toward sustainability. International Energy Agency, Paris

McKinsey (2010) Impact of the financial crisis on carbon economics. http://www.mckinsey.com/clientservice/sustainability/pdf/Impact_Financial_Crisis_Carbon_Economics_GHGcostcurveV2.1.pdf. Last accessed Mar 2011

Ostrom E (2009) A polycentric approach for coping with climate change. Background paper to the 2010 world development report, Policy Research Working Paper 5095, The World Bank, Washington, DC

Parry IWH, Fischer C, Harrington W (2005) Do market failures justify tightening corporate average fuel economy (CAFE) standards? Resources for the future, Washington, DC

Parry IWH, Margaret W, Winston H (2007) Automobile externalities and policies. J Economic Literature 45:374–400

Proost S (2008) Full account of the costs and benefits of reducing CO_2 emissions in transport. Discussion Paper No. 2008-3, OECD/International Transport Forum, Paris

Sallee JM (2010) The taxation of fuel economy. NBER Working Paper No. 16466, National Bureau of Economic Research, Cambridge, MA

Sallee JM, Slemrod J (2010) Car notches: strategic automaker responses to fuel economy policy. NBER Working Paper No. 16604, National Bureau of Economic Research, Cambridge, MA

Schäfer A, Heywood JB, Jacoby HD, Waitz IA (2009) Transportation in a climate-constrained world. MIT Press, Cambridge, MA
Sperling D, Gordon D (2010) Two billion cars – driving toward sustainability. Oxford University Press, Oxford
Tavoni M, Tol RSJ (2010) Counting only the hits? The risk of underestimating the costs of stringent climate policy. Climatic Change 100:769–778
Weitzman ML (2009) Reactions to the Nordhaus critique. Discussion Paper 2009-11, Harvard Environmental Economics Program, Cambridge, MA

Part I
Background – Automobiles and Climate Change

Human society faces a dilemma. People want cheap and easy mobility, and they want to travel in comfort and style. But giving free rein to these desires means more oil consumption and more greenhouse gas emissions; global tensions over scarce oil supplies and a rapidly altering climate; and potential devastation for many regions, many businesses, and many people. The challenge is to reconcile the tensions between private desires and the public interest.

Daniel Sperling and Deborah Gordon, Two Billion Cars – Transforming a Culture. TR News, Issue 259, November–December 2008, p. 9. Transportation Research Board, Washington, DC. http://onlinepubs.trb.org/onlinepubs/trnews/trnews259billioncars.pdf. *Last Accessed Mar 2011*

Chapter 2
The Attractiveness of Car Use

Arie Bleijenberg

Abstract Understanding the driving forces behind car use is necessary for the development of effective transport policies. The high door-to-door speed of the car in comparison with other travel modes forms its main attractiveness. And speed is the main engine for mobility growth, which is not easy to curb. Public transport and urban planning can only modestly influence the growth in car use. Urbanization creates short travel distances and is therefore a complementary way to achieve good accessibility. However, congestion remains an inevitable part of economically prosperous urban areas. Car growth in industrialized countries will gradually decline to zero in the coming decades, because saturation levels will be reached, and aviation will probably take over the dominant role in passenger travel for the European population before 2050. Clean technology is the most promising route toward reducing the impact of car use on climate change. However, fuel efficient cars and low carbon fuels will only conquer the roads if strict policy measures are taken.

2.1 Reduction of Greenhouse Gasses

Global emissions of greenhouse gasses need to be reduced to avoid costly and dangerous changes in our climate. It is generally accepted that industrialized countries need to cut their emissions 50–80% below 1990 levels by 2050. This is a tremendous challenge, not the least for the transport sector where CO_2 emissions have been growing for more than a century and are expected to increase further. Global emissions of CO_2 from cars are projected to double in 50 years. Road transport – cars, trucks and vans – and aviation together will account in 2050 for almost 85% of all transport CO_2 emissions (OECD/ITF 2010).

A major cause of the increasing emissions from cars is the growth in car kilometers. It is expected that car use in the EU-27 will increase by 26% in the coming 20 years to 2030, corresponding with an average growth of 1.2% per year

A. Bleijenberg (✉)
TNO, Schoemakerstraat 97, 2628VK Delft, The Netherlands
e-mail: arie.bleijenberg@tno.nl

(Capros et al. 2008). A second reason for the growing CO_2 emissions is the trend toward larger, heavier and more powerful cars. This development in the car market roughly cancels out any technical achievements of the car industry toward improving the fuel efficiency of cars. As a result, greenhouse gas emissions from cars grow at more or less the same rate as the number of kilometers driven. Driving style and speed also have some impact on the environmental performance of car driving. This results in four potential ways to curb the CO_2 emissions from cars:

- Technology: better energy efficiency and low-carbon fuels, such as electricity and bio fuels, both with low well-to-wheel CO_2 emissions.
- Car performance: reduced size, weight and/or power.
- Car use: lower speeds and a fuel efficient driving style with, for example, proper gear shifting and constant speed.
- Car use: fewer kilometers driven.

This chapter focuses only on the last option: the volume of car traffic. However, the last section, on Sustainable Mobility, will discuss the balance between reduced car use and clean technology.

Many studies and policy documents state the inevitability of a reduction in at least the growth in car use, in order to achieve the required reduction in greenhouse gasses. Some even argue that the industrialized world needs an absolute reduction in the volume of car traffic. However, attempts to reduce the growth in total mobility or to achieve a shift to other modes of transport have thus far had only a limited effect. It appears that strong driving forces lie behind current mobility trends, which make it hard to curb those trends. This chapter examines the fundamental driving forces behind the continued mobility growth and the apparent attractiveness of car use. Proper insight into these driving forces is needed to judge the feasibility and effectiveness of policies aimed at reducing the growth in car traffic. Attempts to change mobility patterns are frequently more the result of wishful thinking than of a sufficient understanding of the forces and of the system dynamics behind current trends. This hinders the development of realistic policies to reduce greenhouse gas emissions from transport.

This chapter identifies increasing speed as the main driving force behind the growth in mobility and car use. Faster travel makes it feasible to travel longer distances in a fixed time frame, making it possible to undertake activities in ever more distant locations. A complementary way to increase our accessibility is to locate our activities close to each other, that is, in cities. Continued urbanization paradoxically yields at the same time both better accessibility and increased congestion and therefore influences mobility patterns and growth. These two topics, speed and urbanization, will be discussed and elaborated on in the following sections. The consequences for climate policy of the given analyses are presented in the final section of this chapter: Toward Sustainable Mobility.

2.2 Speed

2.2.1 Historic Mobility Growth

Mobility is of all ages and Homo sapiens has been moving around since its appearance. In the evolution of species the ability to run was important to catch prey, to escape from enemies and also to move to more favorable territories. Mankind inherited these abilities from its ancestors. Some argue that the 'decision' to walk on two legs instead of four increased the mobility of Homo sapiens substantially, giving it an advantage over other animals. So, the first mobility revolution of mankind can be dated to around 6 million years ago. In the era of hunters and gatherers walking and running were essential for survival, and after the agricultural revolution mobility was needed for the emerging trade required for the first cities. The drive for mobility may have found a way into our genes, however, until the Industrial Revolution mobility was dependent on the energy of people, animals and the wind. Major historic improvements were the invention of the wheel (c. 3,500 BC) and the domestication of horses (c. 2,000 BC). Speeds were low: walking covers around 5 km/h, horses and boats between 8 and 15 km/h. The associated mobility volumes stayed small.

The Industrial Revolution had a tremendous impact on man's ability to travel. In a short period of time the main current transport technologies were developed. Within a few decades around 1900 the car, electric train and aeroplane were all invented. These inventions were possible due to prior development of the internal combustion engine and the electric engine in the period 1880–1890. Surprisingly, after 1910 only a few major inventions followed: the jet engine (c. 1940), the sea container, and the manned space ship (both around 1960). Although current transport technologies are more than a century old, their full deployment has not yet been reached. Car use might be nearing saturation levels in Europe, but air travel still has a long way to go before it is fully incorporated into our society.

It is not surprising that these improved transport technologies had a great impact on mobility. Figure 2.1 presents an excellent summary of the mobility growth induced by the Industrial Revolution.

It is a striking aspect of this figure that the aggregate mobility for all modes together closely follows an exponential growth path for almost two centuries. The average distance travelled per day increases from 20 m in 1800 (walking excluded) to 30 km in 1990. This corresponds with an average annual growth rate of close to 4%.

Furthermore, the figure reveals the rise and fall of transport modes. Horse carriages were still dominant up to 1850 and kept market share until motorized road transport replaced them around 1910. Trains were in the lead for almost a century between 1850 and 1930. Since then the car has been the dominant transport mode. Car ownership and use increased fast because the price of cars decreased strongly and paved roads, and later motorways, made car driving faster and more comfortable. The price of the well-known T-Ford dropped by two thirds in the

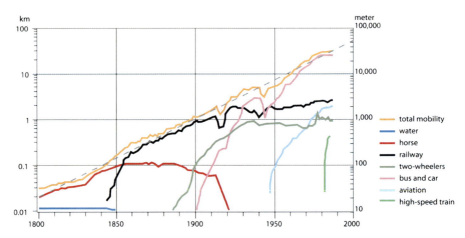

Fig. 2.1 Average mobility (km/day) per person 1800–1990 in France (Grübler 1999; reproduced with permission from the International Institute for Applied System Analysis, IIASA)

period 1908–1920, during which time mass production started. Mass production – also known as fordism – is another milestone in the process of industrialization.

It is obvious that the observed shift in modes of transport corresponds with a shift to ever faster modes. The average speed of the horse-drawn carriage was around 10 km/h. The steam railway was a big improvement at 30 km/h. Electric trains increased the average speed further, but door-to-door speed for trains is constrained by the journey to and from railway stations. The car does not have this disadvantage, and nowadays the average speed of the car in the Netherlands lies around 45 km/h (Verkeer en Waterstaat 2002). So, technological progress in combination with a willingness to use the new transport technologies resulted in faster transport. This in turn led to strong mobility growth. This strong link between travel speed and mobility growth will be further analyzed in the next section.

2.2.2 Faster and Further

One of the most intriguing features of mobility is the notion of constant travel time. People spend on average 1.1 h a day travelling. This constant does not hold for individuals, but is roughly valid when applied to large groups. Yacov Zahavi (e.g. 1974) is one of the pioneers of this concept of constant travel time and he used this as the basis for traffic forecast modelling. In the years since, his approach has been used by a variety of researchers in long-term mobility studies. The validity of this constant has also been disputed by scientists. A reasonable amount of empirical evidence, however, supports this thesis. In 1997 the Massachusetts Institute of Technology (MIT) presented an overview of the main empirical evidence, which is summarized in Fig. 2.2.

2 The Attractiveness of Car Use

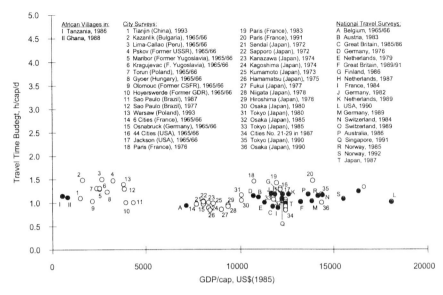

Fig. 2.2 Average travel time (hour/day) per person related to income level (Schäfer and Victor 2000)

Figure 2.2 gives an overview of 58 empirically found average travel times. The investigated studies range from African villages, via many cities, to the USA as a whole, with a related range in income level from close to zero to $18,000 US (1985) per person. The time span covered is from 1965 to 1993. All observed average travel times lie between 0.8 and 1.5 h/day per person. Surprisingly, this is completely independent of income level. So, the daily time spent on travel does not increase (or decrease) with income, and since income level certainly does influence the choice of transport technology, it is safe to assume that the average daily travel time does not depend on the used transport mode.

The notion of constant travel time has far-reaching consequences for understanding mobility growth. If the average time spent on travelling does not increase, growth in mobility per person can only be the result of faster mobility. The formula below reflects this relation.

$$\text{Total mobility} \left[\frac{pkm}{day}\right] = \text{Population size} \left[p\right] \times \text{Average travel time} \left[\frac{h}{day}\right] \times \text{Average speed} \left[\frac{km}{h}\right] \quad (2.1)$$

The above equation is self evident. So, growth in total mobility can only be caused by population growth and by faster mobility, where average travel time is constant.

This way of explaining mobility growth seems rather unconventional, but the usual factors explaining mobility growth determine, in the above formula, the growth in average speed. Thus, for example, income level and prices determine whether people can afford to use a fast transport mode. Also, it takes some years or decades before new transport possibilities are fully integrated into the daily life of a whole population. So, economic, demographic, social, and cultural factors determine the average speed and consequently the growth in mobility in the approach followed here.

Taking speed as the dominant factor in explaining mobility growth is especially suitable for investigating long-term developments, because changes in spatial behavior caused by faster travel require a long time period. Other factors, such as income growth, can change more rapidly and cause fluctuations around the long-term trends.

Another major consequence of constant average travel time is that better transport services do not, in the long run, lead to time savings. While faster transport, for example, a new motorway or a High Speed Train, may lead to short-run time savings, in the longer run people will adapt their choice of location of their activities. Faster transport allows for a longer commuting distance, making it possible for people to live in a rural village but work in the city. It also generates opportunities to travel greater distances for shopping, cultural activities, leisure and so on. It is evident that the spatial adaptation to faster transport takes years, and for some activities probably decades. A change of job or a move to another town are decisions with a long time span and transport possibilities are only one of the factors influencing these decisions. A change of destination for recreation or shopping can happen much sooner.

So, faster transport generates time savings in the short run and longer travel distances in the long run. This, among other things, is reflected in a sustained growth in average commuting distance in industrialized countries. The Netherlands, for instance, witnessed an annual growth of the commuting distance of 2% in the period 1995–2005 (Jorritsma et al. 2008). Another illustration is the first High Speed Train in France between Paris and Lyon, which generated substantial commuting between these two cities, something not feasible before with conventional train or car travel.

The temporary gain in travel time can explain part of the variation in travel time as represented by Fig. 2.2. Shortly after a comprehensive improvement of the transport system the average travel time will be lower. In the longer run travel times will increase again.

The presented overview of average travel times from MIT covers evidence up to 1993. Other studies confirm these findings (e.g. Ausubel et al. 1998). Also, more recent empirical results support the thesis of constant travel time. The presented overview of the MIT study has been updated with empirical data to 2004 (Schäfer et al. 2009). Data for the UK show that in the 30 years between 1979 and 2009 average travel time fluctuated a little around the same level, while in the same period the distance travelled increased by more than 50% (Department for Transport 2006 and 2010).

Finally, a recent study for the city of The Hague in the Netherlands concluded that in 2006 people on average travelled 66 min a day (Dienst Stedelijke Ontwikkeling Den Haag 2008). This happens to correspond exactly with the 1.1 h/day this section started with.

2.2.3 Limits to Public Transport and Urban Planning

The dominant role of travel speed in spatial developments and mobility generates unconventional insights into the effectiveness of policy instruments aimed at changing mobility patterns. Public transport and land use planning are often advocated, but their impact on mobility growth is limited, as will be discussed hereafter.

Speed is crucial in the competition between transport modes. The historic overview above reveals a continuous shift to faster modes. In this speed competition, public transport has the disadvantage that travelling to and from stations and stops reduces the average door-to-door speed. The speed competition between cars and public transport is convincingly shown in Fig. 2.3. The average trip by public transport in the Netherlands takes twice as long as the same trip by car (solid line). Only for a small percentage of trips is public transport faster than the car. If a trip takes twice as long by public transport as by car, only 20% percent of the population – mainly people without a car, or captives – is willing to take public transport (Fig. 2.3 dashed line). If public transport offers the same travel time as the car, its market share rises to around 50%. People apparently do not opt for car use as a matter of principle but make a rather rational choice in which trip time is important. Because public transport can offer a travel time comparable to that of the car only for a small percentage of trips the aggregate share of public transport is limited to somewhere around 10 to 15%. This is an average for entire countries; for large cities it can be higher, as will be discussed hereafter.

Public transport could increase its market share by offering shorter travel times, but this is rather costly and would require an even higher level of public funding.

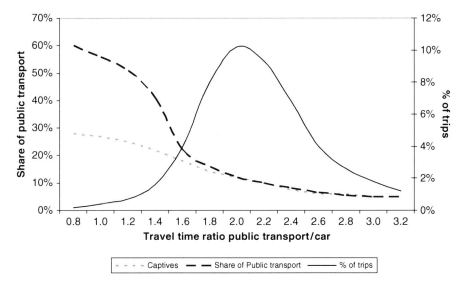

Fig. 2.3 Share of public transport (*dashed*) and percentage of trips (*solid*) related to the travel time ratio public transport/car (van den Heuvel 1997)

A more promising strategy for public transport is to focus on areas where the car is relatively slow, mainly in large urban areas.

Starting from the notion that door-to-door travel time is a crucial element in mode choice, it is also evident that an increase in the average speed of the car will lower the market share of public transport. As a consequence, any transport policy aimed at increasing the share of public transport needs to be consistent with the chosen approach for car travel, for example, road capacity, pricing and parking policy.

The notion of constant travel time also limits the impact of urban planning on mobility. It is often argued that 'good' urban planning is an effective policy for reducing mobility growth and car use. Sometimes, 'bad' urban planning is even blamed for current congestion and the dominance of the car. At first sight urban planning seems effective, because planners and authorities determine locations for houses, offices and industries. By locating houses and work places close to each other, commuting distances are expected to be short. However, this line of argument does not take into account the crucial difference between the spatial structure on the one hand and the spatial behavior of people and firms within this structure on the other hand. If houses and work places are located close to each other, this does not mean that the houses will be occupied by people working nearby. People might commute to other towns for their work and out-of-towners might occupy the work places. That commuters travel both ways in urban areas can be illustrated with data for the Metropolitan Area Amsterdam (Gemeente Amsterdam 2008). The number of jobs in this region is more than one million and although the labor force is insufficient to occupy all jobs, resulting in a net inflow of almost 100,000 commuters, at the same time more than 10% of the labor force has a job outside the Metropolitan Area Amsterdam. They work mainly in the areas around the nearby cities of Utrecht and The Hague, while many other people living in those cities commute to Amsterdam. So, commuting between the cities in the Randstad goes both ways.

This is consistent with the thesis of constant travel time. Faster travel will lead to longer distances for commuting and other activities and therefore has a strong impact on spatial behavior. Longer trip distances even accounted for 70% of the growth in car use per person in the period 2000–2008 in the Netherlands (Olde Kalter et al. 2010). The spatial structure on the other hand only has a small influence on mobility behavior. So, urban planners do determine to a large extent the spatial structure, but this does not fix spatial behavior and mobility growth.

Although the impact of urban planning on total mobility is small, building compact cities is relevant for achieving sustainable transport. In large cities with a high density of the built environment a greater variety of activities can be undertaken within a short distance than in small towns or rural villages. This is what makes urban areas attractive for people and firms. The combination of shorter distances and more attractive places to go results, on average, in more activities per day. The extra activities are a welfare gain for citizens. Total travel time, however, is roughly the same in cities as in rural areas (Maat 2009). The extra trips and the somewhat lower speeds are balanced by the shorter distances in urban areas. Total mileage per person is therefore a little lower.

Furthermore, dense urban areas offer better opportunities for public transport. The average speed of the car is lower in urban areas, making it easier for public transport to compete on speed with the car. In addition, the high volume of potential passengers in large cities makes it affordable to offer public transport of good quality, regarding number of stops, speed and frequency. In particular, a higher density of work locations leads to fewer car kilometers and a higher use of public transport (Maat 2009). The name 'mass transit' reflects adequately the role of subways and trains in urban areas. Without mass transit it would be hard to make our large urban areas accessible and prosperous. This can be illustrated by the high share of public transport in big cities such as New York (56%), London (37%) and Berlin (27%) (Urban Age City Data 2009).

Given the complexity of the impact of urban density on mobility, it is hard to estimate the potential reduction in transport-related CO_2 emissions that would result from building more compact cities. For the Netherlands, for example, the best professional guess is that building more compact cities than currently planned, could reduce CO_2 from mobility by around 5%. This contribution is modest because of the already densely-built Dutch environment. In European countries with a less densely built environment, the potential for CO_2 reduction of building compact cities might lie around 10%.

2.2.4 Future Mobility: The Sky is the Limit

The history of transport can be summarized as a continuous decline in the 'friction of distance'. Travelling became faster through successive improvements in transport technology and their use, as discussed earlier in this paragraph. What does this driving force imply for the future of mobility? It is safe to assume that average speeds will continue to rise. Aviation and, perhaps, high speed trains will gain market share and allow us to travel further in the 1.1 h/day we spend on mobility, while car use in industrialized countries will reach its level of saturation in the coming decades.

The earlier-mentioned MIT study used the thesis of constant travel time to forecast worldwide mobility in 2050 (Schäfer and Victor 2000). In addition, a constant money budget for travel is assumed, so expenditure for travel will increase at the same rate as income growth. This constant money budget is also based on empirical evidence and lies around 11–14% of personal expenditure. Combined with assumptions for population and income growth, the average speed of modes, and price changes of transport, travel forecasts were made for 11 world regions separately. The resulting forecast for Western Europe indicated an annual growth rate of 5.3% for air travel and High Speed Train combined, which seems rather high according to current views. Aircraft builder Boeing (2010) expects an annual growth of airline traffic (passenger kilometers) in Europe of 4.4% in the 20 years to 2029. By lowering the annual growth of high speed modes from 5.3 to 4.4% and allocating the reduced time spent in high speed modes to car travel, a revised forecast is calculated and presented in Fig. 2.4.

According to the presented projection, total mobility in Western Europe is expected to increase further, from 40 km in 2000 to over 70 km/person per day

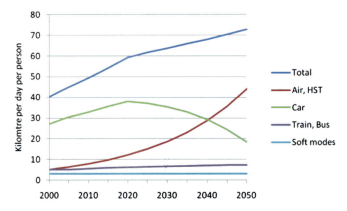

Fig. 2.4 Mobility forecast in kilometers per person per day for Western Europe. HST: high-speed train

in 2050. The average growth rate of 1.2% per year is, however, much smaller than the historic growth rate of 4% in the past two centuries. Taking speed as the main driving force for mobility growth, it is not surprising that fast modes are growing fast. According to this forecast aviation will take over the dominant position from the car shortly after 2040.

The most surprising outcome of the presented forecast is that car mileage will reach a maximum of around 40 km/day per person in Western Europe shortly after the year 2020. After this, car use per person will decline. This forecast conflicts with conventional wisdom. Most projections indicate a further increase in car use. For the EU-27 a further growth in car use per person of 10% is projected for the period 2020–2030 (Capros et al. 2008). However, the study for the EU-27 foresees an annual growth in aviation of only 3%, which seems too low. The Netherlands expects an annual increase in car use per person of 0.7–1.1% in the period 2005–2030 (Harms et al. 2010). This expected growth is, however, only half the level of the preceding 25 years and will mainly take place before 2020. The forecast presented in Fig. 2.4 indicates a 0.6% annual increase in the same 25 years and is therefore close to the lower bound of the Dutch projection.

Although, according to conventional wisdom, we should expect a sustained growth in car use, the thesis of constant travel time in combination with the actual and projected strong growth in aviation are reasons to give it a second thought. The increasing amount of time people will spend using air travel will no longer be available for car use. It is expected that the daily time spend in car travel in the USA will decrease from 60 min in 2005 to somewhere between 44 and 51 min in 2050 (Schäfer et al. 2009). Most of the business-as-usual forecasts for car use do not incorporate aviation as a transport mode, which makes these studies rather unrealistic for long time horizons. So, the future of car use might be different from that commonly expected.

Another argument in favor of the expectation that growth in car use per person will decrease is that the average speed of cars is no longer increasing. The building of motorways – mainly in the 1960s and 1970s – and their intensive use allowed car speeds in the Netherlands to rise from an average of 35 km/h in 1960 to 45 km/h (Verkeer en Waterstaat 2002). It is not likely that average car speeds will rise much further in Western Europe. Only a few new motorways will be built and the capacity increase of existing roads will be insufficient to eliminate congestion. The car system is close to achieving its maximum potential in industrialized countries. And, as discussed before, faster travel is the engine for mobility growth. So, if car travel is no longer becoming faster, the growth in car mileage per person will gradually decrease to zero after a few decades. If the full 1.1 h a day were used for car driving and the average car speed did not increase above 50 km/h, total car mobility would reach a limit of 55 km/person per day. Figure 2.5 below shows that the increase in average speed since 1985 is a result only of a greater use of aviation for passenger travel. The average speed of all land modes taken together is not increasing. This indicates that the growth in land-based travel modes will come to an end in the near future, at least in the urban areas of industrialized countries.

The projected mobility growth to 2050, as presented above in Fig. 2.4, combined aviation and High Speed Trains as 'fast modes'. However, it seems unlikely that HST will achieve a large market share in long-distance travel. The construction costs of high speed rail links (HSR) are high and are only economically justified when used by many passengers. Emerging economic insights indicate that the travel time between cities connected by HST should be around 3 h maximum and that the new rail service should attract at least 9 million passengers in the first year. In addition, building new high speed rail links seems only economically viable if the capacity of conventional rail lines is insufficient and needs to be expanded (Nash 2009). These conditions limit the potential for building new economically feasible HSR links.

Furthermore, building HSR links does not substantially contribute to the reduction of green house gasses from transport (Kågeson 2009). The net environmental

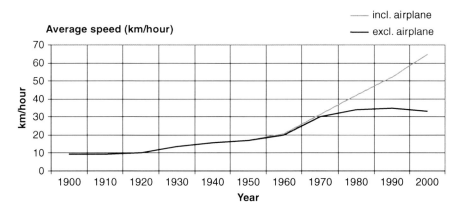

Fig. 2.5 Average door-to-door speed of passenger travel, with and without aviation (Dutch population; Verkeer en Waterstaat 2002)

impact of HST results from a modal shift from aviation and car travel, which reduces emissions, and a shift from conventional rail and extra travellers, which increases emissions. These opposing effects are in the same order of magnitude, so building new high speed rail links is not the answer to the challenge of reducing CO_2 emissions from transport.

2.2.5 ICT for Mobility

The previous section explained the strong mobility growth of the last centuries as one of the results of the Industrial Revolution. New technologies made it feasible to travel at higher speeds and thus cover larger distances in the same amount of time. Increasing speed has been the main driving force behind the growth in mobility, with the car becoming the dominant transport mode because its high door-to-door speed made it faster than other modes of passenger travel. The fruits of the industrial revolution for travelling are not fully incorporated into our society yet. Car use is still growing, though at a declining rate, mainly as a result of the higher average speed generated by the building of motorway networks. Aviation still has a long way to go before saturation is reached.

Now we witness the emergence of a new so-called General Purpose Technology: Information and Communication Technologies (ICT). Within a few decades major technological inventions took place: the personal computer (1981), the Internet (1983), and wireless communication technologies (GSM 1993, Wi-Fi 1996, UMTS 2004). The invention of the computer chip in 1958 was the backbone of these innovations. New information and communication technologies are growing into our society and will continue to do so, including in transport. Intelligent cars, smart traffic lights, traffic forecasts and multimodal travel assistants will improve our mobility: safer, cleaner and more predictable. The main innovations during the coming decades in the transport system will most likely be driven by ICT rather than by the technologies of the Industrial Revolution.

Some argue that the ability to communicate without travelling, through ICT use, will reduce mobility and hence might reduce the amount of time we spend on travelling. Others argue that ICT will allow us to perform other tasks while we are travelling such as working, reading and communicating, and that this might increase the time we spend travelling. Looking at current empirical evidence, it does not seem likely that the average 1.1 h a day for travelling is changing much as a result of the ICT revolution.

2.3 Urbanization

2.3.1 The Attractiveness of Cities

The previous section identified the increasing speed of travel as the main driving force behind mobility growth. Faster travel creates the opportunity to reach distant places in the limited time available for travelling, thus improving accessibility. In addition to speed, urbanization is another successful way to improve accessibility. By concentrating locations for housing, work and leisure in cities, trip distances and travel times will be shortened. So, better accessibility can be achieved in two different and complementary ways: short distances and high speed, as shown in the following formula.

$$\text{Acessibility} \left[\frac{1}{\text{hour}}\right] = \frac{\text{Speed} \left[\frac{\text{km}}{\text{h}}\right]}{\text{Distance [km]}} \quad (2.2)$$

Both increasing speed and urbanization are strong historic trends that are expected to continue in the future. It is obvious that urbanization forms a challenge for mobility. If we want our homes, work and recreation close to each other this will generate a lot of traffic in a limited area. This leads to lower average travel speeds in cities compared with interurban travel.

Urbanization started around 10,000 BC and has continued since then. The beginning of agriculture increased yields and allowed people to live together in somewhat higher densities. Villages and towns came into existence, though the size of ancient and medieval cities remained rather small, mostly, at maximum, 50,000 inhabitants and 20 km^2 (Rodrigue et al. 2009). The limited means of transport and trade, together with the very slow growth of agricultural productivity, were important factors preventing the growth of cities. In 1520 only 3% of the European population lived in cities and this share gradually increased to 18% by 1750 (de Vries 1984).

The Industrial Revolution boosted the process of urbanization. Industrial production required the concentration of labor and materials in factories. At the same time motorized transport increased travel speed and loading capacity. This allowed for longer distances to be covered and for heavier loads to be transported. As a consequence urbanization speeded up and by 1950 more than half of the European population lived in cities. This process of urbanization still continues and it is expected that by 2050 close to 85% of all Europeans will live in a city (Fig. 2.6). For the United Kingdom the estimate is as high as 94%. So, the urban Homo sapiens has become a dominant species.

Why do people and firms cluster together in cities and large urban areas? What makes living in high densities so attractive, despite the related congestion, nuisance and environmental problems? The main benefits lie in greater accessibility. Firms and people profit from the nearness of other firms and people. A large market for consumer goods generates economies of scale in production and distribution, resulting in lower prices. A large demand also generates a higher quality and variety of services, such as theatres, leisure and shopping centers, while a large labor market allows for specialization and a better match between firms and employees. Another important benefit of a greater concentration of population is that knowledge and new ideas spread more easily. These and similar benefits of proximity are often summarized as agglomeration economies. Cities are drivers of productivity growth and innovation mainly takes place in urban areas. So, cities are one of the cornerstones of past and present economic successes (ter Weel et al. 2010). People and firms want to be part of that.

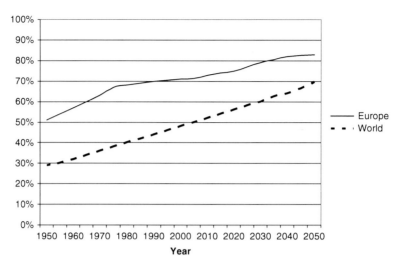

Fig. 2.6 Share of urban population 1950–2050, World and Europe (UNPD 2008)

The growth of the city is closely linked to improvements in transport. Through history few cities exceeded a size corresponding with 1 h of travel. At a walking speed of 5 km/h city size was limited to around 20 km^2. The largest cities prior to the Industrial Revolution, such as Rome, Beijing, or Constantinople never surpassed this size (Rodrigue et al. 2009). The impact of faster transport on city size is demonstrated for the city of Berlin (Marchetti 1994). In 1800 Berlin was a compact city with a diameter of 5 km, consistent with the speed of walking. After the subsequent introduction of faster means of transport – horse tramway, bus, electric tram, subway and, finally, the car – the city grew to its current size with a diameter of 40 km. This is consistent with average car speeds in larger cities.

Another way to illustrate the impact of the transport infrastructure on urban development is to look at the sequence of infrastructure building and the expansion of cities, as shown in Fig. 2.7. The first railway stations were built outside the old towns. This improved accessibility and generated urban development in the zone around the station. As a consequence of this development most central stations are now located within the city, just outside the old town centre. This process was repeated almost a century later when ring motorways were built. These improved the accessibility of the then outskirts of cities and stimulated the development of new sub centers, because these were more accessible for cars than the old centre. These ring roads now belong to the most congested part of the motorway network.

The growth of cities is both facilitated and limited by the transport system. Urban areas are the paradox of accessibility. Good, accessible locations attract people and firms, thus increasing urban density in those areas. This in turn leads to more urban traffic and, mostly, to congestion, which reduces accessibility, and this limits the further growth of the city. In this way, congestion plays a balancing role in the concentration and dispersal of spatial developments. Congestion, together with rising

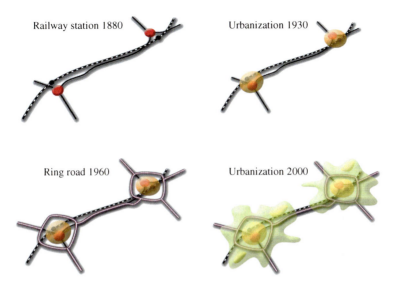

Fig. 2.7 Sequence in new infrastructures and urban development. Railway stations were built just outside the old towns followed by city growth around the stations. A century later ring roads were built just outside the cities and generated urban growth along the ring roads (Kwantes and Govers 2007)

costs in urban concentrations, puts a limit to the growth of cities. Rents for land, offices and houses, as well as wages, are higher in urban areas, especially in city centers. Without these counter forces of higher costs and congestion, cities would grow indefinitely. But, cities have their limits.

2.3.2 Urban Mobility

The average travel speed in urban areas is lower than for interurban traffic. Cars, busses and tramways use the same, often congested, roads. And public transport has frequent stops. The average travel speed in large cities is rather low, for instance 25 km/h in London and 18 km/h in Paris (Verkeer en Waterstaat 2002). In contrast, the average travel speed in the Randstad area, with its much lower density, lies around 35 km/h. Cities in the USA even reach an average travel speed of 40 km/h.

The lower travel speed in metropolitan areas, in combination with the notion of constant travel time, lead to a somewhat reduced growth in mobility measured in passenger kilometers. This does not mean, however, that accessibility in urban areas is worse than in rural areas. As discussed before, accessibility can be achieved both by short distances and by high travel speeds. In large cities the advantage of nearness outweighs the disadvantage of congestion, but, of course, only until congestion gets so bad that locations outside the city are preferable.

Next, urban areas offer good opportunities for mass transit. The high density generates a large volume of passengers, required for an economically viable exploitation of public transport services. And because car speeds are rather low, it is easier for mass transit, such as trains and subways, to offer competitive travel times in large cities (see also Section 2.2.3).

Another major consequence of the described interaction between urbanization and transport is that it is nearly impossible to solve congestion in metropolitan areas. Building new urban roads will in the short run reduce congestion and increase car speeds. In the longer run, however, this will generate both longer trip distances and a higher urban density. The longer distances follow from higher speed combined with the thesis of constant travel time, and the additional urban growth results from the improved accessibility, as discussed in the previous section. These two mechanisms lead to extra urban traffic, which reduces the initial gain in average speed achieved by additional road capacity. This phenomenon is called induced traffic and has been studied since the late 1980s. Calculations using the Dutch national traffic model give an impression of the relevance of this negative feedback loop in urban transport. A politically desired increase in the required speed on motorways during rush hours from 67 to 80 km/h, would generate 1.4 to 2.3% extra car kilometers in 2028 on motorways (Snelder et al. 2010). And building a second motorway between Rotterdam and The Hague is expected to generate 25% extra car trips between these two cities. Almost half of the new road capacity is needed for this extra traffic, leaving only the other half for reducing congestion and coping with general traffic growth.

So, congestion will most likely remain part of urban life. This can be witnessed in all big cities in the world. As discussed before, congestion puts a limit on further urbanization and creates a balance between spatial concentration and dispersal of our activities.

2.4 Toward Sustainable Mobility

Speed is the fundamental driver behind the popularity of the car. The favorable door-to-door speed of the car makes it by far the most attractive means of passenger travel in industrialized countries. The comparable high average speed determines to a large extent the high level of mobility achieved. Urbanization is a second major trend influencing mobility patterns and growth. People and firms like to cluster in urban areas to profit from the benefits of proximity. But urbanization also creates congestion, which puts a limit on city size and also to car use. It is close to impossible to have prosperous urban areas without congestion, because increasing car speed will lead to extra traffic.

What do the combined trends of faster travel and ongoing urbanization imply for policies aimed at achieving sustainable mobility? Some main lines will be sketched in this last section. These will be discussed in more detail – and sometimes disputed – in following chapters of this book.

2.4.1 Mobility Growth

Following from the insight that increasing travel speed is the driving force behind mobility growth, reducing the volume of car use can be achieved by reducing the average speed of cars. So, at least in theory, it is possible to reduce the volume of car traffic, if this is needed in order to achieve the required reduction in greenhouse gas emissions from transport. Several policy instruments are available to reduce the speed of the car, e.g., imposing tight speed limits and not expanding road capacity. In addition, pricing car use – and other fast transport modes – will encourage a shift to slower modes and thus reduce mobility growth. Although these policies are effective in reducing car use, they face two problems.

The first is the lack of public and political support for pricing policies, accepting congestion and tight speed limits. Of course, public support might rise, but 20 years of experience in the Netherlands with plans for different forms of road pricing indicates this is not an easy solution. On the contrary, policies facilitating car use are popular. Combating congestion and thus increasing car speed is a political priority in many urban areas.

The second problem with these policies aiming to effectively reduce car use is more fundamental. A lower car speed also reduces the economic and social benefits of car use. The urbanization process, with its economic benefits, will be slowed down or even reversed. And the benefits of covering larger distances in the 1.1 h travelling time per day will be reduced. The friction of distance will increase and halt further growth in the distance between our activities. This raises questions about finding the right balance. What is the best size of cities, taking into account both the benefits and the costs of transport? How far apart do we want the activities that we undertake to be, looking at the benefits of distance and at the social costs of transport? The compass for urban transport policy, aimed at achieving the right balance, will be discussed in the next section.

Before turning to this search for the right balance, a popular political response with respect to mobility growth is pointed out. Some politicians state that a reduction in the growth of car use is needed to combat climate change, and maybe for other reasons too. But these politicians generally do not favor the necessary effective – but unattractive – policies as sketched above. A usual way out of this dilemma is to create illusions surrounding the effectiveness of policy measures. This has resulted in persistent misunderstandings in the debate on transport policy. Subsidies for public transport, investment in high speed rail links and better urban planning can be useful in specific circumstances, but they will not have a substantial impact on the growth of mobility and on car use. Another popular misunderstanding is that building new roads will solve congestion in metropolitan areas. The lack of effectiveness of such alternative transport policies is illustrated by a review of past projections (de Jong and Annema 2010). Mobility growth and modal split have quickly followed the business-as-usual scenarios, alternative scenarios have remained largely wishful thinking, and congestion has increased more than most studies predicted, despite the aim to reduce it. In order to achieve realistic transport policies it is necessary to face the real dilemma: either accept current trends in mobility growth, car use, and congestion, or have the courage to implement unpopular measures. Wishful thinking has its appeal, but does not help achieve political goals in the real world.

2.4.2 Compass for Urban Transport

A major focus in urban transport policy lies currently on reducing congestion. Congestion in urban areas is a daily nuisance for car drivers, truckers and firms. In many countries the economic loss caused by congestion is calculated and published each year. Congestion costs in the EU are estimated at around 1% of GDP (EC 2006). Promises to combat congestion make politicians popular with both entrepreneurs and many car drivers. However, the analyses presented in this chapter show that it is practically impossible to get rid of congestion in large urban areas. Every gain will be followed by extra urban traffic that reduces the expected gain. In addition, building new infrastructure becomes more expensive with increasing urban density, as a result of the lack of space and required environmental measures. For instance, building new motorway capacity in the Randstad area costs almost twice as much as in the rural parts of the Netherlands. So, a certain level of congestion is part of every prosperous urban area and forms one of the balancing mechanisms between concentration and scattering of human activities. In fact, one should worry if there is no congestion, because in that case the urban economy is probably in bad shape. So, combating congestion is not the right compass for urban transport policy. Is a better compass available?

For the achievement of the ideal size for our cities and the best spatial range of our activities, economic principles form the suitable compass. These help us find the right balance between the economic benefits of cities and the costs of keeping the traffic flowing. The two cornerstones of this compass are charging prices for transport services that reflect the marginal social costs and a proper social Cost Benefit Analysis for infrastructure investments. Book shelves of studies have been published about the benefits of using these two standard economic instruments for transport policy, and several authors of this book have elaborated this economic approach.

In summary, user charges for cars need to be increased, while in several countries fixed taxes should be reduced. User charges for trucks and vans need to be increased substantially. New roads in congested urban areas are only economically justified if the road users are willing to cover the full infrastructure costs for building and maintenance (Dings et al. 2002). If extra motorway capacity is built that is needed only during rush hours, the costs for building and maintaining the road can amount to as much as around €0.20 to €0.25 per car kilometer in densely populated areas. A congestion charge at this level is economically justified, reducing congestion and generating funds for building new infrastructures with a positive socio-economic rate of return.

Using these two economic principles as the compass for urban transport policy would strongly contribute to the vitality and competitiveness of urban areas, and would, furthermore, lower somewhat the volume of car travel. For the EU it is estimated that a 2 to 3% reduction in car use would result from economic pricing (van Essen et al. 2008). The Netherlands expected a stronger reduction of 11% in car use from their earlier intended pricing policy (MuConsult 2009).

When taking a closer look at congestion, it is necessary to distinguish between the time loss caused by traffic jams and the unpredictability of the trip time. The variable journey time caused by congestion seems to be a greater nuisance than the average time loss itself. If we know that a certain trip always takes, say, 10 min longer during rush hour than during quiet hours, we can adapt our travel behavior. But we cannot anticipate unexpected delays and these are the main reason for complaints about congestion. The predictability of journey times needs to become part of the economic compass for urban transport. It is expected that the economic loss due to unpredictable travel times will be higher than the time loss alone and will triple in the Randstad in the period 2008–2030 (Snelder et al. 2008). Transport policy in congested cities needs to focus on improving the predictability of trip times, instead of on reducing congestion in general. Positive results can be achieved, because predictability is not subject to the negative feedback loops caused by the constant travel time and enhanced urbanization, while time loss is.

This is not the appropriate place to discuss urban transport policy at length. Only a few main lines will be touched on, following from the sketched compass for urban transport. Compact cities have greater accessibility than a dispersed built environment and will emit somewhat less transport related CO_2. Mass transit is needed in addition to car traffic, in order to keep large urban areas accessible. Mass transit can be economically viable in dense urban areas because of the large volume of passengers and the relatively low speed of the car. Biking forms an attractive contribution to the accessibility of cities. In addition, the robustness of road and rail networks need to be improved, to prevent congestion spreading as an oil splash over large areas of the networks. And the utilization of the scarce infrastructure capacity needs to be improved by congestion pricing and smart traffic management, supported by new applications of ICT in mobility. Both national governments and city authorities have effective policy instruments at their disposal to optimize urban transport. Parking policy, providing good mass transit, and biking facilities, are the main instruments at the local level.

Using the outlined compass for urban transport policy – instead of an oversimplified focus on combating congestion – will lead to better accessibility for urban areas and a somewhat lower growth in car traffic. This will, however, only result in a modest reduction in the growth of CO_2 emissions from transport. The next section will discuss clean technology as a more promising route.

2.4.3 Clean Technology

Cars can become much more fuel efficient and low carbon fuels are available to reduce CO_2 emissions further. The efficiency of the global car fleet can be improved by 50% by the year 2050, according to four international organizations (IEA et al. 2009). Summarizing international research they conclude: 'The technologies required to improve the efficiency of new cars 30% by 2020 and 50% by 2030, mainly involve incremental change to conventional internal combustion engines and drive systems, along with weight reduction and better aerodynamics. To achieve a 50% improvement by 2030, the main additional measures would be full hybridization of a much wider range of vehicles'. So, deployment of currently available technologies could achieve a drastic reduction in CO_2 emissions from cars.

In line with these promises of clean technology the International Energy Agency developed a scenario for a reduction of global greenhouse gas emission from transport in 2050 to 40% below 2005 levels (IEA 2009). This reduction can be achieved through efficiency and the use of alternative fuels, despite the projected growth in transport volume. Chapter 3 of this book will further elaborate on the technical developments and on scenarios for transport-related greenhouse gas emissions.

Although technology can substantially reduce the production of greenhouse gasses from transport, the main question is whether these technological promises will become reality. From past experiences it is clear that powerful policies are needed to get clean technologies indeed on the road. The following chapters of this book will discuss at length policies at the EU and national level. Three types of policy instruments have thus far proven to be effective:

- Setting and tightening *standards* for the specific fuel consumption or CO_2 emissions from vehicles. This is an important element of current EU policy and by 1975 had already been introduced in the USA as standard for the Corporate Average Fuel Economy (CAFE).
- *Fuel taxes* increase the fuel price, which generates an incentive to buy fuel efficient cars. National governments can set the level of fuel taxes to anywhere above the European minimum level, but due to tank tourism, countries can in practice not raise taxes much higher than their neighbors. International cooperation is thus needed to increase fuel taxes.
- *Fiscal and financial incentives* to stimulate the purchase of fuel efficient cars. National governments can use sales taxes, vehicle taxes and the tax regime for company cars to stimulate fuel efficiency. Many European countries have successfully done so.

It is evident that public and political support is needed to implement these policies in the firm way required to realize the technical promise of cars that are 50% more fuel efficient. The technology is available and the policy instruments are available too. Now, the political will is needed to implement these policy instruments. Policies to drastically improve the fuel efficiency of cars will probably gain public and political support with greater ease than policies aimed at reducing car use.

2.4.4 Aviation

The trend toward ever faster travel will lead to an increasing market share for aviation. Current annual growth rates in air travel are higher than in car use, at least in Europe and other industrialized countries. According to the projection presented before in Fig. 2.4, shortly after the year 2040 we will travel more kilometers by plane than by car. This seems hard to imagine at this moment. However, it is important to recall that the question is not which of our current trips are suited for aviation. Instead, we have to look at the share of our 1.1 h/day that we will spend flying. In 2040 Europeans will on average fly only around 4% of their available travel time, corresponding with 17 h/year. This seems realistic. The implication is that we shift from shorter trips by car to long distance travel by airplane. So, people will visit distant places more frequently. This is consistent with current trends in holiday travel and business trips. Social trips and commuting will probably also take their share of air travel.

It is hard to predict how air travel services will look 40 years from now; the current market is very dynamic. Low-cost carriers are fiercely competing with traditional airline companies. Which balance will emerge between hub-and-spoke operations on the one hand and direct point-to-point services on the other hand? Will air taxies, using small aircrafts or helicopters, have a commercial future, as several start ups indicate? Will small, regional airports have a healthy economic future or will large airports keep their dominant position? Will the private jet become popular among the rich? Such questions cannot be answered yet. It is clear, however, that the expected strong growth in air travel will bring changes in the industry and in the air travel services they provide. It is time that policy makers realize that aviation will become the dominant mode of travelling in a few decades and that these developments need to be guided in an economically, spatially and environmentally balanced way.

The strong growth in aviation will cause additional greenhouse gas emissions. Because emissions per passenger kilometer of air travel are higher than for cars, total CO_2 emissions from European aviation could become larger than from car driving. This creates a new challenge for climate change policy. Most likely a similar combination of policy instruments will be effective to reduce the emissions from aviation as has been developed for cars. Proper pricing of aviation in combination with standards for the fuel efficiency of aircraft and the carbon content of jet fuels will reduce the growth of emissions. It is expected that the well-to-wheel CO_2 emissions from air travel per passenger kilometer can be halved by 2050, due to technical improvements (IEA 2009). And again, this will only happen if firm policy measures are taken soon.

2.5 Conclusions

This chapter ends with the main conclusions related to the required strong reduction in greenhouse gasses from car use. The focus lies on the industrialized countries of Europe, but a few recommendations are added for emerging economies such as Brazil, Russia, India, China and many other countries.

Although the volume of car traffic can be reduced in theory, in practice this is not likely to happen for both economical and political reasons. Increasing speed is a strong driving force behind the growth in mobility and car use. This trend will not easily be reversed. Following conventional wisdom, CO_2 emissions from cars in Europe will increase by a quarter in the period 2010–2030. Setting the right prices for car use is economically beneficial and could reduce CO_2 from cars by several percentage points if implemented properly. And building compact cities instead of urban sprawl can reduce transport related CO_2 emissions by another 5–10%. The related urban transport policy should follow economic principles and focus at better predictability of trip times, instead of the over-simplified approach of combating congestion. Economically justified prices and infrastructure building in combination with compact cities, could limit the projected increase in CO_2 from car use to somewhere around 10%, but only if the required policies are implemented.

After 2030 the growth in European car traffic will gradually stabilize and maybe even decline. The car system has matured and shaped the current spatial behavior of people and firms. Aviation will probably take over the dominant role of the car before 2050. A small but growing part of the available daily travel time will be used flying, at the expense of the time used for car driving. The continued growth in aviation asks for stringent policy measures to reduce its impact on climate change. Emission standards and proper pricing are the key elements of an effective climate policy for aviation, just as for cars.

The largest contribution to a strong reduction in the emissions of greenhouse gasses from cars can be achieved by clean technology. The combination of very fuel efficient cars with low carbon fuels has the potential to reduce CO_2 per car kilometer around 50% by 2050. No new technological breakthroughs are required to achieve this tremendous gain. However, the potential technological progress will only become reality if strict policy measures are introduced. The needed policy instruments are available – standards and economic incentives. So, achieving a strong reduction in CO_2 emissions from cars is mainly a matter of the political will to turn the knobs. Such stringent policies will also stimulate further innovations in fuel efficiency and low carbon fuels. These will allow for additional reductions in CO_2 from cars after 2050.

Setting tight standards and creating financial incentives for low carbon propulsion will also counterbalance the current market trend toward heavier and more powerful cars. Some downsizing of the vehicle fleet might even be the outcome, which could add a CO_2 reduction of around 10% in addition to that resulting from technical improvements.

Information and communication technology (ICT) can also contribute to a reduction in CO_2 from cars by stimulating a fuel efficient driving style with indicators in the car advising the driver to shift gears or to change driving style to reduce fuel consumption. This so-called eco-driving could reduce CO_2 per kilometer by around 15% (Klunder et al. 2009).

The conclusions about achieving sustainable mobility presented above apply to industrialized countries, with a focus on Europe. But the world-wide growth in car use will mainly take place in emerging economies such as China, India, Brazil, Russia, and many others (see next chapter). Because mass motorization is still in an early phase, these countries have more scope to guide developments in a sustainable way from the outset. Many emerging economies have introduced CO_2 standards for cars comparable to or even tighter than in Europe. This stimulates a fuel efficient car fleet, reduces oil dependency and lowers the strong growth in CO_2 emissions from transport.

Another easy-to-implement measure is a gradual increase in fuel taxes. This creates an extra incentive for fuel efficient cars and is a first step toward proper economic prices for car driving. Higher fuel taxes generate state income, which can be balanced by lower taxes on labor and capital. From a macro-economic point of view it is better for most emerging economies to make car users pay fuel taxes to the state, than to have a higher national bill for oil imports.

Not only does car use grow fastest in emerging economies, urbanization also takes place rapidly and many mega-cities will be built in the coming decades. Now is the time to guide this urbanization process in the direction of compact cities, and transport policy is one of the main keys to steering urban development, as has been emphasized before in this chapter.

References

Ausubel JH, Marchetti C, Meyer P (1998) Towards green mobility: the evolution of transport. Eur Rev 6(2):137–156

Boeing (2010) Current market outlook 2010–2029. Boeing

Capros P, Mantzos L, Papandreou V, Tasios N (2008) European energy and transport – trends to 2030 – update 2007. European Commission, Brussels

de Jong M, Annema JA (2010) De geschiedenis van de toekomst – Verkeer- en vervoerscenario's geanalyseerd [History of the future – analysis of traffic and transport scenarios]. Kennisinstituut voor Mobiliteitsbeleid. Den Haag

Department for Transport United Kingdom. National Travel Surveys: 2005 and 2009. 2006 and 2010

de Vries J (1984) European urbanization, 1500–1800. MIT Press, Cambridge, MA

Dienst Stedelijke Ontwikkeling Den Haag (2008) Mobiliteit van de Hagenaar, 2006 [Mobility of the population of The Hague 2006]

Dings JMW, Leurs BA, Hof AF, Bakker DM, Mejjer PH, Verhoef ET (2002) Returns on roads – optimizing road investments and use with the 'user pays principle'. CE Delft and 4Cast

EC (2006) Keep Europe moving – sustainable mobility for our continent – Mid-term review of the European Commission's 2001 Transport White Paper. Commission of the European Communities. COM (2006) 314 final

Gemeente Amsterdam (2008) Metropoolregio Amsterdam in beeld 2007 [Image of the Metropolitan Area Amsterdam]

Grübler A (1999) The rise and fall of infrastructures – dynamics of evolution and technological change in transport. International Institute for Applied Systems Analysis and Arnulf Grübler (originally published in 1990)

Harms L, Olde Kalter MJ, Jorritsma P (2010) Krimp en Mobiliteit – Gevolgen van demografische veranderingen voor mobiliteit [Shrinkage and mobility – impact of demographic changes for mobility]. Kennisinstituut voor Mobiliteitsbeleid. Den Haag

IEA (2009) Transport, energy and CO_2 – moving towards sustainability. International Energy Agency, Paris

IEA et al (2009) 50 by 50 report – global fuel economy initiative. International Energy Agency, International Transport Forum, United Nations Environment Programme and FIA Foundation, Paris

Jorritsma P, Berveling J, Harms L, Kolkman J, Koopmans C, Lijesen M, van der Loop H, Olde Kalter MJ, van Ooststroom H, Visser J, Warffemius P (2008) Mobiliteitsbalans 2008 [Mobility review 2008]. Kennisinstituut voor Mobiliteitsbeleid

Kågeson P (2009) Environmental aspects of inter-city passenger transport. OECD/ITF Discussion Paper No. 2009-28

Klunder GA, Malone K, Mak J, Wilmink IR, Schirokoff A, Sihvola N, Holmén C, Berger A, de Lange R, Roeterdink W, Kosmatopoulos E (2009) Impact of information and communication technologies on energy efficiency in road transport. European Communities

Kwantes C, Govers B (2007) Trends: opeenvolgende levenscycli van vervoerssystemen [Trends: successive life cycles of transport systems]. Bureau Goudappel Coffeng

Maat K (2009) Built environment and car travel – analyses of interdependencies. Thesis. TU Delft

Marchetti C (1994) Anthropological invariants in travel behaviour. Technol Forecasting Social Change 47:57–88

MuConsult (2009) Effecten milieudifferentiatie basistarieven kilometerprijs [Effects of differentiated kilometre charges]. MuConsult

Nash C (2009) When to invest in high-speed rail links and networks? OECD/ITF Discussion Paper No. 2009-16

OECD/ITF (2010) Transport outlook 2010 – the potential for innovation

Olde Kalter MJ, Loop H van der, Harms L (2010) Verklaring mobiliteit en bereikbaarheid 1985–2008 [Explaining mobility and accessibility 1985 – 2008]. Kennisinstituut voor Mobiliteitsbeleid

Rodrigue, JP, Comtois C, Slack B (2009) The geography of transport systems. Routledge, New York

Schäfer A, Victor DG (2000) The future mobility of the world population. Transportation Res Part A 34:171–205

Schäfer A, Heywood JB, Jacoby HD, Waitz IA (2009) Transportation in a climate-constrained world. Massachusetts Institute of Technology, Cambridge, MA

Snelder M, Schrijver J, Landman R, Mak J, Minderhoud M (2008) De kwetsbaarheid van Randstedelijke vervoernetwerken uit verkeerskundig perspectief [Vulnerability of transport networks in the Randstad from a traffic engineering perspective]. TNO

Snelder M, Schrijver J, Smokers R, Wilmink I, Ligterink N, Sharpe R, Swinkels V, van de Wall Bake D (2010) Effecten van de Mobiliteitsaanpak op de uitstoot van CO_2 [Impact of the Mobility Policy on emissions of CO_2]. TNO

ter Weel B, van der Horst A, Gelauff G (2010) The Netherlands of 2040. CPB – Netherlands Bureau for Economic Policy Analysis

United Nations Population Division (UNPD) (2008) World urbanization prospects: the 2007 revision population database. http://esa.un.org/unup/. Accessed 14 Nov 2010

Urban Age City Data (2009) Transport and mobility. http://www.urban-age.net/cities/istanbul/data/2009/. Accessed 14 Nov 2010

van den Heuvel MG (1997) Openbaar vervoer in de Randstad – een systematische aanpak [Public Transport in the Randstad – A Systematic Approach]. Thesis. TU Delft

van Essen HP, Boon BH, Schroten A, Otten M, Maibach M, Schreyer C, Doll, C, Jochem P, Bak M, Pawlowska B (2008) Internalization measures and policy for the external cost of transport. CE Delft

Verkeer en Waterstaat (2002) Perspectief op Mobiliteit – sneller, goedkoper en verder [View on mobility – faster, cheaper and further]. Ministerie van Verkeer en Waterstaat. Den Haag

Zahavi Y (1974) Travel time budgets and mobility in urban areas. United States Federal Highway Administration

Chapter 3
The Importance of Passenger Cars for Global Greenhouse Gas Emissions – Today and Tomorrow

Lew Fulton

Abstract Worldwide energy use and CO_2 emissions are on a trajectory to double by 2050. Transport is on a similar trajectory. The stock of light duty vehicles could triple. This paper explores scenarios to cut the energy and CO_2 emissions of light-duty vehicles. It is found that deep reductions will require both the widespread adoption of current best available technology, e.g. via measures to maximize their use to improve fuel economy, and the longer-term development and deployment of a range of new technologies such as electric and plug-in hybrid vehicles. A combination of doubling fuel economy (halving fuel intensity) from 2005 levels and strong rates of adoption of new technology vehicles and fuels (e.g. electric and plug-in hybrid vehicles accounting for more than half the vehicles on the road by 2050) could cut oil use and CO_2 emissions by well more than half in 2050, compared to Baseline 2050 levels (and to half of 2005 levels in Europe). However, the changes needed will be dramatic including unprecedented penetration rates for certain key technologies. At the same time, some other trends must stop: those toward ever larger, more powerful cars, and trends in some countries toward ever-greater dependence on the car for all types of trips. And while the emergence of low cost cars can provide mobility to millions of people, society must ask if this is the best way to provide such mobility, rather than (for example) via advanced bus and train systems, and with land use patterns supporting a bigger role for non-motorized transport.

3.1 Introduction

The global energy economy is on a path roughly to double its energy use and CO_2 emissions by 2050 (IEA 2010). To meet IPCC emission reduction targets (e.g. for achieving 450 or even 550 ppm CO_2 levels in the atmosphere), deep reductions will be needed in the 2050 time frame. The IEA ETP 2010 report (IEA 2010) shows that a 50% reduction globally across all energy sectors will be needed to have a chance to hit a 450 ppm target. If developing countries can eliminate all future growth in

L. Fulton (✉)
International Energy Agency, Paris 75015, France
e-mail: lew.fulton@iea.org

CO_2 (a difficult challenge in itself), then currently developed countries will need to achieve an 80% reduction in order to achieve an overall 50% global reduction target. In other words, Europe and other developed parts of the world must strive for very deep emissions reductions in order to have a chance of a 450 ppm scenario, which is believed necessary to stabilize global temperatures at 2°C warmer than today's level.

In 2008, the transport sector worldwide accounted for about 23% of energy-related CO_2 emissions. This share is likely to grow slowly in the future as transport CO_2 is expected to rise faster than average across energy end-use sectors (IEA 2010). Even if other sectors such as industry and buildings cut emissions dramatically, it will be necessary for transport to deeply cut its emissions in order to achieve the overall targets. Globally this means achieving 2050 CO_2 emissions from transport that are below today's levels. For Europe it means achieving emissions that are half of today's levels or less.

Deeply cutting CO_2 emissions will mean deeply cutting fossil energy use in transport. This will be extremely challenging. But the IEA ETP 2010 report shows that transport CO_2 emissions in 2050 could be reduced to about 25% below their 2005 levels through a combination of measures and adoption of new technologies.

Worldwide, transport sector energy and CO_2 trends are very strongly linked to rising population and incomes. Transport continues to rely primarily on oil. Given these strong connections, decoupling transport growth from income growth and shifting away from oil will be a slow and difficult process. In projecting trends, this inertia must be taken into account. Large reductions in greenhouse gas (GHG) emissions by 2050 can only be achieved if some of the elements contributing to the inertia of income-driven transport growth are overcome, so that change can happen much more quickly in the future than it has in the past. For example, improvements in vehicle and system efficiencies of 2–4% per year will need to replace past improvement rates of 0.5–1%. New technologies and fuels will need to be adopted at unprecedented rates. The baseline trends and the technology and system efficiency improvement needs are explored through the rest of this chapter.

3.2 Worldwide Mobility and Energy Use Trends

From 1971 to 2006, global transport energy use rose steadily at between 2 and 2.5% a year, closely paralleling growth in economic activity around the world (Fig. 3.1). The road transport sector (including both light-duty vehicles and trucks) used the most energy and grew the most in absolute terms. Aviation was the second largest user of energy and grew the most in relative terms.

Within the transport sector, light-duty vehicles (LDVs) use about half of the travel related energy used world-wide and perhaps two thirds of energy use within national borders (i.e. excluding international shipping and air travel). These relative shares of energy use have remained fairly stable over the past 15 years as underlying growth

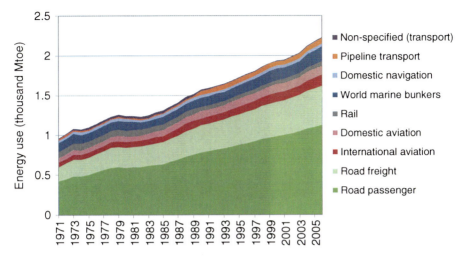

Fig. 3.1 World transport energy use by mode, 1971–2006

in activity, such as the rapid growth in air transport, has been offset by increases in efficiency. Nearly 97% of energy use in LDVs around the world is petroleum-based.

3.2.1 Recent Trends in Passenger Travel

Though data for many countries are unavailable, the IEA Mobility Modelling (MoMo) project has developed a large data base on vehicle sales and stock numbers, travel mode shares, and related data for many countries and on a regional level, also providing global totals (IEA 2009). Regional averages for the shares of travel undertaken by different motorized modes in 2005 are shown in Fig. 3.2. This excludes non-motorized modes of travel such as walking and bicycling because there are little data on these modes and they do not use fuel. OECD countries rely on 4-wheel LDVs far more than non-OECD countries. And people in OECD countries also undertake far more air travel per person. Developing countries show far higher modal shares for buses and, in some regions, for motorized two-wheelers, i.e. scooters and motorcycles. Section 3.3 explores the data and policy related issues in respect to passenger travel.

The total worldwide stock of passenger LDVs has grown steadily, reaching over 800 million worldwide in 2008. From 1990 to 2008, the stock of LDVs grew by about 70%, or about 3% per year, dominated by gasoline vehicles in most countries. In the same period, world population grew by 25%, from 5.2 to 6.6 billion.

In wealthy countries, the rate of growth in passenger LDV ownership has declined in recent years. This reflects slowing growth in population and may indicate saturation in terms of cars per capita, at least in some countries. It is also possible that cultural shifts are occurring – more people may be choosing not to own an

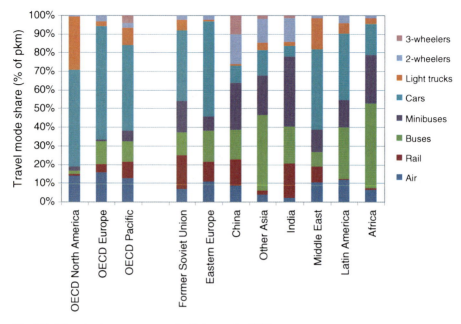

Fig. 3.2 Motorized passenger travel split by mode, 2005
Source: IEA Mobility Model database estimates

automobile or a family may be choosing to own only one instead of two or more, when there is increased access to mass transit options. But in developing countries, rates of LDV ownership are growing rapidly, suggesting that mass transit options are insufficient; many families purchase LDVs as soon as they can afford them. The emergence of low-cost LDVs such as the Tata Nano in India will probably further accelerate LDV ownership rates. The number of motorized two-wheelers continues also to grow rapidly.

3.2.2 Energy Efficiency by Mode

Estimates of recent average vehicle efficiency by mode are shown in Fig. 3.3 in grams of CO_2 per tonne-km for freight modes and in grams of CO_2 per passenger-km for passenger modes. The same pattern would emerge if the x-axis was in energy units rather than grams of CO_2. The figures reveal a wide range of values for each mode of transport, the range corresponding to the lower and higher boundary of the geographical zones considered in MoMo and the average value being shown as a vertical line. Some modes are generally more efficient than other modes: for example, rail is more efficient than air in both freight and passenger movement. But the most efficient mode can depend on the range of travel: for example passenger air travel is generally less efficient than passenger LDV travel, except for very long distances. These efficiency values can be heavily influenced by average loads or

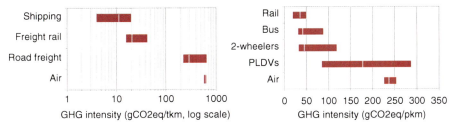

Fig. 3.3 GHG efficiency of different modes, freight and passenger, 2005
Note: The clear line indicates world's average, the bar representing MoMo regions' discrepancy
Source: IEA (2009)

ridership. For example, buses in the United States have significantly higher CO_2 emissions per passenger-km than those in most other parts of the world where buses tend to be fuller.

3.3 Projections and Scenarios

Here we present a look into the future (to 2050), using a range of scenarios to do so. The Baseline scenario represents a projection of where things appear to be headed in the absence of strong new policies to change expected future trends. Using the IEA MoMo, a number of additional scenarios have been developed to show how the transport sector might evolve differently to 2050. These scenarios represent just a few of very many possible futures, selected to illustrate the impacts of specific policy and technology developments. They are not predictions.

Two main scenarios are covered in this chapter (others are available in IEA 2009). The scenarios are outlined below. The key assumptions for each of these scenarios are summarized in Table 3.1.

Baseline – Vehicle ownership and travel per vehicle for LDVs, trucks and other modes are consistent with the IEA's World Energy Outlook (WEO) (IEA 2008) and a world oil price of 100 US dollars per barrel (USD 100/bbl) rising to USD 120/bbl by 2030. This scenario implies somewhat lower passenger LDV ownership in the developing world, at a given level of income, than has occurred historically in many OECD countries. This could be caused by a number of factors including greater urbanization in developing countries and lower suburbanization than in OECD countries, greater income disparities between the wealthy and the poor in non-OECD countries, and limits on the infrastructure needed to support large numbers of vehicles. This scenario also assumes a continuation of the decoupling of freight travel growth from GDP growth around the world, which has clearly begun in OECD countries. Note that a High Baseline scenario has also been developed, though it is not considered here. Information on this scenario is available in IEA (2010).

BLUE Map – This scenario reflects the uptake of technologies and alternative fuels across transport modes that can help to cut CO_2 emissions at up to USD

Table 3.1 Scenario descriptions and main assumptions

	Baseline	BLUE map	BLUE shifts	BLUE map/shifts
Scenario definition	Baseline projection	Greater use of biofuels, deployment of EVs, FCVs	No advanced technology deployment, gain through modal shifting only	BLUE Map+ BLUE Shifts
Passenger light-duty vehicles	Total vehicle travel more than doubles by 2050; fuel economy of new vehicles 30% better than 2005	FCVs each reach 40% of market share in 2050, so do EVs/PHEVs	Passenger travel in LDVs 25% lower than Baseline in 2050. Ownership and travel per vehicle reduced	BLUE Map+ BLUE Shifts
Biofuels	Reaches 260 Mtoe in 2050 (6% of transport fuel) mostly 1st generation	Reaches 850 Mtoe in 2050 (33%), mostly 2nd generation biofuels growth after 2020	Reaches 200 Mtoe in 2050 (6%), mostly 1st generation	Reaches 670 Mtoe in 2050 (32%) mostly 2nd generation biofuels growth after 2020
Low GHG hydrogen	No H_2	220 Mtoe in 2050	No H_2	170 Mtoe in 2050
Electricity demand for transport	25 Mtoe (mainly for rail)	390 Mtoe in 2050 primarily for EVs and PHEVs	40 Mtoe (mainly for rail)	500 Mtoe in 2050 primarily for EVs and PHEVs

Mtoe, million tonnes of oil equivalent

200/tonne of CO_2 saved by 2050. New powertrain technologies such as hybrids, plug-in hybrids vehicles (PHEVs), electric vehicles (EVs) and fuel cell vehicles (FCVs) start to penetrate the LDV and truck markets. Strong energy efficiency gains occur for all modes. Very low GHG alternative fuels such as H_2, electricity and advanced biofuels achieve large market shares.

BLUE Shifts – This scenario envisages that travel is shifted toward more efficient modes and total travel growth is modestly reduced as a result of better land use, greater use of non-motorized modes and substitution by telecommunications technologies. The scenario envisages that this has happened by 2050, with passenger travel in LDVs and aircraft approximately 25% below 2050 Baseline scenario levels as a result. The 25% reduction is simply a construction allowing analysis of the effects.

BLUE Map/Shifts – Finally, there is a scenario called BLUE Map/Shifts, which combines the technology gains in BLUE Map with the modal shifts in BLUE Shifts. This naturally provides the deepest energy and CO_2 reduction of any of the scenarios.

Population and GDP growth trends are assumed to be the same in all scenarios, matching recent UN and OECD projections (IEA 2008). The current global economic downturn is not fully reflected in these GDP projections. This will cause near-term projections, e.g. for 2010, to diverge from the assumed trend lines. But over the long term, e.g. to 2050, the impacts are likely to be minor assuming that the world economic system returns to its projected growth track within a few years.

The future oil prices assumed in this analysis are also based on IEA WEO (IEA 2008), rising from USD 80/bbl in the near term, after the recovery from the current economic downturn, to USD 120/bbl in 2030, in 2006 real USD. We assume prices stay at that level in real terms through 2050, although this implies a nominal oil price of over USD 300/bbl in that year. This price forms the basis for the transport and efficiency trends in the Baseline scenario.

3.3.1 Global Scenario Results

The overall picture that emerges from the projections and scenarios is that OECD countries are nearing or have reached saturation levels in many aspects of travel, whereas non-OECD countries and especially rapidly developing countries such as China and India are likely to continue to experience strong growth rates into the future through to at least 2050. In OECD countries, the biggest increases in travel appear likely to come from long distance travel, mainly by air. In non-OECD countries, passenger LDV ownership and motorized two-wheeler travel are likely to grow rapidly in the decades to come, although two-wheeler travel may eventually give way to passenger LDV travel as countries get richer. Freight movement, especially trucking, is also likely to grow rapidly in non-OECD regions. In all regions of the world, international shipping and aviation are likely to increase quickly.

In the Baseline scenario, travel growth will be triggered by strong growth in the number of households around the world that gain access to individual motorized transport modes. This will in turn lead to a rise in average travel speeds and increased travel distances, and reinforce land use changes such as suburbanization. Increasing wealth will also trigger more frequent and longer-distance leisure-related trips, in particular through increased tourism generating considerable amounts of long-distance travel. Figure 3.4 shows the projected evolution of motorized passenger mobility by mode to 2050 for the Baseline scenarios, as well as the effect of the modal shift policies adopted in the BLUE Shifts scenario. Estimated motorized passenger travel was about 40 trillion kilometers in 2005; this is projected to double by 2050 in the Baseline scenario.

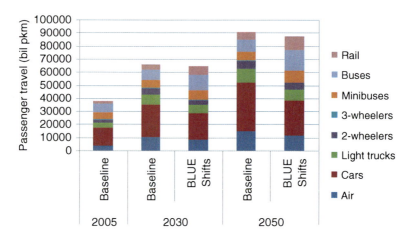

Fig. 3.4 Passenger mobility (trillion passenger kilometers) by mode, year and scenario

The BLUE Shifts scenario projects a different sort of future travel. Although it only reduces overall travel slightly on a worldwide basis compared to the Baseline scenario, the composition of that travel changes significantly, with much greater travel shares being undertaken by buses and rail, the most efficient travel modes. It is assumed that strong investments in, and expansion of, bus and rail services in the developing world induce a significant increase in motorized travel. For most non-OECD countries, travel by bus and rail is so much higher in the BLUE Shifts scenario than in the Baseline scenario that it results in net increases in the total amount of travel worldwide, more than offsetting decreases in travel in OECD regions where the use of telematics and changes in land use result in a net reduction in travel compared to the Baseline scenario.

3.3.2 Energy and GHG Intensity

The future energy intensities of different transport modes will play an important role in determining overall energy use and CO_2 emissions. Taking into account both efficiency improvements and GHG intensity of fuels, Fig. 3.5 shows the projected GHG intensity by passenger transport mode in the Baseline and BLUE Map scenarios. Given the relatively high oil price assumptions in WEO (IEA 2008) and existing policies such as the fuel economy standards in many OECD countries, the GHG intensity of LDVs decreases by 30% between 2005 and 2050 in the Baseline scenario. This is a substantial improvement. The GHG intensity of all other modes except motorized two-wheelers decreases as well, typically by about 15%. In the BLUE Map scenario, all modes reduce their GHG intensity by at least 50% by 2050, and FCVs, EVs, two-wheelers and rail help to cut modal CO_2 emissions by 80% or more, due to the widespread availability of very low carbon H_2

3 The Importance of Passenger Cars for Global Greenhouse Gas Emissions... 51

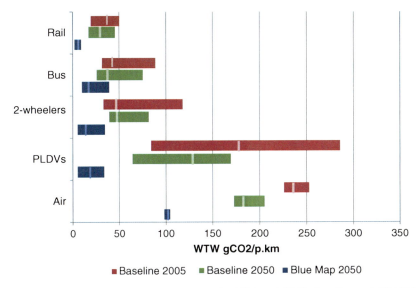

Fig. 3.5 GHG intensity of passenger transport in 2005 and 2050, Baseline and BLUE Map scenarios
Note: The clear line indicates world's average, the bar representing the range of values across regions

and/or electricity by 2050; thus all passenger modes except air travel reach less than 50 g of CO_2 per kilometer.

3.3.3 Energy Use Scenarios

The net impacts on energy use in each of the scenarios are shown in Fig. 3.6. In the Baseline scenario, energy use grows substantially to 2050 as efficiency improvements are outweighed by growth in transport activity. In the BLUE Shifts scenario, energy use in 2050 is around the same level as in the Baseline scenario in 2030, suggesting a degree of stabilization. In the BLUE Map scenario, energy use returns to the 2005 level, and if the BLUE Shifts scenario is combined with the BLUE Map scenario, energy use drops to a level lower than that in 2005.

There are also important differences between scenarios in the composition of fuels used. In the Baseline, little non-petroleum fuel is used even in 2050. As a result, fossil fuel use increases by 50% in the Baseline scenario. By contrast, in the BLUE Map scenario, the need for fossil energy for transport halves, given very large shifts to low-CO_2 alternative fuels such as low-CO_2 electricity and hydrogen and advanced biofuels. In the BLUE scenarios, most conventional gasoline and diesel powered LDVs have disappeared by 2050, being replaced by largely hydrogen and electricity powered vehicles. But for heavier long-distance modes such as trucks, planes and ships, diesel fuel, jet fuel and heavy fuel oil or marine diesel still

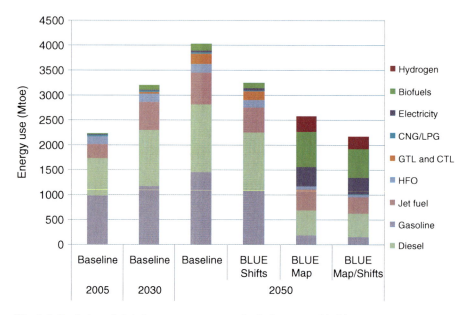

Fig. 3.6 Evolution of global transport energy use by fuel type, worldwide

dominate. Biofuels, which comprise mainly biodiesel rather than ethanol in 2050, play an important role in displacing liquid fossil fuels in these long-distance modes. Biofuels reach about 33% of total transport fuel use in BLUE Map in 2050, including about 30% of truck, aircraft and shipping fuel and 40% of LDV fuel. For LDVs, nearly all the rest is electricity and hydrogen whereas for trucks, ships and aircraft most of the rest remains petroleum fuel.

3.3.4 Focus on LDVs

Driven by income and population growth, in the Baseline scenario sales of LDVs around the world nearly triple to 150 million vehicles a year by 2050, from about 60 million a year in 2005 (Fig. 3.7). In the BLUE Shifts scenario, the shift from private to public transport constrains the growth in LDV sales well below that in the Baseline scenario, rising to about 110 million in 2050. Even this lower level of growth represents nearly a doubling of world LDV sales from today's levels.

In addition, the types of vehicle sold vary considerably between scenarios. In the Baseline and BLUE Shifts scenarios, two thirds of the new LDVs sold in 2050 are still conventional internal combustion engine (ICE) vehicles, with the remaining third hybrids. In the BLUE Map scenario, by 2050 over half of the vehicles sold are EVs and FCVs, powered by electricity and hydrogen respectively.

In the BLUE Map scenario, changes over time are based on the projected evolution of technology potential and cost, described later in the chapter. Strong policies will be needed to bring about this scenario. As shown in Fig. 3.8, after 2010 the rate

3 The Importance of Passenger Cars for Global Greenhouse Gas Emissions ...   53

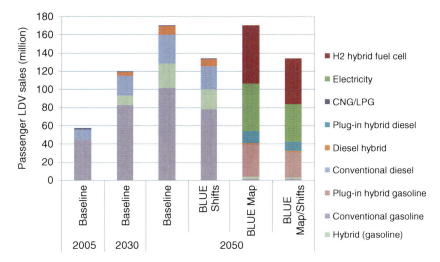

Fig. 3.7 LDV annual sales by technology and scenario
Source: IEA Mobility Model

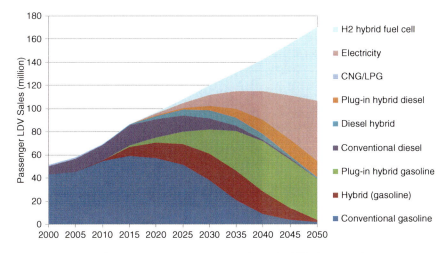

Fig. 3.8 Evolution of LDV sales by technology type in the BLUE Map scenario

of growth in conventional gasoline and diesel LDV sales begins to be trimmed by the sale of hybrids and PHEVs, with EV sales increasing after 2015. By 2020 PHEV sales reach 5 million and EV sales 2 million worldwide. Around 2020, commercial FCV sales begin. Through 2030 EV and FCV sales increase significantly, taking a progressively higher proportion of the overall growth in LDV sales. From 2030 onward, demand for non-PHEV ICEs declines rapidly in absolute terms. By 2040, more EVs and FCVs are sold than any ICE vehicle. By 2050, LDV sales are equally split between FCVs, EVs, and PHEVs.

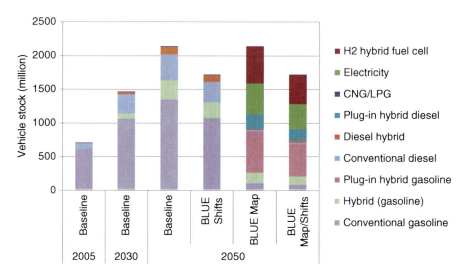

Fig. 3.9 Vehicle stocks by technology and scenario

Most LDVs stay in use for 15–20 years, hence the changes in vehicle sales take time to affect the total stock of vehicles (Fig. 3.9). In the BLUE Map scenario in 2050, for example, EVs and FCVs only account for about 45% of all LDVs on the road. It would take until perhaps 2065 for these vehicle types to represent 99% of all vehicles in use.

3.3.4.1 LDV Fuel Economy

Average LDV fuel economy is expected to improve over time. The rate of improvement is likely to be driven by technological improvements and their costs, by consumer choices over vehicle performance and size, by fuel costs and by policies to help achieve GHG targets. Not all these factors point in the same direction in all circumstances.

In the Baseline scenario, average new LDV test fuel economy is expected to improve by about 25% (reduction in fuel intensity) by 2030 in both OECD and non-OECD countries (Fig. 3.10). This is driven mainly by current (and in some cases very recent) efficiency policies in OECD countries such as the United States, EU countries and Japan, described in more detail in Chapters 4 and 15 of this book. Most of these policies are set to run through 2015. After 2015, if such policies are not renewed and strengthened, increases in vehicle size, weight and power may start to reverse the benefits of higher efficiencies.

The improvements in non-OECD countries are expected broadly to parallel those in OECD countries, since most vehicles around the world are produced in the latter. Eventually more vehicles will be produced in non-OECD than in OECD countries. As this happens, it will be very important for non-OECD countries to have

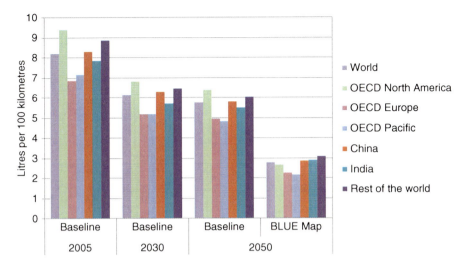

Fig. 3.10 New LDV tested fuel economy for selected regions

in place fuel economy policies that ensure efficiency technologies are adopted and fully exploited, and limit increases in average vehicle size, weight and power.

Beyond about 2020, electric motors and fuel cells will become increasingly important to support continuing improvements in LDV fuel economy. In the BLUE Map scenario, moving away from conventional gasoline and diesel vehicles toward PHEVs, EVs and FCVs improves new LDV fuel efficiency by a factor of two between 2030 and 2050 (Fig. 3.10). While LDV fuel economy in OECD countries remains slightly better than in non-OECD countries, new LDVs in all regions use less than three liters gasoline-equivalent of fuel per kilometer by 2050, compared to about eight today.

As discussed in Section 3.4, actual in-use fuel economy is generally worse than tested fuel economy, due to in-use conditions such as traffic congestion, use of auxiliary equipment etc. The gap between them may be as high as 25% in some countries, though it appears to average about 15–20%. This gap could increase further in the future if, for example, traffic congestion worsens around the planet. On the other hand it could shrink with the introduction of better technologies, such as vehicle start-stop systems that stop the engine while a vehicle is at idle. Hybrids, PHEVs, EVs and FCVs all experience less deterioration in fuel economy in congested traffic than today's conventional ICEs. In the Baseline scenario, the gap between tested and in-use fuel economy for most regions remains around 15–20% in the future. In the BLUE scenarios, it improves to 10% by 2050 due to improved component efficiency, the introduction of advanced technology vehicles and the use of policies to improve traffic flow and (in BLUE Shifts) cut the growth in car travel.

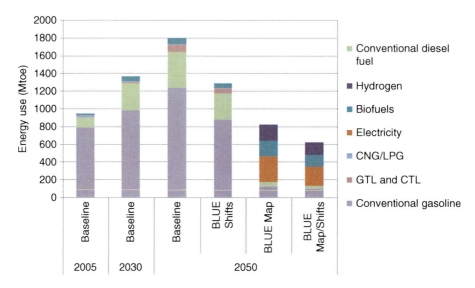

Fig. 3.11 LDV energy use by scenario

3.3.4.2 Energy Use and CO$_2$ Emissions

In the Baseline scenario, the relative shares of different energy sources remain broadly constant as total energy use doubles or more by 2050 (Fig. 3.11). In BLUE scenarios, total energy use is far lower. In the BLUE Shifts scenario, with 25% lower car travel (as described in Section 3.5), the shares of different sources are very similar to those in the Baseline scenario. In the BLUE Map scenario there is both strong fuel economy improvement and a major shift to biofuels, electricity and H$_2$ by 2050. Combining the BLUE Map and BLUE Shifts assumptions achieves a total fuel use of slightly more than half of the 2005 level.

3.3.5 Scenarios for OECD Europe

OECD Europe has a high average travel per capita. With little expected population growth over the next 40 years, total transport activity is unlikely to grow significantly in Europe. It is also unlikely that energy use per passenger kilometer and per tonne-kilometer for freight will improve significantly without strong policy interventions.

In the Baseline scenario, transport energy use in OECD Europe remains fairly flat, reflecting the impact of a wide range of initiatives around Europe that are expected to help cut energy intensity over the next 5–10 years. Without further significant expansion of these initiatives, energy intensity is projected to improve little if at all after 2020, especially in LDVs.

In 2007, the transport sector in OECD Europe used about 450 million tonnes of oil equivalent (Mtoe), or around 20% of global transport energy use (Fig. 3.12). By

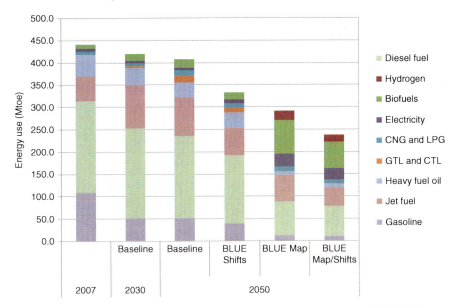

Fig. 3.12 Transport energy use in the baseline and BLUE scenarios by fuel type in OECD Europe
Source: IEA (2009); IEA analysis

2050 this share is likely to drop to about 10%, as the transport sector energy use of developing economies grows very quickly over the next 40 years.

Different European countries, with cultural differences, transport system differences, climate differences and a range of different commitments on CO_2, will adopt different approaches to ensuring that their transport sectors make the contributions they need to make to the outcomes implicit in the BLUE Map scenario. Some countries will rely heavily on biofuels, others more on electrification. Some countries may have particular opportunities to deploy EVs, for example because they have a proportion of LDVs used exclusively within large cities. Cold and biomass-rich Scandinavian countries may be more likely to go toward compressed (and eventually bio-synthetic) natural gas or biomass-to-liquids fuel options. In most of the big passenger LDV markets such as the United Kingdom, France, Germany, Spain and Italy, the electrification of vehicles is now high on the agenda.

The projected OECD Europe GHG emissions in each of the transport scenarios are set out in Fig. 3.13. The emissions for individual modes depend on a combination of efficiency improvements and the use of low carbon fuels. Modal shifts to the most efficient modes account for the remaining reductions.

In particular, reductions depend on:

- Achieving a 50% improvement in new LDV fuel efficiency by 2030 compared to 2005.
- Achieving efficiency improvements in the stock of trucks, ships, trains and aircraft of the order of 40–50% by 2050.

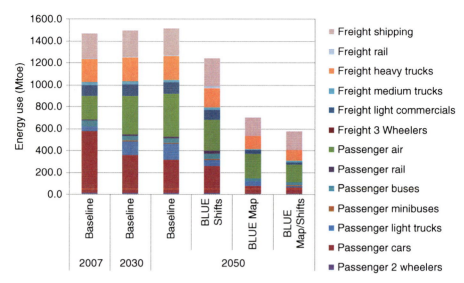

Fig. 3.13 OECD Europe's GHG emissions evolution by transport mode
Note: WTW (well-to-wheel) GHG emissions comprise the emissions of the vehicle (tank-to-wheel) and the upstream emissions along the fuel production pathway (well-to-tank)
Source: IEA (2009); IEA analysis

- Reaching substantial sales of EVs and plug-in hybrid electric vehicles (PHEVs) by 2030 (6 million) and 2050 (10 million).
- Biofuel being about 12% of transport fuel by 2030 and 25% by 2050. This assumes that most of the biofuel is imported into OECD Europe.

In the BLUE Shifts scenario, travel by rail and bus in 2050 increases by 50–100% compared to the Baseline scenario in that year. This, together with other changes such as improvements in land use planning and investment in non-motorized transport infrastructure, results in a 25% cut in the growth of car and air travel in OECD Europe.

Decarbonization of power generation will also play an important part in reducing GHG emissions in the transport sector as EVs start to play a larger role. Europe starts from a relatively good position, producing on average 345 g of CO_2 per kilowatt-hour (kWh) of generation in 2007. This is expected to reduce to 208 g CO_2/kWh in 2050 in the Baseline scenario and to 15 g CO_2/kWh in the BLUE Map scenario. In the BLUE Map scenario, CO_2 emissions reductions benefit not only from there being many more EVs than in the Baseline scenario, but also from the much lower carbon footprint of the electricity that runs them.

In the BLUE Map scenario, transport GHG emissions are reduced by around 60% in OECD Europe, with the aggressive promotion of low-GHG technologies into the market. The cost of such GHG emissions reductions over the life time of a vehicle

3 The Importance of Passenger Cars for Global Greenhouse Gas Emissions ...

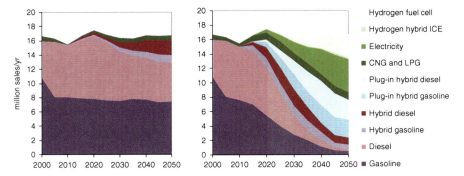

Fig. 3.14 Passenger light duty vehicles sales by technology type in OECD Europe in the baseline and BLUE Map scenarios
Source: IEA (2009); IEA analysis

depends on energy prices, but it will often be negative as energy savings exceed the extra investment cost in new technologies.

In the BLUE Map scenario, PHEV and EV technologies dominate new LDV sales after 2030 (Fig. 3.14). Sales of EVs and PHEVs begin in earnest in 2015; by 2030 they reach nearly 80% of sales; and by 2050 nearly all new vehicles are electric.

Transport volumes in OECD Europe are relatively stable. They may also decline during periods of slow economic growth or when energy prices increase, as in 2008. Deep cuts in GHG emissions can be achieved by adopting an aggressive strategy toward efficiency. This has already begun for passenger LDVs. Further big reductions will come from shifting toward electricity and advanced biofuels. Natural gas can also play a significant role in European transport for cars and perhaps especially for trucks. Over time there must be a transition to biogas and bio-synthesized natural gas in order to reach very low CO_2 intensities by 2050. Pursuing a growth strategy for the most efficient transit and non-motorized modes, and dampening demand growth for the least efficient single-occupant passenger LDVs can also contribute to substantial energy savings and GHG reductions by 2050 or even earlier.

3.4 LDV Technologies and Market Uptake – The Need for Market Transformation

Achieving a low fossil fuel, low-CO_2 future for LDVs will require a combination of strong fuel economy improvements and adoptions of advanced technologies and new propulsion systems, like electric and plug-in hybrid vehicles. In each case, it will be critical to develop policies that incentivize rapid uptake, thus essentially transforming the market.

The first of these (fuel economy) represents an area where the key transformations have already begun, but must continue over the next 20 years in order to reach key sustainability targets. The second area (EVs/PHEVs) represents a transformation that is just beginning, but must also be successful in order to help reach sustainability targets.

There are other transformations and transitions that must also play an important role in achieving sustainable LDV transport. These include adoption of new low-carbon liquid fuels such as advanced biofuels that very clearly provide low net GHG emissions without damaging ecosystems or threaten food security. Another transformation relates to the role of the car in society – and toward much more sustainable configurations of cities and transport systems, to allow people to easily travel by the most efficient modes, and travel less when desirable. But here the focus is on the twin examples of fuel economy improvement and adoption of EVs/PHEVs.

3.4.1 Fuel Economy Improvement as a Market Transformation Area

Vehicle fuel economy, measured either as miles per gallon (MPG) or liters per hundred kilometers (L/100 km), has the potential for substantial improvements around the world over the next 20 years and beyond. The Global Fuel Economy Initiative, for example, calls for a target of a 50% reduction in new car fuel intensity compared to 2005 levels around the world by 2030 (GFEI 2009).

However, during the past two decades, especially during the late 1980s and most of the 1990s, little progress was made in almost any region. Many new technologies were adopted by manufacturers but most often these were used effectively to maintain fuel economy levels while allowing increases in vehicle size, weight and power. The US fuel economy standards that ramped up rapidly between 1978 and 1985 had, however, shown that it was possible to achieve rapid improvements in fuel economy. But after 1985, the US standards have not significantly changed. However around the mid-1990s both the EU and Japan adopted new strategies to promote fuel economy.

In Japan the Top Runner approach was adopted, which set up requirements for a range of LDV market classes based on weight, and identified the top performer(s) in each class. The results have been strong fuel economy improvements in Japan over the past 15 years, with Japan now having among the most fuel efficient vehicles in the OECD.

In the EU, voluntary commitments were made by the three major European auto manufacturing groups (ACEA, JAMA and KAMA) in the late 1990s to achieve a 25% improvement in fuel economy between 1995 and 2008 (IEA 2009). After fairly good progress for several years, EU manufacturers began falling behind the pace of improvement needed to reach the 2008 target. One reason was that dieselization, which had accounted for a large share of the fuel economy improvements, had reached high levels in many EU countries, so new strategies and technology approaches were needed, but manufacturers did not seem to be adopting these quickly enough. Trends toward larger, heavier vehicles also offset much of the technology uptake. Finally, in 2005 the EU concluded that the voluntary system wasn't working and would not reach the targets, and moved to adopt a mandatory system. Based on the EU's expected future sales mix of LDVs, the requirement is consistent with achieving 135 g/km by 2015, compared to 154 g/km in 2008. More details on the EU legislation are provided in Chapter 4.

The European system gained a place at the forefront of fuel economy improvements around the world, alongside Japan with its Top Runner program (Fig. 3.15). With the standards in place through 2015, Europe and Japan will have significantly more efficient vehicles than almost anywhere else on the planet.

These initiatives have clearly changed the course of fuel economy trends, and have helped give impetus to other countries. Within the past 4 years the United States, Canada, Korea and China have adopted or significantly tightened their LDV fuel economy regulations.

In the *United States*, the Obama Administration recently finalized a rulemaking on new car and light-truck fuel economy standards, expected to increase MPG by about 25%, to 35.5 miles per gallon by model year 2016. As shown in Fig. 3.15, in terms of carbon emissions per kilometer, this puts the US on a strongly downward track, equal to or even exceeding the pace set by other OECD countries (though

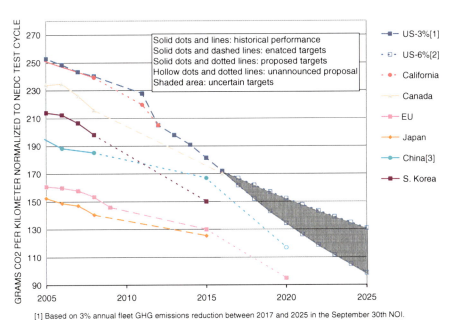

[1] Based on 3% annual fleet GHG emissions reduction between 2017 and 2025 in the September 30th NOI.
[2] Based on 6% annual fleet GHG emissions reduction between 2017 and 2025 in the September 30th NOI.
[3] China's target reflects gasoline fleet scenario. If including other fuel types, the target will be lower.

Fig. 3.15 Average CO_2 emissions trends through 2008 with targets enacted or proposed thereafter by region
Note: Enacted standards in *dashed lines*, proposed with *dotted lines*
Source: ICCT (2009)

starting from a higher fuel intensity position). The Administration estimates that this will save the equivalent of 1.8 billion barrels of oil. It has also launched a process to develop efficiency standards for heavy duty vehicles (freight trucks), which to date no other country except Japan has implemented. More details are provided in Chapter 15 of this book.

China imposed fuel economy standards in 2005 on LDVs, using a weight-class based system. The standards set a minimum fuel economy for vehicles in each weight class. The 2005 standards affected few vehicles since the targets for smaller vehicles were all set above 8 L/100 km for automatic transmission vehicles and were set at over 11 L/100 km for 1,500 kg curb weight vehicles. The standards for 2008 were tightened by 1 L/100 km for the lightest vehicles and by almost 2 L/100 km for heavier vehicles. Standards were planned to become more stringent in 2012. The Chinese government has also been concerned about the trend to larger and more powerful vehicles and has taken two steps in this regard. First, the fuel economy standards that it has imposed are more stringent (relative to current average vehicles) for heavier vehicles than lighter vehicles. Second, it has lowered the tax rate on vehicles with engines smaller than 1.6 L from 3 to 1%, while increasing taxes on vehicles with engines over 3 L from 15 to 25%.

In the US and China, the world's two largest car markets, the adoption and tightening of fuel economy standards will save literally billions of barrels of oil and billions of tonnes of CO_2 over the next 20 years. In all countries adopting standards, fuel cost savings to drivers will be substantial, and will likely pay for most or all of the costs of fuel economy technology, based on recent cost/benefit studies conducted by the IEA and others (IEA 2009, Bandivadekar et al. 2008).

In the ETP BLUE Map scenario, a range of measures to cut oil use is explored, including on-going tightening of standards (and other approaches) to reach a 50% reduction in LDV stock-average fuel intensity between 2005 and 2050 (and by 2030 for new vehicles). Fuel economy improvements across all vehicle types (including ships and aircraft) account for more than half the total oil savings in BLUE Map, which is about 2.5 billion tonnes oil equivalent (50 million barrels per day) in 2050. Vehicle efficiency also plays a key role in reducing the cost of oil in BLUE Map, to USD 70/bbl in 2050 compared to USD 120/bbl in 2050 in the Baseline scenario (IEA 2010).

In summary, what Fig. 3.15 shows is essentially a market transformation in progress – a new direction for fuel economy, now clearly on a path toward much more efficient vehicles in the future in contrast to the stagnation in fuel economy levels that prevailed in recent decades. Fuel economy in OECD countries is now on track toward the Global Fuel Economy Initiative target of a 50% reduction in new car fuel intensity by 2030, compared to 2005. This initiative sets targets that the IEA considers a key part of achieving a low CO_2 future for transport. However, in order to reach these targets (also including a 30% improvement by 2020 and a total stock improvement of 50% by 2050, i.e. '50 by 50'), four key outcomes must be secured:

- OECD countries must continue to tighten their fuel economy standards over time, and complement the targets with vehicle tax-related policies (such as differential

3 The Importance of Passenger Cars for Global Greenhouse Gas Emissions ...

vehicle purchase taxes as a function of fuel economy or CO_2 emission) to help spur demand for more efficient vehicles – another key element in this market transformation.
- Non-OECD countries must also begin to take up bold policies toward improving fuel economy. Large markets like India and Brazil would likely benefit from adopting standards; smaller markets and vehicle importing countries may find that vehicle tax incentives may be more appropriate. Getting the fuel prices 'right' is also very important, by removing subsidies to gasoline and diesel fuel and setting taxes for all fuels at levels that at least offset the societal/environmental costs of these fuels.
- All countries must work to ensure that actual on-road fuel efficiency improvements match the improvements in tested fuel economy, across the entire stock of vehicles. Measures such as promoting eco-driving, reducing traffic congestion, and revising test procedures to more closely reflect real-world conditions can all play an important role in this regard.
- Standards and other efficiency measures are also needed for trucks, which account for up to half of road fuel use in some countries. Until early 2011, only Japan had developed fuel economy standards for trucks, but they were under development in other countries such as the US and should be implemented as soon as is feasible.

In these ways, the fuel economy transformation that has begun can be continued, and can reach targets such as a 50% reduction in new LDV fuel intensity by 2030. Probably by this date, in order to continue cutting fuel intensity as well as CO_2 emissions, the focus will need to change to mass adoption of different vehicle propulsion systems and fuels, such as plug-in electric vehicles and possibly fuel cell vehicles. In particular, EVs and PHEVs are expected to play a key role. The next section covers the need for this second market transformation.

Fuel cell vehicles may also play a key role in the future, but appear likely to need another decade for research and development to reduce costs and improve performance before they are ready for mass market introduction. The need for extensive new hydrogen refuelling structure is another concern. But fuel cell vehicles may ultimately offer significant advantages to electric vehicles, particularly in terms of vehicle driving range. Or they may be combined with EVs, for example, with the ICE in plug-in hybrids eventually being replaced by a fuel cell system.

3.4.2 EVs/PHEVs as a Key Future Market Transformation Area

Even a 50% reduction in the energy intensity of cars will only cut fuel use by half, and in the IEA Baseline projection fuel use doubles worldwide by 2050, so the targeted fuel economy improvements will basically help avoid large future increases in vehicle CO_2 emissions. But in order to move toward outright reductions in CO_2, we must then turn to new fuels and new vehicle propulsion systems. Of these, perhaps the most important over the next 20–40 years will be electricity and electric vehicles.

Electric vehicles will provide two important benefits in this regard: the possibility to run on a near-zero GHG energy source (along with zero vehicle emissions of any kind), and also continued efficiency gains compared to internal-combustion engine vehicles. For example, highly efficient gasoline vehicles (probably hybridized) may achieve 4 L/100 km of fuel economy (60 MPG); EVs today can achieve the equivalent of 2 L/100 km gasoline equivalent, running on electricity and electric motors.

PHEVs offer a compromise between pure EVs and ICE vehicles – keeping both systems on board. These vehicles are more complex than EVs, and overall likely to be less efficient, but offer excellent flexibility (able to use both electricity and liquid fuels) and preserve the long driving range offered by ICEs.

However, for a number of reasons, it will take considerable time and very strong policies to bring EVs and even PHEVs into the market in substantial numbers. The effort must begin – and is beginning – now. For even with very aggressive increases in vehicle production and sales, it will be 10 years before these vehicles begin to approach a significant share of total sales around the OECD, and perhaps 20 years before they obtain a significant share of global vehicle stocks (see Fig. 3.8). On the other hand, this also buys some time – over the next 5 years, for example, it should be possible to work out many of the key technologies and systems that will then need to be built up over the following 5–10 year period as EV production goes large scale.

In 2009 the IEA published a 'roadmap' for countries around the world, collectively, to adopt EVs and PHEVs and ramp up sales over the next 10 years and beyond, on a path consistent with the IEA BLUE Map scenario (Fig. 3.16). The target by 2050 is over 50% of total LDV sales around the world – over 100 million vehicles per year. Backing out from this, sales will need to be on the order of 20 million per year by 2030, and probably at least 5 million by 2020. Even to reach this level will be quite challenging, since it suggests that, starting from perhaps a few thousand EVs and PHEVs produced around the OECD in 2010, there will need to be more than a doubling in vehicle production each year over the next 10 years. In fact growth in some countries will need to be even faster, if they are to hit their own targets that in some cases reach into the hundreds of thousands of vehicles by 2015.

What needs to happen in the next 5–10 years to achieve the ambitious targets? It breaks down roughly as follows:

2010–2012: Limited production of most new EV/PHEV models, many as part of demonstration projects in key cities; initial, well-coordinated investments in private and public recharging infrastructure; determination and harmonization of key codes and standards. Substantial vehicle purchase incentives will also be needed.

2012–2015: Lessons learned during 2010–2012 applied; fully optimized vehicle designs and batteries are introduced, production runs increase toward

3 The Importance of Passenger Cars for Global Greenhouse Gas Emissions ... 65

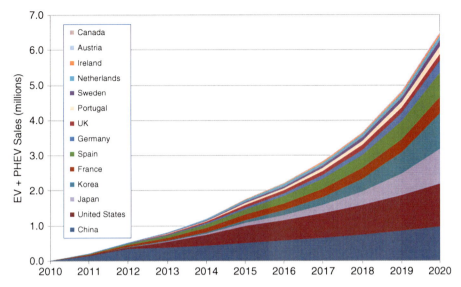

Fig. 3.16 Country targets for EV/PHEV annual sales as of December 2010
Notes: Based on government announcements and reports where available, otherwise from news reports; for targets set before 2020 (e.g. China 2012), a 10% annual sales growth is assumed thereafter

commercial scale (e.g. in the tens of thousands per year for some models). Purchase incentives will need to continue.

2015–2020: Rapid increases in EV/PHEV production reaching over 100,000 per year for many models (for example, in order to reach sales of 5 million per year by 2020, this could require sales of 100,000 units for 50 different models). Vehicle purchase incentives and other government support should begin to decline.

These three time periods, and the actions described, amount to a process of rapid market transformation. What specific actions should governments be taking to bring on this market transformation? The key steps are outlined below. In most cases, major economy governments are taking the correct steps at this point (France sets a good example, see Box 3.1), but must continue to do so and follow through on many fronts over the coming years.

Box 3.1: France's Plan to Launch EVs

In October 2009, the French government committed 1.5 billion euros ($2.2 billion) to a 10-year plan to help put 2 million electric (plug-in) cars on the road by 2020. The funds will help pay for manufacturer and buyer subsidies, a nationwide network of more than 4 million EV charging stations (one million by 2015), and subsidies for battery manufacturing and industrial research. Other elements include:

- Requiring all new apartment developments in the country to install charging stations, beginning in 2012
- Introducing purchase 'bonuses' of up to €5,000 to support consumers buying EVs
- Calling for public and private tenders for fleets of EVs to generate demand – with a target for these fleets to account for 100,000 EVs by 2015
- Providing €125 million for an EV battery manufacturing plant near Paris.

The two major French car manufacturers, Renault and PSA Peugeot Citroen have pledged to begin selling electric vehicles in France by 2012.

The national government has named an EV coordinator between ministries, who is also working closely with cities, electric utilities, vehicle manufacturers and other stakeholders to coordinate all aspects of EV developments.

Source: French Ministry of Ecology press release, 1 October 2009, and news reports.

3.4.3 Ensuring EVs/PHEVs are Cost-Competitive

The cost of EVs and PHEVs will likely be substantially higher than the cost of similarly sized ICE vehicles, for at least the next 5–10 years, and probably longer. Battery costs are the main reason for this (and for PHEVs, the cost of the hybrid system).

The IEA estimates that in the near term batteries are likely to cost in the range of USD 600–800 per kWh even at large volume which means for a 30 kWh EV (typical to provide 150 km of range), the batteries alone will cost USD 18,000 to USD 24,000 per vehicle. Through on-going research and development, deployment activities to encourage larger scale production, and learning by doing (including optimization of production processes), this cost is expected to drop with a target of USD 300 or less, in the time frame of 2015–2020. This will help bring EV battery costs down under USD 10,000 per vehicle. Along with savings from removing the ICE system, the energy cost savings associated with EVs (likely to be in the range of USD 3,000–5,000 per vehicle over vehicle life) may then pay for most of the battery costs over the vehicle life. But in the very near term, lifetime fuel cost savings will not nearly cover the higher costs of EVs.

A range of studies suggests that this target of USD 300 per kWh will be attainable, but only if large scale production (e.g. 100,000 units/year) can be achieved and cumulative production achieves much higher level – through which manufacturing optimizations can be achieved. This is precisely what occurred to bring down the cost of small appliance Li-ion batteries over the past 10 years.

3 The Importance of Passenger Cars for Global Greenhouse Gas Emissions ...

To achieve this, therefore, it will be necessary to achieve both large scale production of batteries and vehicles and a growing market that helps yield the learning and optimization that ultimately will drive costs down. And to achieve these conditions, there will be a need to provide incentives to make EVs and PHEVs have a chance to succeed in the market. The amount that will be necessary to spur demand is unclear, since few EVs had reached the market by early 2011. But if one assumes that consumers will not pay a net life-cycle cost premium for EVs, the market incentives in the near term may need to be on the order of USD 7,500–10,000 per car (this could be set on the basis of performance, such as for very low GHG vehicles). This type of approach is in fact about what has been put in place in a number of countries (Table 3.2). This approach appears sensible, but must be carefully managed to avoid excessive costs to treasuries over time. Reducing this level of subsidy as battery and EV costs drop will be needed, with a full phase-out of subsidies perhaps by 2020, as production volumes rise rapidly at that point.

Table 3.2 Comparison of EV/PHEV-related policies in several countries as of October 2010

Country	Sales target	Fiscal incentives	Other comments
China	Production of 500,000 cars by end of 2011	Up to about USD 8800 per vehicle	Available in 12 Chinese cities
France	Up to 2 million stock by 2020; 50,000 purchase order for gov't fleets	Eur 5000 (USD 6300) tax credit per vehicle	Total funding of EUR 1.5 billion (USD 1.9 billion), includes funding for 4 million recharging points by 2020, battery production
Japan	Maximum 1 million sales by 2020 (based on 20% share target)	Up to Y 1.3 million (USD 14,000) per vehicle	Fiscal incentives can change frequently
Germany	1 million total stock by 2020	No direct incentives at this time	Eur 285 million (USD 350 million) for infrastructure development and battery R&D
Spain	250,000 sales by 2014	Up to EUR 6000 (USD 7500) per vehicle	Primary focus on Madrid, Barcelona, Sevilla
United Kingdom	1.5 million stock by 2020	Up to 5000 GBP (USD 7500) per vehicle	Total funding of GBP 250 million (USD 375 million) for low-CO_2 transport
United States	1 million total stock by 2015	Up to USD 7500 per vehicle	US DOE providing R&D funding and grants of over USD 2 billion

Sources: Based on government announcements and reports where possible, otherwise from news reports

Over the next 10 years, the IEA BLUE Map scenario includes total (cumulative) worldwide sales of EVs and PHEVs totalling about 20 million. At an average subsidy level of USD 5,000 per car (reflecting higher early subsidies with a phase-down over time), this would cost USD 100 billion over 10 years across a range of countries. While this is a large figure, in comparison to the USD 10 trillion or more[1] that the world will spend on new LDVs over the next 10 years, this amount is less than a 2% incremental cost, which could be considered a bargain if it succeeds in generating the market transformation to EVs and PHEVs. Put another way, these estimates suggest that an average 2% tax on new cars around the world would pay the average subsidy cost of EVs.

3.4.4 Coordination of EV/PHEV Infrastructure Development

There are a range of important actions that will be needed by the year 2020 to ensure that recharging infrastructure is developed alongside the market uptake of EVs and PHEVs. These include:

- Ensuring that basic charging stations are available at reasonable cost for residential locations for EV owners and providing public recharging options for urban dwellers that don't have home recharging possibilities. This also involves coordination of investments in public and private retail recharging stations, and in some cases direct provision of this infrastructure – especially in areas where there is a fairly high density of EVs.
- Coordinating the ramp up in EV/PHEV sales with recharging availability, working closely with cities in this respect.
- Ensuring common systems and standards for charging, at least within each country and preferably across regions (such as North America, Europe etc.). Deciding how and when fast charging stations will be rolled out will be a critical aspect, as fast charging will provide important benefits but will be expensive, and may have adverse impacts on electricity generation profiles. Managing the introduction of smart metering and vehicle-to-grid services will also be a critical area for coordination.
- Providing other types of support such as information to the public, funding for research projects etc.

This type of support is beginning to occur in some countries (as indicated in Table 3.1), but it will be critical in the coming years that national governments pay close attention to each of these aspects, while also working closely with other stakeholders trying to make progress in each area. Key stakeholders include city governments (who clearly will play a critical role in coordinating developments in their cities), electric utilities, 3rd party providers of various goods and services,

---

[1] Based on average global car sales of 80 million per year over 10 years at USD 12,500 per vehicle.

and automobile and parts (such as batteries) manufacturers. All of these stakeholders will have critical roles to play, but it is up to national governments to ensure coordination of all actors and activities.

3.5 Conclusions

Worldwide energy use and CO_2 emissions are on a trajectory to double by 2050. Transport is on a similar trajectory. The stock of light duty vehicles could triple. To cut energy use and CO_2 emissions of vehicles will require both the widespread adoption of current best available technology, e.g. via measures to maximize their use to improve fuel economy, and the longer-term development and deployment of a range of new technologies such as electric and plug-in hybrid vehicles. It will also require strong policies to encourage sustainable, low carbon fuels and sensible changes in travel patterns. It will involve industry, governments and consumers, but government policies will need to lead the way.

As shown in the scenarios presented in this chapter, the combination of maximizing fuel economy and wide-spread adoption of new technology vehicles and fuels, along with changes in travel patterns, could cut oil use and CO_2 emissions from cars around the world by well more than half in 2050, compared to 2005 levels.

However, as also shown, the changes needed will be dramatic and need to happen quickly with fast penetration rates for a range of new technologies. Improvements in fuel economy on the order of 3% per year must replace typical past rates of 1–2% (where it improved at all). Electric and plug-in hybrid vehicles, perhaps representing a bigger change in vehicle technology than anything that has occurred in the past 50 years, must be adopted rapidly, in order to reach commercial scale and maturity as quickly as possible. At the same time, some other trends must stop: those toward ever larger, more powerful cars; trends in some countries toward ever-greater dependence on the car for all types of trips. And while low cost cars can provide mobility to millions of people, society must ask if this is the best way to provide such mobility, rather than (for example) via advanced bus and train systems.

While the costs of key advanced technologies appear high, their potential benefits are also high. It appears worth the 'gamble' that if production at commercial scales occur, so will the benefits of that scale, including learning and optimization, such that costs will come down to the point where eventually electric and perhaps even fuel cell vehicles are on a par (at least on a life-cycle basis with low discount rates) with the cost of owning today's ICE vehicles. And even if costs remain higher than for today's vehicles, these must be compared to the costs spent annually on cars and on transport in general. Through 2020, cars will likely cost more than USD 10 trillion; through 2050 the number could rise to well over USD 100 trillion. The incremental cost of advanced technologies could add a few percent to these already very large numbers. But on the other hand, there will be fuel savings that could offset most or even all of these costs, depending on the price of oil and other fuels. Overall, the costs of taking action and following the sustainable pathway appear likely to be manageable, and may be far lower than the costs of not taking action.

References

Bandivadekar A, Bodek K, Cheah L, Evans C, Groode T, Heywood J, Kasseris E, Kromer M, Weiss M (2008) On the Road in 2035: Reducing Transportation's Petroleum Consumption and GHG Emissions, Report No. LFEE 2008-05 RP, MIT Laboratory for Energy and the Environment, July 2008.

GFEI (2009) 50BY50 – making cars 50% more fuel efficient by 2050 worldwide. Global Fuel Economy Initiative, London

ICCT (2009) Passenger vehicle greenhouse gas and fuel economy standards. 2009 Update. International Council on Clean Transportation, Washington, DC

IEA (2008) World energy outlook 2008. International Energy Agency, Paris

IEA (2009) Transport, energy and CO_2 emissions: moving toward sustainability. International Energy Agency, Paris

IEA (2010) Energy technology perspectives 2010. International Energy Agency, Paris

Part II
European Union Policies

The EU will never be a superpower, but could be a model power of regional cooperation. For success, the EU must be open to ideas, trade and people... Any model power in the 21st century must be a low carbon power, so the European Union must become an Environmental Union.

From a speech of David Miliband, former British Foreign Secretary, at College of Europe, Bruges, Belgium, 15 November 2007. http://news.bbc.co.uk/2/hi/uk_news/politics/7097162.stm. *Last Accessed Mar 2011*

Neither energy security nor climate change can be solved at the national level. Both are at least European in domain, and both require an ability to negotiate at the global level... Europe's policies on energy security and climate change have so far not matched the scale of the challenges it faces.

Dieter Helm, Energy and Environmental Policy: Options for the Future. In Tsoukalis, L (ed) (2009), 'The EU in a world in transition: Fit for what purpose?', pp. 141–152. Policy Network, London. www.policy-network.net/publications_download.aspx?ID=3328. *Last Accessed Mar 2011*

Chapter 4
The Past and the Future of EU Regulatory Policies to Reduce Road Transport Carbon Emissions

Karl-Heinz Zierock

Abstract The paper describes the road decarbonization policy of the European Union from its initial phase in the early nineties to the state-of-play in late 2010. It shows that this policy has been developed from a voluntary approach, aimed at reducing the specific emissions of passenger cars, to a complex integrated strategy. Legally binding efficiency requirements for passenger cars have been set in the meantime and efficiency requirements for other road vehicles are in preparation. More important for the long-term success of the decarbonization strategy, however, are strategic decisions on alternative fuels and alternative power-train technologies. But the key for the long-term success of transport decarbonization lies outside the transport sector: It is the production of renewable electricity, be it as future transport fuel or as energy source for the generation of low carbon transport fuels. Currently no other approach seems to be able to deliver the emission reductions necessary to meet the goal of limiting global temperature increase to maximum 2°C. Most likely it also requires to set-up new regulatory approaches. However, the way toward the realization of this vision is full of obstacles. In fact, it requires re-inventing motorized transport, thus entailing significant changes in the automobile industry, but even more drastic changes in the oil industry.

4.1 Introduction

Behind power generation with a share of 37%, road transport is the second largest carbon dioxide (CO_2) emitting sector in the European Union, with a contribution of about 22%. Moreover, while total EU emissions fell 7.1% from 1990 to 2005, transport emissions rose by about 26% over the same period due to large increases in traffic (European Environment Agency 2010a). Hence, road transport emissions remain one of the few sectors whose emissions keep rising. The sub-sector passenger car usage contributes with about 12% to the overall EU emissions of CO_2.

K.-H. Zierock (✉)
EnviCon – Environmental Consultancy, Berlin 41, Germany
e-mail: Dr.Karl-Heinz_Zierock@t-online.de

Currently road transport in the EU is nearly completely oil-powered; other energy carriers contribute with only about 6% to the total. The EU consumes about 700 million tonnes of oil per year but has a production of somewhat above 100 million tonnes only. About two thirds of the oil products are consumed by the transport sector. Thus, the EU depends heavily on imports to keep motorized road transport (but also transport in general) going (European Commission 2008).

These aspects highlight the two main drivers for the EU policy to reduce car carbon emissions:

- the fight against climate change, and
- the security of energy supply.

In addition, and linked to these two drivers, it is the general objective to promote sustainability, i.e. to develop a more resource-efficient, greener and more competitive economy.

The economic and industrial boundary conditions are at the same time drivers and obstacles for the transport decarbonization policy:

- Although the EU oil reserves are very small in a global context, the petroleum exploration, production, manufacturing and marketing industry makes a significant contribution to the EU economy.
- At the same time the EU is the world's largest producer of motor vehicles. The automotive industry is central for prosperity in a number of Member States. In these countries it is a large employer of skilled workforce and a key driver of innovation.

Looking at the EU in total the economic importance of the automotive industry, expressed in turnover, profits and capitalization at the stock markets, is significantly smaller than the one of the oil industry. In fact, oil firms, including Europe-based companies like British Petroleum, Royal Dutch Shell and Total, are holding since years top rankings on top ten lists of the world's largest companies. (Economy.one GmbH 2010).

In addition to the private international companies, national oil companies play an important role at global level; they own with a share of about 85% the majority of petroleum reserves and hold exclusive rights to exploration of the resources within their home country (Pirog 2007).

4.2 First Steps Toward Regulating Car Carbon Emissions

The issue of CO_2 emission reduction from passenger cars, the largest sub-sector of road transport, appeared on the political agenda of the European Union in the end of the eighties as part of the more general discussion on climate change and possible counter policies. In fact, Article 6 of Directive 89/458/EC requires that

'Acting by a qualified majority on a proposal from the Commission, which will take account of the results of the work in progress on the greenhouse effect, the Council shall decide on measures designed to limit CO_2 emissions from motor vehicles' (Official Journal of the European Communities 1989). In reaction to discussions in established EU automotive expert circles, aiming at identifying an appropriate methodological approach to reduce CO_2 emissions from cars, the European car industry, represented by the European Automobile Manufacturers' Association (ACEA), offered a voluntary agreement as an alternative to legislation. In the light of the fruitless results of the discussions in the expert circles, the Commission reacted positively to ACEA's offer, integrating it as a key corner stone into a more comprehensive Community strategy.

Although Members of the European Parliament (EP), but also non-institutional policy circles, were quite skeptical due to mixed experiences with voluntary approaches and the fact that such agreements partly by-pass the power given to the EP, it finally accepted this way forward, mainly due to the fact that the regulation of CO_2 did not fit into the methodological approaches applied for conventional pollutants so far. The reduction of CO_2 emissions from cars is more complex, encompassing the whole vehicle and not just the power-train. Moreover, in the light of the differences in manufacturers' product portfolio, it was obvious from the beginning that the issue is highly sensitive to competition. Based on the Commission's proposal, the Council (i.e. the EU Environment Ministers) and the European Parliament specified, as an objective for the Community strategy, that an average CO_2 emission figure for new passenger cars of 120 grams of CO_2 per kilometer (gCO_2/km) was to be met by 2010 at the latest (European Commission 1995).

The target of 120 gCO_2/km was a political one, not based on any sort of impact assessment. In fact, it derived from the so-called '3 Liter-Car' discussion which was based on a single Greenpeace demonstration vehicle. Looking backward, this demonstration vehicle can be considered as a very successful public relation campaign, seeing that this vehicle – in market terms – could hardly be considered as a competitive passenger car.

In the following years the European Commission, in co-operation with the Council and the European Parliament, negotiated with ACEA in order to define the details of the agreement. The key issues were, of course, the level of the target to be achieved by technical measures, e.g. those the car manufacturers considered to be under their direct influence, and the year of achievement. In July 1998, the European Commission and ACEA reached an agreement on the reduction of CO_2 emissions from passenger cars (Official Journal of the European Communities 1999). In this agreement, ACEA committed itself to achieve an average CO_2 emission figure of 140 gCO_2/km by 2008 for all its new cars registered in the European Union by technical measures and related market changes. Subsequently, Commitments have also been concluded with the Japanese and the Korean Automobile Manufacturers' Associations (JAMA and KAMA respectively), according to which they had to achieve 140 gCO_2/km by 2009 (Official Journal of the European Communities 2000a, b). In addition, 'estimated interim target ranges' were set for 2003/2004

(For ACEA 165–170 gCO_2/km in 2003; for JAMA 165–175 gCO_2/km in 2003; for KAMA 165–170 gCO_2/km in 2004) in order to guarantee that sufficient progress is made.

As mentioned, the 1995 strategy covered more than just the voluntary agreement. It included measures aimed at influencing consumer behavior, addressing the fact that it is the consumer who finally makes the choice. Therefore two additional pillars were defined: Consumer information by means of car efficiency labelling (Official Journal of the European Communities 2000c) and car taxation (European Commission 2002). According to the strategy, these two pillars were supposed to deliver the missing 20 gCO_2/km between the 140 gCO_2/km target of the voluntary agreements and the Community target of 120 gCO_2/km.

Hence, about 10 years after the issue was raised, the EU policies to reduce car carbon emissions were in place and ready for implementation.

The policy finally proposed and adopted reflects the understanding of the issue at that point of time. In practical terms it was a rather isolated sectorial policy approach. Although it was part of the wider climate change policy, its links to this policy were quite limited and there were no real links to discussions on the general transport policy and the fuel and energy policy. The general understanding was rather that the 'CO_2 & cars' policy will make a needed contribution since it addresses the most important road transport source and does not disturb developments in other policy fields. This was reflected by the First European Climate Change Program (ECCP), which aimed at listing and identifying measures and action necessary to achieve the 8% Kyoto reduction target, but where the 'CO_2 & cars' policy was simply added to the stock rather than integrated in any sort of consistent CO_2 reduction strategy (European Commission 2001a). Nevertheless, this deficit of the EU policy became obvious with the First ECCP and created the drive and background for the search of a better and more integrated decarbonization policy.

4.3 The Implementation and Enlargement of the Strategy and the Search for Alternative Options

In the following years the focus of the EU policy on the decarbonization of road transport was put on four issues:

- To implement the voluntary agreements properly
- To develop further the other two pillars of the 'CO_2 & Cars' strategy, as well as other related measures
- To develop policies for the case that the car industry could not live up to its commitments
- To develop policies for a more sustainable long-term solution.

The main practical task to be tackled in the initial phase with regard to the implementation of the voluntary agreements was the establishment of a reliable monitoring and reporting scheme, based on the so called Monitoring Decision (Official Journal of the European Communities 2000d). This was accomplished in 2002 after a number of improvements in Member States' registration and type approval data banks had been implemented. The crucial check on the interim target ranges to be carried out in 2003 could therefore be based on sound data sets. The monitoring up to 2003 showed that the car industry was on track, e.g. all applicable undertakings specified in ACEA's and JAMA's CO_2 Commitment had been met, and in some cases over-achieved (European Commission 2005a). The 'Major Review' of the Commission, carried out in parallel, came to the conclusions that the reduction in specific CO_2 emissions (i.e. emissions per kilometer) had been overwhelmingly achieved by technological developments, and observed market changes did not significantly influence the CO_2 emission reductions achieved (Mehlin et al. 2004, Commission 2006a).

It is important to note that these findings highlighted also another aspect of the implementation of the strategy: the negligible impact of measures taken under the other two pillars: labelling and fiscal measures. On the contrary: While remarkable changes in the use of technologies for the reduction of the fuel consumption for new passenger cars had been observed within the period between 1996 and 2003, leading to fuel efficiency improvements of the power-trains of individual models in the range of 10–30%, the overall new fleet reduction remained at about 10%. The reason was simple: In a time period of increasing wealth and low fuel prices, but also driven by the features of cars presented by the car industry to the market, consumers tended to purchase heavier and more powerful cars. And this was not an isolated EU trend but applied also to other key markets, e.g. in the United States due to the absence of stricter fuel consumption (CAFE) standards. Thus, the technical improvement was partly used up for other purposes than increasing fuel efficiency.

The Commission's attempts to stop this development by increasing the taxes on fuels (Official Journal of the European Communities 2003a), by improving the car efficiency labelling Directive (Gärtner 2005) and by forwarding a proposal on the approximation of car taxation (European Commission 2005b) did not change the picture, mostly because the attempts were either of limited practical relevance (fuel taxation), half-hearted (labelling) or rejected by the Council (taxation).

In parallel, work on related issues was initiated by the Commission:

- Studies on the reduction of emissions from mobile air conditioning was launched (Rijkeboer et al. 2002)
- Preparatory work on the reduction of the emissions of fluorinated gases from mobile air conditioning system (MAC) was carried out (Callaghan et al. 2003)
- Options to integrate light commercial vehicles into the strategy were studied (Elst et al. 2004).

The work on fluorinated gases was the first to come up with results: In 2006 the Directive 2006/40/EC was adopted, which aimed at controlling the leakage of fluorinated greenhouse gases with a global warming potential (GWP) higher than 150 in MACs and at the prohibition of MACs using those gases (Official Journal of the European Union 2006). This Directive, amended and further improved in the meantime (Official Journal of the European Union 2007a, b), results in $CO_{2equivalent}$ emission reductions of about 80–90%, to be compared with the estimated unregulated case of about 16–28 g/km in 2010 and 19–32 g/km in 2020.

Another preparatory outflow of the work on light duty vehicles was the integration of CO_2 values into the type-approval legislation, which paved the way for future CO_2-related measures for this vehicle segment (Official Journal of the European Union 2004).

In 2004, however, it became also very clear that the car industry had no internal mechanism that could force individual manufacturers to take over their responsibility within the voluntary agreement, finally confirming one of the key objections expressed by some Members of the European Parliament in the very beginning of the process. Consequently, the car industry could not assure anymore that the target would be met in 2008/9 (European Commission 2006a). The Commission concluded that the likelihood that the car industry would live up to its Commitment was very limited. Therefore, the Commission decided to carry out in 2005–2006 a review of the Community's strategy to reduce CO_2 emissions from passenger cars and light commercial vehicles, including legislative options. The 2008 monitoring report of the Commission confirmed the Commission's view: It showed that the average CO_2 emissions from new passenger cars in 2008 was 153.5 gCO_2/km, far away from the 140 gCO_2/km target (European Commission 2009a). Nevertheless, compared to the 1995 situation a progress in specific CO_2-efficiency of 17.5% was achieved, providing evidence that the voluntary approach was at least partially successful.

Fortunately, the Commission was not unprepared for such a situation. Already in 2002 it started studying alternative options (ten Brink 2003, 2005, Elst et al. 2004) and had some ready-for-use policy approaches at hand for the redesign of the strategy (Smokers et al. 2006, Zierock et al. 2007a).

In parallel to this process, focusing more on the two other drivers mentioned in the introduction – sustainable use of resources and security of supply – the more general problem of how to reduce the carbon dependency of transport has been addressed. The key question is: How to decarbonize motorized road transport in an economically sustainable way, e.g. keeping transport affordable and improving energy security. The general answer given is:

- Improvement of efficiency of energy use, and
- Phasing-in of non-carbonaceous or emission-neutral carbonaceous fuels or energy carriers that can be produced in the EU.

Thus, intensive work has been launched in order to identify the most appropriate options for the realization of these goals.

The progress in efficiency of fuel until about 2003 was mainly caused by direct-injection turbocharged (TDI) diesels and the continued market penetration of numerous incremental efficiency technologies. Other technical options like engine downsizing and weight reduction of cars was not on the list of priorities of automobiles companies. On the opposite: cars became heavier and larger and downsizing of engines took place out in only a few cases. Other technologies like hybridization were considered as too expensive. In these years nobody thought that Toyota, for example, made any profit with its hybrid model. And electric battery vehicles were also considered to be no option due to the high price for relatively low-performing batteries – e.g. Ford took its model ('Th!nk') from the market in 2003.

Obviously, there was a need to force the automobile industry to make use of the existing technological potential.

The search for the most appropriate alternative fuel was intensified in these years as well. In its 2001 Communication on alternative fuels for road transport the Commission identified biofuels, natural gas and hydrogen as possible future energy sources for transport (European Commission 2001b). Two years later, under Directive 2003/30/EC on the promotion of biofuels, the EU established the goal of reaching a 2% share of biofuels in the transport sector in 2005 and a 5.75% share by 2010 (Official Journal of the European Communities 2003b). Since biofuels can be produced at least to some degree in Europe, unlike oil and gas, which have to be imported, they can contribute also to the security of supply. As biofuels are more expensive than traditional fuels, the EU also allowed in Directive 2003/96/EC Member States to apply a total or partial tax exemption for biofuels (Official Journal of the European Communities 2003c).

In parallel to this action, the Commission established the 'European Hydrogen and Fuel Cell Technology Partnership', aiming at the development of a broad and far-reaching hydrogen and fuel cell strategy at the EU level in order to secure the EU's position as a leading world-wide player in the supply and deployment of hydrogen technologies (European Commission 2003a).

While attempts to establish natural gas as a key alternative transport fuel for the EU failed (European Commission 2003b), intensive work on the overall energy and greenhouse gas efficiency of alternative fuels was launched by these early initiatives. Eucar,[1] Concawe[2] and the JRC[3] started at that point of time their still ongoing joint evaluation of the Well-to-Wheels energy use and greenhouse gas emissions for a wide range of potential future fuels and power-train options (Concawe, Eucar, JCR 2003). This work has been playing a major role in the development of the Commission's decarbonization strategy.

Although the EU policies on the phasing-in of non-carbonaceous or emission-neutral carbonaceous fuels was based on an overall target, in practical terms it concentrated on the blending of conventional fuels with biofuels. This has the advantage that the existing distribution infrastructure can be used. The mineral oil industry – somewhat hesitating in the beginning – identified quickly the advantages of this approach compared to the marketing of pure alternative fuels: It stretches the mineral oil reserves and keeps other players more or less out of the market.

---

[1] European Council for Automotive Research and Development.
[2] Conservation of clean air and water in Europe. The oil companies' European association for environment, health and safety in refining and distribution.
[3] Joint Research Centre of the European Commission.

However, not only the voluntary car efficiency improvement approach fell short. In the 2007 biofuels progress report the Commission had to state that biofuels reached only 1% of the 2005 market and that it was unlikely that the EU would meet its 5.75% target for 2010.

Another development to be mentioned in this respect is the general transport policy of the European Union. In the White paper of the Commission 'European transport policy for 2010: time to decide' of 2001 and the mid-term review of 2006 the Commission proposed and reviewed a large number of measures aimed at improving the efficiency of the European transport system (European Commission 2001c, 2006b). A more efficient, sustainable transport system can contribute significantly to the strategic goals mentioned in the introduction of this chapter. However, the list of possible measures is long and often difficult to implement, and requires sometimes to take unpopular decisions (see for example Zierock et al. 2006). Nevertheless, the general transport policy is another important line to be considered in the decarbonization of road transport, which sets a frame for the efficiency improvement and fuel policy.

In summary, in 2006/7 the EU's policies on the decarbonization of transport showed some results but did not fully match the goals set at the turn of the century. While the actual level of integration of policies was still unsatisfying, intensive discussions and research took place in order to find answers to the obvious question: What will be the best future vehicle technology/alternative fuel combination?

This may be the right occasion to mention that – although the history of the EU decarbonization policy is described in this chapter mainly on the basis of EU legislation – the EU policy is always a result of an in-depth exchange of views between the Commission, Member States, the European Parliament, stake holders and the scientific community. Thus, the search for the best way forward took place and is taking place throughout the EU.

With regard to the scientific work the evaluation of Eucar, Concawe and JRC on the overall efficiency of different energy/power-train combination showed the importance of the whole energy pathway, including primary energy generation, for the solution of the problem (Concawe, Eucar, JCR 2007, 2008). It widened the scope of the political discussions and opened the way toward a cross-sectoral, more integrated policy.

Another aspect that influenced discussion on the decarbonization of transport was the general climate change policy. The post-Kyoto discussions had developed to a common understanding that very large Greenhouse Gas (GHG) emission reductions are needed in order to meet the goals. For the EU the climate change policy became the top policy driver and moved as a permanent topic on the agenda of European Council meetings (i.e. meetings of the EU Heads of State). It became obvious to all sides that this policy already now, and even more in future, will have a major impact on the economy. This fact has brought the focus – more than in the past – to questions of burden sharing and competition. While in the first and second phase of policy development measures could still be considered as no-regret and low-hanging fruit issues, the next phase could not be developed without significant repercussions for one or the other side concerned.

An issue of importance to be tackled with regard to the decarbonization of transport was the question whether the re-design of the failed 'CO$_2$ & cars' policy should be carried out under the umbrella of the emission trading policy or not. It was finally decided to develop a separate strategic approach for the reduction of CO$_2$ emissions from cars instead of integrating this sector into the emissions trading system. Apart from some practical aspects like monitoring and transaction costs (which are, however, also part of all other policy approaches) the key driver behind this decision was certainly the fact that a car-specific policy allows to achieve higher emission reductions. The integration into the emissions trading approach, although considered by the industry as more cost-effective, has the risk that the full technological potential would not be used. Instead of using available emission reduction technology, emissions allowance certificates would be bought by the car industry and the costs would be passed to the consumer. Since emission reductions in other sectors, e.g. power generation, appear less costly, the price for the emissions allowance certificate would be lower than many of the possible technical measures available to the car sector. In particular Member States without car industry were opposed to this approach. They saw advantages in an approach that puts more burden on sectors which contribute little to their national economies. This decision, however, entailed that Member States with a significant car industry, as well as associated industries, had and have to take over more economic risks than others.

4.4 The New Decarbonization Strategy

It was not by coincidence, but a result of the more comprehensive understanding of the problem, that the Commission forwarded in parallel to the new car emissions strategy (European Commission 2007a) a proposal (European Commission 2007b) on the introduction of compulsory requirements aimed at the gradual decarbonization of road fuels.

In addition the Commission published a Renewable Energy Roadmap, calling for a mandatory target to satisfy 20% of Europe's energy demand and 10% of transport energy demand from renewable sources by 2020 (European Commission 2007c). The target was endorsed by EU leaders in March 2007.

Based on supporting studies (Fergusson et al. 2007), the Commission's revised strategy on the reduction of carbon dioxide emissions from new cars and vans sold in the European Union aims at limiting average CO$_2$ emissions from new cars to 120 gCO$_2$/km by 2012 – a reduction of around 25% from 2006 levels (European Commission 2007a). Due to the experience gained under the Voluntary Agreement, the Commission proposed a legislative framework to achieve the target in order to ensure that progress is made.

A key element is a mandatory reduction of the emissions of CO$_2$ to reach the objective of 130 gCO$_2$/km for the average new car fleet by means of improvements in vehicle motor technology. As part of the Regulation a burden sharing proposal was forwarded, which requires the small fraction of heavier cars to achieve significantly larger reductions of emissions per kilometer than the large fraction of light cars. This proposal of the Commission was quite controversial, firstly because it was far away from competition neutrality and secondly because it took insufficiently into account the issue of car ownership and future fleet composition (see for example Papagiannaki and Diakoulaki (2007) that shows the important impact of these parameters) and therefore underestimated the reduction potential of smaller cars. The burden sharing proposal was finally accepted, after a number of regulatory details were added that increased the flexibility of the approach.

As part of the strategy a further reduction of 10 gCO_2/km had to be realized by other technological improvements and by an increased use of biofuels, more specifically (Zierock 2009):

- the compulsory fitting of accurate tire pressure monitoring systems;
- setting maximum tire rolling resistance limits in the EU for tires fitted on passenger cars and light commercial vehicles;
- the use of gear shift indicators;
- fuel efficiency progress in light-commercial vehicles (vans);
- the setting of minimum efficiency requirements for air-conditioning systems; and
- increased use of renewable fuels.

In this way the increase of the biofuel content in fuels became one element of the integrated approach to reduce CO_2 from cars, bringing together the issues of vehicle technology and fuels.

In addition, as part of the revised strategy, the Commission announced a revision of the fuel efficiency labelling Directive 1999/94/EC. Furthermore, it explained that new and additional attention will be paid to the definition of the Light Environmentally Enhanced Vehicles (LEEV). Finally, the Commission announced that a voluntary agreement on an EU-wide code of good practice regarding car marketing and advertising should be signed with the automotive industry. The key elements of the new strategy are shown in Table 4.1.

By the beginning of year 2011, the implementation of the strategy was well on track:

- Regulation (EC) No 443/2009, which sets emission performance standards for new passenger cars, has been adopted (Official Journal of the European Union 2009a).
- Measures to increase the biofuel content in fuels have been adopted (Official Journal of the European Union 2009b).
- The low carbon fuel standard established in the fuel quality Directive (Official Journal of the European Union 2009b).
- The new Regulation on the general safety of motor vehicles contains requirements on the rolling resistance of tires and on gear shift indicators (Official Journal of the European Union 2009c).
- In an additional regulation rules on tire labelling, including the rolling resistance have been laid down (Official Journal of the European Union 2009d).

4   The Past and the Future of EU Regulatory Policies to Reduce Road Transport ...   83

Table 4.1 Elements of European commission's 2007 strategy on CO_2 reduction from passenger cars

Pillar	Measure	Target	Target year
Technical measures	– Forcing EU regulation	130 gCO_2/km	2012
Demand related measure	– Improved energy efficiency labelling – Codex for advertising	Not specified	About 2009/10
Taxation measure	– Approximation of passenger car taxation – Definition of LEEV	Not specified	Not specified
Other measures	– CO_2 reduction measures for light commercial vehicles – Extended use of alternative fuels – Tire rolling resistance – Tire pressure control – Gear shift indicator – Improvement of energy efficiency of mobile air conditioning equipment	Further reduction by 10 gCO_2/km	About 2012

- Details concerning tire pressure monitoring are currently under discussion in expert circles, and preparatory steps to introduce fuel efficiency requirements for mobile air conditioning are being taken.
- Finally, the Commission proposed a Regulation to reduce CO_2 emission from light commercial vehicles (European Commission 2009b), which was adopted by the Council and the European Parliament in January 2011.

All these measures mainly aim at implementing the existing technical potential. No progress however, could be made with regard to the advertising codex and car taxation.

With regard to car taxation, measures are blocked since years due to principal reservations of a number of Member States concerning EU legislation on taxation in general. However, the concept as such penetrates policies in Member States: The number of countries basing car taxation either wholly or partially on a vehicle's CO_2 emissions is steadily increasing, and currently 17 of the EU's 27 Member States apply CO_2-based car taxation (Bastard 2010, ACEA 2010; see also Chapter 8 of this book).

The tendency that Member States take national action in policy areas which in theory should better be regulated at EU level, but for which no majority can be found internationally, applies also to car efficiency labelling. Many Member States used the frame given in Directive 1999/94/EC to develop more efficient national approaches, most recently Germany (Zierock et al. 2007b, Zierock 2010).

Related to the strategy, the Directive on the promotion of clean and energy-efficient road transport vehicles has been adopted (Official Journal of the European

Union 2009e). It requires that public authorities should take into account energy and environmental impacts linked to the operation of vehicles over their lifetime for public procurement. This gives a competitive advantage to green vehicles and provides strong support to their broad market introduction.

With regard to the general climate change policy a number of milestone decisions have been taken in 2009, which also provide important boundary conditions for the future decarbonization of road transport policy:

- The so called energy and climate package (Official Journal of the European Union 2009f–2009i) has been adopted, which is embedded in a set of Climate and Energy directives, guidelines and decisions with, inter alia, the following objectives for 2020:
   - Binding 20% GHG reduction until 2020 and independent EU commitment of 30% GHG reduction compared to 1990 in context and under condition of an international agreement,
   - 20% renewable share of final energy consumption,
   - 20% improvement of the energy efficiency,
   - 10% renewable share in transport, with focus on production being sustainable, linked to a priority for second-generation biofuels commercially available,
   - 6% reduction in life cycle GHG intensity of energy used in road transport and in the non-road mobile machinery sector,
   - Use of renewable electricity for transport.
- Under the two Directives on fuel quality and renewable energy (Official Journal of the European Union 2009b, 2009f), biofuels can only be accounted for if they produce emissions savings of at least 35% compared to fossil fuels. It aims at ensuring that only sustainable biofuels, which generate a clear and net GHG saving and have no negative impact on biodiversity and land use, should be used by establishing sustainability criteria for biofuels and bioliquids. The Commission was still working in early 2011 to develop a methodology for calculating the lifecycle greenhouse gas emissions of fossil fuels under the fuel quality directive (Official Journal of the European Communities 1998) to make the low carbon fuel standard (LCFS) operational and to provide a benchmark against which renewable fuels will be benchmarked to deem whether they produce sufficient emissions savings.
- The renewable energy Directive establishes a common framework for the promotion of energy from renewable sources. Member States have to meet mandatory national targets for the overall share of energy from renewable sources in gross final consumption of energy and for the share of energy from renewable sources in transport.
- At the level of Heads of States the EU has announced at the European Council meeting of October 2009 its support for an ambitious long term objective of reducing CO_2 emissions by 80–95%, compared to 1990 levels, by 2050, in the context of action by developed countries as a group (European Council 2009).

The decisions mentioned above reflect the further improved understanding of the road decarbonization problem and take into account technological decisions taken by industrial stakeholders: On one hand the identified difficulties associated with a large volume use of biofuels, e.g. to comply with sustainability criteria, showed the current limits of this option, on the other hand the car industries' decision in favor of battery-supported power-trains for the years to come, in addition to a further significant improvement of the conventional power trains, paved the way for an increased use of electricity as key energy carrier for road transport.

In fact, with the adoption of Regulation 443/2009 the time to decide could not be moved further into the future anymore by the automobile industry. Practically the battery-supported power-train is currently the only option to achieve significant reduction of the specific emissions of passenger cars and light duty vehicles. It has the additional benefit that electrification of power-trains can be achieved incrementally (e.g. from stop-start systems, through increasing degrees of hybridization, to plug-in hybrids and ultimately battery electric vehicles) The other main option, fuel cell technology, although in technological power-train terms nearly as well developed as battery technology, proved to be – for the time being – less realistic for many reasons, in particular due to generally higher power-train costs of this technology and the lack of a hydrogen distribution system, resulting in hesitation of potential players in this field to move into this direction.

The increase of oil prices since the turn of the century, culminating in an oil price shock in 2008, accelerated this decision as well. With huge market chances in the rapidly growing less industrialized countries in vision, the automotive industry needed urgently a convincing technical concept for future individual motoring at acceptable prices. And in doing so, as a desired side effect, it plays the CO_2 emission ball into the yard of the primary energy providers.

The climate change and energy package reacted, not only for this reason, by setting targets for the generation of electricity from renewable sources. Directives on the liberalization of the electricity market provided another step stone toward an increase of the overall efficiency of primary energy supply as well as the use of alternative primary energy sources. In addition, the European Directive for the promotion of renewable energy asks Member States to define national sectoral targets helping to reach the binding overall national targets. This sets the scene for the discussions on the next steps to be taken within the decarbonization strategy of road transport.

4.5 Options for the Future

For the near future, up to about 2020, the general targets as well as the way forward seems to be defined: Under the umbrella of the agreed greenhouse gas emission reduction target of 80–95% until 2050, compared to 1990, and in order to achieve the goal of 2°C as the maximum acceptable average temperature increase (European Council 2009), the target of 95 gCO_2/km for passenger cars has to be met as well as the targets for biofuels and renewable primary energy generation.

The policy concentrates on the implementation of the decarbonization strategy (Barroso 2009). In the Commission's Communications on Europe 2020 (European Commission 2010a), on 'A sustainable future for transport: Toward an integrated, technology-led and user friendly system' (European Commission 2009c) and 'A European strategy on clean and energy efficient vehicles' (European Commission 2010b) a number of political, organizational as well as practical issues have been addressed, which need to be tackled in the coming years:

- to increase consumer acceptance, e.g. the safety, standardization, electricity consumption, environmental aspects and affordability of electric vehicles,
- to enable synergies with smart grids, promote renewable sources of energy and offer a possibility of energy storage,
- to speed up the market uptake of clean and energy-efficient vehicles, in compliance with existing State aid rules and the principle of subsidiarity,
- to create business confidence by a well-timed and well-tailored public policy,
- to focus on research excellence in order to ensure that alternative power-trains receive targeted research financing,
- to ensure access to, recycling and recovery of indispensable materials, including rare earth elements and notably lithium reflecting their importance for the production of alternative power-train components, inter alia, batteries,
- to take all necessary measures so that a skilled and qualified workforce is available for alternative power-train and energy-efficient technologies,
- to accelerate the standardization of interfaces in view of the interoperability between electric vehicles and the charging infrastructure, in order to ensure that electric vehicles can be recharged, domestically or at public station points, without difficulty within the territory of the EU and with the use of any electric vehicle charger,
- to re-launch earlier stakeholder consultation processes, in order to allow stakeholders to contribute to strategic regulatory policies.

The last point underlines that Commission aims at a high level coordination across relevant policy areas in order to ensure internal coordination and co-operation with industry and Member States. Inside the Commission, in organizational terms, the strategy to reduce CO_2 emissions from road vehicles is coordinated under the European Climate Change Programme (ECCP) and the integration of this strategy into the overall EU transport policy with the March 2011 White Paper on the European Transport Policy has to be ensured and intensified (European Commission 2009c). This includes also the link to the sustainable development policy, which shows first signs of decoupling transport volumes from economic growth (European Commission 2009d, Eurostat 2009).

It can be assumed that this as well as other, national action programs, will pave the way for an increased electrification of power trains. However, pure battery-electric vehicles still have a limited use, e.g. mainly inside cities where they compete with public transport, and are quite expensive. Thus, their market success is far from being granted. Innovative business models are needed in order to boost consumer acceptance and overcome the remaining barriers, such as high battery costs, green electricity supply and charging infrastructure.

In a number of studies the needed contribution of electric or other zero-emission or nearly zero-emission vehicles to the achievement of the agreed GHG emission reduction target have been studied, assuming that the transport sector has to reduce emission in the same order of magnitude (e.g. European Environment Agency 2010b, Skinner et al. 2010, McKinsey 2010a). All studies come to the conclusion that these technologies have to achieve significant market shares of 60% to 80% or more in order to get closer to the goal. However, the economic development and customer acceptance define the demand for more fuel efficient cars; these influencing parameters cannot be changed by legislation. Therefore, some studies came to the conclusion that efficiency improvements and zero emission vehicles alone will not be enough to meet the EU's reduction goals of 2050 (European Environment Agency 2010b, Skinner et al. 2010). The scenarios showed that the greatest savings potential arises from combined packages, in which technological improvements that reduce fuel consumption are used alongside measures to shift journeys to lower emission modes and to avoid the need to travel altogether, e.g. by measures like high density, mixed use land planning. Scenarios of the International Energy Agency presented in Chapter 3 of this book confirm this finding. It should be mentioned that all these studies have to be interpreted with care since their results depend heavily on assumptions with regard to the general economic development, e.g. GDP increases, the world oil supply and its impact on fuel prices, the economic viability of alternative transport fuels, and to the R&D outcomes in several transport-related technologies. Moreover, mainly linked to GDP, the needed reduction percentages depend heavily on the forecasts of transport volumes.

In any case, electric mobility can actually contribute only little to the achievement of the 95 gCO_2/km target, laid down in Regulation 443/2009 due to low market penetration forecasts up to 2020. Thus, although electric cars may make an important contribution to reducing road transport emissions in the long run, conventional engines will play a predominant role for the foreseeable future and well beyond 2020. Further improvements and important contribution to overall CO_2 reductions are still possible on the primary power-train in the next decades. Therefore, efficiency requirements for passenger cars and light commercial vehicles will have to be reviewed and other types of road vehicles, e.g. two- and three-wheelers and quadricycles as well as heavy-duty vehicles, have to be covered in the coming years. This has also to include a revision of the type approval procedure in order to bring tested values closer to those reported under real world conditions (see for example Johannsen 2010).

Reviews of the current fuel efficiency policy will most likely also have to look at the approach as such again since the introduction of a completely new technology like electric vehicles does not fit easily to the general manufacturer-related approach laid down into Regulation 443/2009: In the transitional phase the manufacturers can hardly be made responsible for the up-take of new technology like electric vehicles by the market. A bonus system for low-emitting vehicles, as currently and time-limited applied in the Regulation, will most likely not be sufficient.

It should be underlined that the EU road decarbonization strategy continues embedding hydrogen fuel and hydrogen vehicles as a medium term option. Hydrogen technology is a quite universal energy carrier and combines different sources of primary energy but, among other things, is far from being cost-effective. Nevertheless hydrogen can offer the opportunity to 'de-carbonize' the transport energy system in the mid-term, maybe starting with a sub-sectorial approach for the heavy duty vehicle sector, and is therefore worth to stay on the list of options to be developed further.

A further increase of the use of biofuels remains, of course, an option if the current sustainability criteria discussions lead to a positive result. The key uncertainty at present is the issue of GHG emissions due to indirect land use change. This raises a real risk that many types of land-based biofuels may result in much smaller or negligible GHG savings than had previously been thought. Under the fuel quality and renewable energy Directives, biofuels can only be accounted for if they produce emissions savings of at least 35% (50% in 2017 and 60% in 2018) compared to fossil fuels (including sources like oil sands and deep sea exploitation). The legislation includes a ban on biofuels planted in protected areas, forests, wetlands and 'highly biodiverse' grasslands. The Commission is now developing a methodology for calculating the life-cycle greenhouse gas emissions of diesel and petrol under the fuel quality Directive, against which renewable fuels will be benchmarked to deem whether they produce sufficient emissions savings.

The huge specific land use and the associated land use change, entailing potentially additional CO_2 emissions, are currently the big disadvantage of biofuels. Moreover, if biofuels are significantly limited in large-scale production, they should better be used in future for air transport and shipping since for these transport modes electric mobility and hydrogen technology are less or not attractive. In addition, direct use for heat and energy production is an energy-efficient alternative for biofuel use and is therefore competing with transport fuel use. Thus, little would be left for the decarbonization of road transport. In any case biofuels, at the current state-of-play, cannot fuel future road transport alone.

This highlights that, although important decisions in Europe with regard to the near future technology mix have been taken, a large number of uncertainties still exist with regard to the optimal future transport approach.

While there is still no single favorite future 'final' technology/fuel mix in sight, it is in any case necessary, next to the further improvement of conventional propulsion technologies, to increase the production of regenerative electricity in order to make electric mobility, but also hydrogen, a valid option from an environmental point of view. The environmental success of these two future fuel options will strongly depend on the GHG emissions of primary energy. This is underlined by the fact that there are physical limits for the specific efficiency gains of marketable conventional vehicles, which might be in the range of 60–80 gCO_2/km for the EU fleet, depending, among other things, on the composition of the fleet.

Therefore, 'green electricity production' is the key to the decarbonization of transport and the decrease of EU's oil dependence.

Scenarios show that in particular wind and solar energy production could cover in future fully the primary energy demand of the EU, leading to a significant amount of residue energy that needs to be used or stored in one or the other way (McKinsey 2010b, Nitsch et al. 2010). The production of hydrogen via electrolysis could become an option, which would change the cost balance of this fuel and make it more attractive. But there are, of course, also other options to make use of the residue energy. The discussions on the future transport policy (Commission 2009c) as well as the development of a new Energy Action Plan for Europe, expected to be presented by the Commission in the beginning of 2011, provide excellent opportunities to give the right signals for a move toward these alternative energy sources and energy carriers.

The Energy Action Plan may also have an impact on the fine-tuning of the current strategic 'decarbonization' approach of the Commission. Up to now it is quite simple: The idea is that in the coming decades the costs of alternative technology (vehicle technology and fuels) should decrease while the costs of conventional technology and in particular fuels should increase. At a particular point of time there would be a mutual shift in technology due to costs advantages of the alternative technology approach. The big players in this game are the governments in so far as they influence, via fuel taxes, the price of fuels and, via vehicle taxes, the price of vehicle technology. However, the governments are significantly affected by such measures as well, since they profit substantially from these taxes. Thus, the timing needs to be fine-tuned. Currently we are far away from the shift point: In cost terms, only CNG and LPG vehicles are able to compete with conventional gasoline and diesel cars. In fact, in Germany for example the costs per kilometer of lower mid-class vehicles are nearly identical in the range of 0.45–0.50 EURO/km, irrespective whether gasoline, diesel, CNG, LPG or an ethanol-gasoline blend (E85) is used as fuel. Thus, cost is not the only decisive factor for the consumer. Already now CNG, LPG and E85 enjoy significant fiscal incentives and are therefore subsidized by the government without making a breakthrough. Other alternative technologies like electric or hydrogen vehicles may need significantly more support in order to become successful.

Therefore the big question whether this strategy will work out remains open for the time being. It requires very sensitive timing of measures and depends on many parameters difficult to predict. Obviously, the future policy development will become an iterative process which might include frequent policy reviews.

4.6 Heading Toward a 'New Deal'

In addition to the European situation, the Commission has to take into account the world-wide developments when defining a successful decarbonization strategy since transport activity is increasing significantly around the world as economies grow (Metz et al. 2007). In fact, the vehicle market is growing and until 2030 the global passenger car fleet is projected to double in size. EU's efforts to meet the 2°C goal might be jeopardized easily be developments in other parts of the world. This does not mean that the EU should slow-down its efforts to decarbonize road transport. However, it has to take care that the so called 'green paradox' is avoided: Efforts to decarbonize transport emissions in the EU might have the effect of increasing CO_2 emissions because the efficiency improvement measures result in a downward pressure on prices for conventional fuels (Sinn 2008).

The fact that energy consumption in transport has also been addressed in other countries and resulted in legislative measures does not solve the problem since the currently adopted approaches lead mainly to a stretching of oil reserves (Delphi 2009). If these measures remain uncoordinated, they might even lead to regionally different technological developments that could increase the reduction costs significantly. Coordination and a joint vision are urgently needed.

The target is clear: the burning of oil has to be slowed down significantly between now and 2050 and may need to be even stopped completely: Limiting cumulative CO_2 emissions to a 25% probability of warming exceeding 2°C means that less than half of the proven economically recoverable oil, gas and coal reserves can still be emitted up to 2050. Emitting all known oil reserves leads to about the same result (Meinshausen et al. 2009). Thus, to meet the 2°C goal means that one cannot allow the last droplet of oil to be burnt.

An alternative regulatory approach, which could help avoiding the 'green paradox', would be the establishment of a fixed global emissions limit, compatible with the 2°C goal, and a global emissions trading system on worldwide fossil fuel consumption, e.g. on the basis of the allocation principle 'one human-one emissions right' (Wicke et al. 2010). Of course, the 'one human-one emissions right' approach needs to be phased-in and based on appropriate national targets, taking into account birth control efforts and other national or regional parameters (Höhne and Moltmann 2009). Calculations show that the average CO_2 emission per capita in 2050 should be in the range of 0.7–2.4 t $CO_{2equivalent}$. While India and Brazil are currently still in that range, the EU, the USA, China and Russia are far away from such values (Olivier and Peters 2010). In particular China, as well as other non-Annex I countries, still increase per capita emissions rapidly.

One could also define appropriate national per capita sub-caps, based on worldwide burden sharing agreements, for road transport. This plays the ball back into the yard of those who are finally responsible: The individuals who decide with their individual lifestyle about their level of CO_2 emissions. They could react in a more complex way than sectorial legislation can achieve, be it by moving closer to their workplace, by buying more CO_2 efficient vehicles, by using more often public transport or non-motorized means of transport and so on. While such options seem to be visible already now for the private transport sector, they would still have to be developed for the goods transport sector. Currently the efficiency improvements made in the motorized individual transport sector are in many countries overcompensated by the increase in the transport volume of goods. Improving conventional vehicles will be insufficient. No decision on the technology/fuel mix is in sight for the moment for the road transport sector, and measures aimed at improving overall transport efficiency do not show the desired effect up to now.

Technically it is less difficult to set up and manage a 'one human-one emissions right' system than often thought. It requires a 'Fuelling Card' – quite comparable to the normal, well established credit card systems – and to allocate one card to each motorized road vehicle owner. The owner and their families, enterprises etc., would then have to obtain the needed CO_2 road transport allowances.

There is another aspect to consider: Currently the production and recycling of a conventional passenger car causes CO_2 emissions of about 4.5 t CO_2, corresponding to around 0.3 t CO_2/year for a car lifetime of about 15 years. The lifecycle balance of alternative vehicles has still to be seen. Thus, in order to achieve such low per capita values, the production, useable lifetime and recycling of vehicles need to be improved as well.

As mentioned in the introduction of this chapter, the automotive and the oil industry, including national oil companies that are, in practical terms, represented by governments at the negotiation tables, are the big players in this game. The automotive industry can cope with reasonable requirements on new technology and energy efficiency improvements since it can build vehicles for nearly any energy carrier available. The risks for the automotive industry are huge R&D budget and marketing requirements for multiple technology developments to be invested into products with unclear market chances. This includes the risk that regional markets decide in favor of different technology/fuel approaches. For the EU with a strong automobile industry this requires to set up a balanced policy framework based on realistic EU but also world-wide market assessments, regarding alternative power-trains as well as conventional internal combustion engines. The oil industry, however, would have to change their business model completely since the future technology has to be carbon-free. The question therefore is: How can the oil industry be transformed into a 'clean fuel' industry? Currently, the oil industry makes little effort to go that way, e.g. by investing heavily into alternative fuel research. On the opposite, the oil industry, be it the private or national, invests the overwhelming part of its profits into the identification and exploitation of new oil and gas reserves (United States Senate 2007, New York Times 2009). This highlights the huge challenge associated with any serious decarbonization policy: It requires rebuilding a significant part of the current economy. Currently, there is no policy in sight which will help or force the oil industry to develop an alternative to its business model. It is time to seriously think about it.

References

ACEA (2010) Tax guide 2010. European Automobile Manufacturers Association. Brussels Belgium

Barroso JM (2009) Political guidelines for the next commission. Document sent to the President of the European Parliament. http://ec.europa.eu/commission_2010-2014/president/about/political/index_en.htm

Bastard L (2010) The impact of economic instruments on the auto industry and the consequences of fragmenting markets – focus on the EU case. OECD/ITF Discussion paper 2010-8

Callaghan P, Vainio M, Zierock K-H (2003) Options to reduce greenhouse gas emissions due to mobile air conditioning. Summary of the discussions. Brussels 10-11 February 2003
Concawe, Eucar, JCR (2003) Well-to-wheels report. Version 1b, December 2003
Concawe, Eucar, JCR (2007) Well-to-wheels report. Version 2c, March 2007
Concawe, Eucar, JCR (2008) Tank-to-wheels report. Version 3, October 2008
Delphi Cooperation (2009) World-wide emission standards: passenger cars & light duty vehicles 2010/2011. www.delphi.com
Economy.one GmbH (2010) Die 500 größten börsennotierten Unternehmen Europas
Elst D, Gense R, Riemersma IJ, van de Burgwal HC, Samaras Z, Fontaras G, Skinner D, Haines D, Fergusson M, ten Brink P (2004) Measuring and preparing reduction measures for CO_2-emissions from N1 vehicles. Final report to EU Study contract B4-3040/2003/364181/MAR/C1
European Commission (1995) The community's strategy to reduce CO_2 emissions from passenger cars and improve fuel. COM (1995) 689 final
European Commission (2001a) European climate change programme report. Brussels, June 2001
European Commission (2001b) On alternative fuels for road transportation and on a set of measures to promote the use of biofuels. COM (2001) 547 final
European Commission (2001c) White paper: European transport policy for 2010: time to decide. COM (2001) 370 final
European Commission (2002) Communication from the commission to the council and the European Parliament: taxation of passenger cars in the European Union – options for action at national and community levels. COM (2002) 431 final
European Commission (2003a) Hydrogen energy and fuel cells – a vision of our future. Final Report of the High Level Group. 20719 EN. Brussels 2003
European Commission (2003b) Market development of alternative fuels. Report of the alternative group fuels contact group. December 2003, Brussels
European Commission (2005a) Communication from the commission to the council and the European Parliament: implementing the community strategy to reduce CO_2 emissions from cars: fifth annual communication on the effectiveness of the strategy. COM (2005) 269 final
European Commission (2005b) Proposal for a council directive on passenger car related taxes. COM (2005) 261 final
European Commission (2006a) Implementing the community strategy to reduce CO_2 emissions from cars: sixth annual communication on the effectiveness of the strategy. COM (2006) 463 final
European Commission (2006b) Keep Europe moving – sustainable mobility for our continent. Mid-term review of the European Commission's 2001 Transport White Paper. COM (2006) 314 final
European Commission (2007a) Results of the review of the community strategy to reduce CO_2 emissions from passenger cars and light-commercial vehicles. COM (2007) 19 final
European Commission (2007b) Proposal for a directive of the European Parliament and of the council amending directive 98/70/EC as regards the specification of petrol, diesel and gas-oil and introducing a mechanism to monitor and the introduction of a mechanism to monitor and reduce greenhouse gas emissions from the use of road transport fuels and amending council directive 1999/32/EC, as regards the specification of fuel used by inland waterway vessels and repealing directive 93/12/EEC. COM (2007) 18 final
European Commission (2007c) Renewable energy road map renewable energies in the 21st century: building a more sustainable future. COM (2006) 848 final
European Commission (2008) European energy and transport, trends to 2030, update 2007. Directorate-General for Energy and Transport. Office for Official Publications of the European Communities, 2008. ISBN 978-92-79-07620-6
European Commission (2009a) Report from the commission to the council and the European Parliament: monitoring the CO_2 emissions from new passenger cars in the EU: data for the year 2008. COM (2009) 713 final

European Commission (2009b) Proposal for a regulation of the European Parliament and of the council: SETTING emission performance standards for new light commercial vehicles as part of the Community's integrated approach to reduce CO_2 emissions from light-duty vehicles. COM (2009) 593 final

European Commission (2009c) A sustainable future of transport: towards an integrated, technology-led and user friendly system. COM (2009) 279 final

European Commission (2009d) Mainstreaming sustainable development into EU policies: 2009 review of the European Union Strategy for Sustainable Development. COM (2009) 400 final

European Commission (2010a) Europe 2020: a strategy for smart, sustainable and inclusive growth. COM (2010) 2020 final

European Commission (2010b) A European strategy on clean and energy efficient vehicles. COM (2010) 186 final

European Council (2009) Conclusions of the council meeting of October 2009

European Environment Agency (2010a) Annual European greenhouse gas inventory 1990–2008 and inventory report 2010. ISBN 978-92-9213-100-5

European Environment Agency (2010b) Towards a resource-efficient transport system – TERM 2009: indicators tracking transport and environment in the European Union 2010. ISBN 978-92-9213-093-0 DOI 10.2800/40099

Eurostat (2009) Sustainable development in the European Union 2009 monitoring report of the EU sustainable development strategy executive summary. Product code: KS-78-09-865, ISBN: 978-92-79-12695-6

Fergusson M, Smokers R, Passier G, ten Brink P, Watkins E, Valsecchi C, Hensema A (2007) Possible regulatory approaches to reducing CO_2 emissions from cars. Final report to EU Study contract 070402/2006/452236/MAR/C3

Gärtner A (2005) Study on the effectiveness of directive 1999/94/EC relating to the availability of consumer information on fuel economy and CO_2 emissions in respect of the marketing of new passenger cars. Final Report to the European Commission, Directorate-General for Environment; Contract No.: 07010401/2004/377013/MAR/C1

Höhne N, Moltmann S (2009) Sharing the effort under a global carbon budget. Ecofys Germany GmbH, Berlin

Johannsen R (2010) Weiterentwicklung der EU-Richtlinie zur Messung der CO_2-Emission von Pkw-Untersuchung der Einflüsse verschiedener Parameter und Verbesserung der Messgenauigkeit. Umweltbundesamt, FE-Vorhaben, FKZ 3709 52 141. In print

McKinsey & Company (2010a) Beitrag der Elektromobilität zu langfristigen Klimaschutzzielen und Implikationen für die Automobilindustrie. Abschlussbericht 30 April 2010. Study for the BMU, Germany. In print http://www.bmu.de/files/pdfs/allgemein/application/pdf/elektromobilitaet_klimaschutz.pdf

McKinsey & Company (2010b) Roadmap 2050: a practical guide to a prosperous, low-carbon Europe, European Climate Foundation, ECF

Mehlin M, Gühnemann A, Aoki R, Schmid H-P (2004) Preparation of the 2003 review of the commitment of car manufacturers to reduce CO_2 emissions from M1 vehicles. Final report to EU Study contract B4 – 3040/2002/343537/MAR/C1 and 070501/2004/377441/MAR/C1

Meinshausen M, Meinshausen N, Hare W, Raper SCB, Frieler K, Knutti R, Frame DJ, Allen MR (2009) Greenhouse-gas emission targets for limiting global warming to 2°C. Letters to nature. Nature 458, 30 April 2009

Metz B, Davidson OR, Bosch PR, Dave R, Meyer LA (eds) (2007) IPPC 2007: climate change 2007: working group III: mitigation of climate change: contribution of working group III to the fourth assessment report of the intergovernmental panel on climate change, 2007 Cambridge University Press, Cambridge, UK and New York, NY, USA

New York Times (2009) Some see Exxon investment into alt energy signaling 'paradigm shift' for big oil. New York Times of October 15th, 2009

Nitsch J, Pregger T, Scholz Y, Naegler T, Sterner M, Gerhardt N, von Oehsen A, Pape C, Saint-Drenan Y-M, Wenzel B (2010) Langfristszenarien und Strategien für den Ausbau der erneuerbaren Energien in Deutschland bei Berücksichtigung der Entwicklung in Europa und global "Leitstudie 2010". Study for the BMU, Germany, BMU FKZ 03MAP146. http://www.bmu.de/erneuerbare_energien/downloads/doc/47034.php

Official Journal of the European Communities (1989) Council directive 89/458/EEC of 18 July 1989 amending with regard to European emission standards for cars below 1,4 litres, directive 70/220/EEC on the approximation of the laws of the Member States relating to measures to be taken against air pollution by emissions from motor vehicles. O.J. L 226 of 03.08.1989

Official Journal of the European Communities (1998) Directive 98/70/EC of the European Parliament and of the council of 13 October 1998 as amended relating to the quality of petrol and diesel fuels and amending council directive 93/12/EEC. O.J. L 350, 28.12.1998

Official Journal of the European Communities (1999) Commission recommendation of 5 February 1999 on the reduction of CO_2 emissions from passenger cars (notified under Document Number C (1999) 107) 1999/125/EC. O.J. L 40/49 of 13.2.1999

Official Journal of the European Communities (2000a) Commission recommendation of 5 February 1999 on the reduction of CO_2 emissions from passenger cars (JAMA) (notified under Document Number C (2000) 803) 2000/304/EC. O.J. L 100/57 of 20.4.2000

Official Journal of the European Communities (2000b) Commission recommendation of 5 February 1999 on the reduction of CO_2 emissions from passenger cars (KAMA) (notified under Document Number C (2000) 801) 2000/303/EC. O.J. L 100/55 of 20.4.2000

Official Journal of the European Communities (2000c) Directive 1999/94/EC of the European Parliament and of the council of 13 December 1999 relating to the availability of consumer information on fuel economy and CO_2 emissions in respect of the marketing of new passenger cars. O.J. L 12/16 of 18.1.2000

Official Journal of the European Communities (2000d) Decision No 1753/2000/EC of the European Parliament and of the council establishing a scheme to monitor the average specific emissions of CO_2 from new passenger cars. O.J. L 202/1 of 10.8.2000

Official Journal of the European Communities (2003a) Council directive 2003/96/EC of 27 October 2003 restructuring the community framework for the taxation of energy products and electricity. O.J. L 283/51 of 31.10.2003

Official Journal of the European Communities (2003b) Directive 2003/30/EC of the European Parliament and of the council of 8 May 2003 on the promotion of the use of biofuels or other renewable fuels for transport. O.J. L 123/42 of 17.5.2003

Official Journal of the European Communities (2003c) Council directive 2003/96/EC of 27 October 2003 restructuring the community framework for the taxation of energy products and electricity. O.J. L 283/51 of 31.10.2003

Official Journal of the European Union (2004) Directive 2004/3/EC of the European Parliament and of the council of 11 February 2004 amending council directives 70/156/EEC and 80/1268/EEC as regards the measurement of carbon dioxide emissions and fuel consumption of N1 vehicles. O.J. L 49/36 of 19.02.2004

Official Journal of the European Union (2006) Directive 2006/40/EC of the European Parliament and of the council of 17 May 2006 relating to emissions from air-conditioning systems in motor vehicles and amending council directive 70/156/EEC which aims at reducing emissions of specific fluorinated greenhouse gases in the air-conditioning systems fitted to passenger cars (vehicles of category M1) and light commercial vehicles (category N1, class 1). O.J. L161/12 of 14.6.2006

Official Journal of the European Union (2007a) Commission regulation (EC) No 706/2007 of 21 June 2007 laying down, pursuant to directive 2006/40/EC of the European Parliament and of the council, administrative provisions for the EC type-approval of vehicles, and a harmonized test for measuring leakages from certain air-conditioning systems. O.J. L 292/1 of 31.10.2008

Official Journal of the European Union (2007b) Directive 2007/37/EC of 21 June 2007 amending annexes I and III to council directive 70/156/EEC on the approximation of the laws of the member states relating to the type-approval of motor vehicles and their trailers. O.J. L 161, 22.06.2007

Official Journal of the European Union (2009a) Regulation (EC) No 443/2009 of the European Parliament and of the council of 23 April 2009 setting emission performance standards for new passenger cars as part of the community's integrated approach to reduce CO_2 emissions from light-duty vehicles. O. J. L 140/1 of 5.6.2009

Official Journal of the European Union (2009b) Directive 2009/30/EC of the European Parliament and of the council of 23 April 2009 amending directive 98/70/EC as regards the specification of petrol, diesel and gas-oil and introducing a mechanism to monitor and reduce greenhouse gas emissions and amending council directive 1999/32/EC as regards the specification of fuel used by inland waterway vessels and repealing directive 93/12/EEC. O.J. L 140/88 of 5.6.2009

Official Journal of the European Union (2009c) Regulation (EC) No 661/2009 of the European Parliament and of the council of 13 July 2009 concerning type-approval requirements for the general safety of motor vehicles, their trailers and systems, components and separate technical units intended therefore. O.J. L 200/1 of 31.7.2009

Official Journal of the European Union (2009d) Regulation (EC) No 1222/2009 of the European Parliament and of the Council of 25 November 2009 on the labelling of tyres with respect to fuel efficiency and other essential parameters. O.J. L 342/46 of 22.12.2009

Official Journal of the European Union (2009e) Directive 2009/33/EC of the European Parliament and the council of 23 April 2009 on the promotion of clean and energy-efficient road transport vehicles. O.J. L 120 of 15.5.2009

Official Journal of the European Union (2009f) Directive 2009/28/EC of the European Parliament and of the council of 23 April 2009 on the promotion of the use of energy from renewable sources and amending and subsequently repealing directives 2001/77/EC and 2003/30/EC. O.J. L 140/16 of 26 March 2009

Official Journal of the European Union (2009 g) European Union: directive 2009/29/EC of the European Parliament and of the council amending directive 2003/87/EC so as to improve and extend the greenhouse gas emission allowance trading scheme of the Community. O.J. L 140/136 of 5.6.2009

Official Journal of the European Union (2009h) European Union: Decision No 406/2009/EC of the European Parliament and of the council on the effort of member states to reduce their greenhouse gas emissions to meet the community's greenhouse gas emission reduction commitment up to 2020. O.J. L 140/136 of 5.6.2009

Official Journal of the European Union (2009i) European Union: directive of the European Parliament and of the council on the geological storage of carbon dioxide and amending council directive 85/337/EEC, directives 2000/60/EC, 2001/80/EC, 2004/35/EC, 2006/12/EC, 2008/1/EC and Regulation (EC) No 1013/2006. O.J. L 140/114 of 5.6.2009

Olivier JGJ, Peters AHW (2010) No growth in total global CO_2 emissions in 2009. Netherlands Environmental Assessment Agency. PBL publication number 500212001

Papagiannaki K, Diakoulaki D (2007) Decomposition analysis of CO_2 emissions from passenger cars: the cases of Greece and Denmark. Energy Policy 37(8):3259–3267, August 2009

Pirog R (2007) The role of national oil companies in the international oil market. CRS Report Order Code RL34137. Washington

Rijkeboer RC, Vermeulen RJ, Gense NLJ (2002) The design of a measurement procedure for the determination of the additional fuel consumption of passenger cars due to the use of mobile air conditioning equipment. Contract no. B4-3040/2003/367487/MAR/C1

Sinn, H-W (2008) Das grüne Paradoxon. Plädoyer für eine illusionsfreie Klimapolitik Econ: Berlin. ISBN-13: 9783430200622

Skinner I, van Essen H, Smokers R, Hill N (2010) Towards the decarbonization of the EU's transport sector by 2050. EU study Contract ENV.C.3/SER/2008/0053

Smokers R, van Mieghem R, Gense R, Skinner I, Fergusson M, MacKay E, ten Brink P, Vermeulen R., Fontaras G, Samaras Z (2006) Review and analysis of the reduction potential and costs of technological and other measures to reduce CO_2 emissions from passenger cars. TNO Report 06.OR.PT.040.2/RSM. Report to the European Commission. EU Study contract SI2.408212 ten Brink P, Skinner I, Fergusson M, Haines D, Smokers R, van der Burgwal E, Gense R, Wells P, Nieuwenhuis (2003) The past and the future of EU regulatory policies to reduce car carbon emissions. Final Report to the Commission. EU study contract. Contract Ref.: B4-3040/2002/343442/MAR/C1

ten Brink P, Skinner I, Fergusson M, Smokers R, van der Burgwal E, Gense R, Wells P, Nieuwenhuis P (2005) Service contract to carry out economic analysis and business impact assessment of CO_2 emissions reduction measures in the automotive sector. Final report to EU Study contract B4-3040/2003/366487/MAR/C2

United States Senate (2007) How much are oil and gas companies investing in clean alternative fuels. Democratic Policy Committee, April 23, 2007

Wicke L, Schellnhuber HJ, Klingenfeld D (2010) Nach Kopenhagen: Neue Strategie zur Realisierung des 2° max. Klimazieles. Potsdam Institute for Climate Impact Research (PIK). Report 116. ISSN 1436-0179

Zierock K-H, Valli R, Pairault O, Hunhammar S, Bridgeland E, Kroon M (2006) Reduction of energy use in transport. Final Report to Working Group Under the Joint Expert Group on Transport and Environment

Zierock K-H, Mehlin M, Köhler K (2007a) Developing a legislative approach for limiting specific CO_2 emissions from passenger cars in the European Union. Report to the Federal Ministry for the Environment, Nature Conservation and Nuclear Safety

Zierock K-H, Mehlin M, Köhler K (2007b) Neugestaltung der Energieverbrauchskennzeichnung von neuen Personenkraftwagen in Deutschland. Report to the Federal Ministry for the Environment, Nature Conservation and Nuclear Safety

Zierock K-H (2009) Die EU-Verordnung zur Verminderung der CO_2 – Emissionen von Personenkraftwagen. BMU publication. http://www.bmu.de/files/pdfs/allgemein/application/pdf/eu_verordnung_co2_emissionen_pkw.pdf

Zierock K-H (2010) Die Neugestaltung der Pkw-Energieverbrauchskennzeichnungs-verordnung. BMU publication

Chapter 5
Fuel Policies in the EU: Lessons Learned from the Past and Outlook for the Future

Sandrine Dixson-Declève

Abstract This chapter will look at the lessons learned from past and present EU decision making in the area of cleaner conventional and renewable fuels as well as the problems confronting policy makers in trying to move toward low carbon fuels. Today we are tasked with finding low carbon substitutes for fossil fuels yet most existing replacements are not issue free nor do they necessarily fit total low carbon requirements. Many would also say that the focus of our low carbon search should not be on existing fuel products and traditional transport modes but rather on new transport solutions and demand management options. Without a doubt the feat ahead is by far more complex than any challenge to the fuels industry or policy makers previously. We are now confronted with the need to take an entirely new approach to liquid fuels and their role in society within a short time horizon. Hence we have to completely re-think the role of liquid fossil fuels in meeting societal demands for transport energy. As in all system changes this must allow for transitional periods and adaptation steps while innovation and scale up is streamlined. It must also go hand and hand with simultaneous efforts to manage demand and implement mandatory efficiency requirements across Europe as no technology or single product will solve the fuel de-carbonization challenge.

5.1 Setting the Stage

5.1.1 Energy Security

Countless studies across the globe have demonstrated that the main source of transport greenhouse gas (GHG) emissions is the combustion of fossil fuels. The problem in the European Union (EU) is that transport's energy consumption and thus resulting GHG emissions are increasing steadily because transport volumes are growing faster than the positive impact of new offsetting vehicle and clean fuel legislation and the energy efficiency of different means of transport.

S. Dixson-Declève (✉)
Cambridge Programme for Sustainability Leadership, B-1050 Brussels, Belgium
e-mail: sandrine.dixson@cpsl.cam.ac.uk

The increase in GHG emissions from transport threatens the European Union's (EU's) progress toward its Kyoto targets (8% reduction by 2012 from 1990 levels) but most importantly its own European target of 80–95% total GHG emission reduction by 2050 as agreed by European Ministers in the spring of 2009. In particular since 90% of the increase in CO_2 emissions between 1990 and 2010 is attributable to transport and transport now accounts for 21% of total GHG emissions (excluding international aviation and maritime transport). Transport is also the sector with the largest predicted growth. In fact the 2010 Business As Usual (BAU) scenarios show an increase of EU transport emissions of 60–70% between 1990 and 2050. By 2050 transport is expected to still be 50% of the total emissions pool even if efforts are made to cut emissions by 50% by that time (Hill et al. 2010).

Recent data gathered by the European Commission, the EU's executive body, shows that 97.5% of the liquid fuel pool (excluding gas) in Europe consists of petrol (motor spirit) used in passenger cars and light commercial vehicles and diesel (gas diesel oil), which is mainly used in heavy duty road vehicles, some railways, inland waterways and maritime vessels. The remaining 2.5% of the liquid fuel pool is split between bioethanol blends in petrol and biodiesel. Liquid petroleum gas (LPG) and natural gas currently makes up a very small percentage of energy used for transport while the main source of energy for railways in Europe is electricity.

Even the most recent International Energy Agency (IEA) World Energy Outlook 2010 data shows that demand for energy for transport in the EU from now until 2030 will continue to be predominantly met by petroleum products (oil) for road and air transport with a small amount of electricity for rail transport. Although biofuel consumption will continue to grow it will remain a small percentage of total fuel consumption in 2030.

With regard to consumption and demand, the European Commission and the IEA warn of Europe's growing oil import dependency. Recent data from the European Commission shows that Europe's import dependency on energy is rising. The European Commission predicts in its report EU Transport GHG: Routes to 2050 – Energy Security and the Transport Sector (2010) that unless EU domestic energy becomes more competitive and new energy outlets are found, the EU is expected to import approximately 70% of its energy requirements by 2030 versus 50% today. In 2006, EU27 oil imports were 545.6 million tonnes, 82.6% of total oil consumption. Today these imports are coming from only a handful of countries thus increasing European dependency on a few key supplying countries. For example, 33% of the oil imported into the EU27 in 2006 came from Russia. In addition to energy dependency, oil prices in the range of 88 US dollars per barrel ($88/bbl) will cost the EU economy approximately €40bn in fuel imports (European Commission 2010). Deutsche Bank forecasts that oil prices could hit $175/bbl by as early as 2016 which could cost €80bn midway through the EU's current 2020 GHG emissions target deadline of 20%. This prediction was made before the continued riots in the Middle East in the winter/spring of 2011, which led forecasts to move closer to $200/bbl in spring 2011.

Therefore, greater energy efficiency and alternative fuels policy are two tools meant to respond to the need for new domestic energy products whilst meeting the 'de-carbonization' challenge.

5.1.2 The De-carbonization Challenge

When addressing the fuel de-carbonization challenge, one must take into consideration that conventional fuels and vehicles are much cleaner than in the 1990s, that energy diversification is increasingly important and that Europe needs to meet a 2020 GHG reduction target of 20% (which could move to 25–30% if agreed by all Member States) and a 2050 GHG reduction target of 80–95%. As a result low carbon fuels and energy sources, in particular renewable fuels and renewable electricity for electrical vehicles are receiving a great deal of attention.

The challenge is to find the alternative fuel or energy solution that not only matches current fuel quality specifications and enables the vehicle to meet its emission requirements and thus air quality targets, but also brings CO_2 reductions and energy diversification options beyond conventional gasoline or diesel fuels. In addition, transport must now be seen as directly linked to changes in our energy systems and our use of energy. Therefore, energy efficiency and demand management must be addressed alongside supply solutions.

This chapter will look at the lessons learned from past and present EU decision making in the area of cleaner conventional and renewable fuels as well as the current and future 'wicked problems'[1] confronting policy makers in trying to move toward low carbon fuels. Without a doubt the feat ahead is by far more complex than any challenge to the fuels industry or policy makers previously. We are now confronted with the need to take an entirely new approach to liquid fuels and their role in society within a short time horizon. Hence we have moved from addressing product environmental, health and safety concerns to a complete re-think of the role of liquid fossil fuels in meeting societal demands for transport energy. As in all system changes this must allow for transitional periods and adaptation steps while innovation and scale up is streamlined. It must also go hand and hand with simultaneous efforts to manage demand and implement mandatory efficiency requirements across Europe as no technology or single product will solve the fuel de-carbonization challenge.

One of the main challenges linked to this complete re-think in our transport and energy systems is that we will need to move away from traditional transport modes and energy products. If we look at de-carbonizing the conventional fuel pool the entire basket of fossil fuels and gases will need to be substantially de-carbonized or replaced by 2050 to assist in meeting Europe's GHG reduction goal of 80–95%.

Recent work by AEA for the European Commission's Directorate General for Climate Action (Hill et al. 2010) has estimated that the average GHG emissions factor per energy carrier including electricity, will need to be substantially cut in some cases by between 50 and 90% as seen under Fig. 5.1. This includes very high levels

---

[1] By their nature, wicked problems have no definitive nature and are difficult to describe. They represent sets of inter-related problems, some of which are actually not problems if considered individually. There are no easy, yes or no, or discreet solutions to wicked problems. Because every potential solution matters a great deal and it is virtually impossible to model the entire system, potential solutions must be tested and feedback assessed in a more or less immediate and dynamic manner. Every wicked problem is typically a system of another (wicked) problem and therefore when taken in totally represent the imbalance and imperfections in large and complex systems.

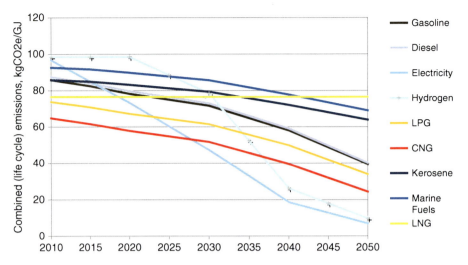

Fig. 5.1 Average GHG emissions factor by fuel
Source: AEA (2010)

of substitution with biofuels for each of the conventional energy carriers (i.e. gasoline, diesel, LPG, CNG, kerosene and marine fuels), only possible with substantial improvements to vehicle efficiency and reduction in forecast baseline demand. It also will entail the use of carbon capture and storage (CCS) once cost-effective and available at a large scale.

Therefore, we are calling upon existing actors in the system to dramatically change their current high fossil fuel products and focus their attention on low carbon options when most of them have stranded investments in high carbon conventional fuels or vehicle technology.

We are also asking the old actors in the system to work with new actors under a fully integrated linked-up approach bringing together stakeholders from different sectors and policy departments. In this respect we would hope that we could learn from the policy lessons of the European Auto Oil programs I and II spanning from 1992 to 2002, (refer to section below) and to a certain degree we can – but the issues confronted by stakeholders at that time were by far simpler than today.

In the area of fuels, the focus under Auto Oil I and II was on changing the existing product quality of fossil fuels to reflect health and environmental concerns. We can safely say that fuel providers rose to the challenge and invested what was necessary to respond to policy demands and product changes to produce ever cleaner fuels over the course of a very short time period. This was not without a fight and extreme lobbying for and against by industry, civil society and governments. But the final result is incredibly positive and all 27 EU Member States now implement stringent vehicle emissions and fuel quality requirements. In fact the EU is the global leader in fuel quality specifications and its legislation has and continues to be replicated across the globe.

Today we are tasked with finding low carbon substitutes for fossil fuels yet most existing replacements are not issue free nor do they necessarily fit total low carbon requirements. For example, biofuels can lead to higher GHG emissions when indirect emissions from land use change are taken into account, natural gas is still fossil based and with today's engine technology is not necessarily a much lower carbon alternative to low sulfur conventional gasoline or diesel, nor is hydrogen if produced from fossil-based fuels. Many would also say that the focus of our low carbon search should not be on existing fuel products and traditional transport modes but rather on new transport solutions and demand management options. This encompasses a brand new group of stakeholders and necessitates cooperation between old stakeholders such as conventional fuel producers and automotive manufacturers with the full gamut of biofuel/biomass producers, electrical vehicle manufacturers, electricity companies, and system operators to name but a few as well as a multitude of government departments from agriculture, transport to energy and environment who have rarely worked together or linked up their policies and in some instances do not always see eye to eye.

5.2 EU Conventional Fuel History

5.2.1 Clean Fuels Policy: The Systems Approach

To date the main focus on fuel quality has been on technical and environmental properties related to engine performance and air pollution. It is not until 2008 with the passage of *The Climate and Energy Package*[2] *including the amendments to fuel quality legislation* (further discussed below) that the carbon content of fuels became important.

The oil and automotive industries in the United States (US), EU and Japan have in each case made considerable efforts to understand the relationship between urban air quality, vehicle emissions, engine technology and fuels, and to work with regulators accordingly. To date, their philosophy has been to adopt a 'systems' approach taking into consideration that the vehicle and the fuel work as a system and transport related air pollutants cannot be reduced unless the vehicle and the fuel are cleaned up together. As a consequence, these countries or regions have led the way, through political will and industry innovation, toward the reduction of ambient air emissions from transport. They have also served as role models for many other countries across the globe.

As shown in Fig. 5.2, it is these three pillars, based on the 'systems approach', which are the foundation for a successful fuel quality strategy and which to date have succeeded in reducing ambient air emissions from the vehicle and fuel as a system. The impact of a vehicle on air quality is directly linked to the type of engine

---

[2] The Climate and Energy Package is a package of EU climate and energy measures approved in December 2008 and implemented in 2011. The package also includes fuel quality amendments.

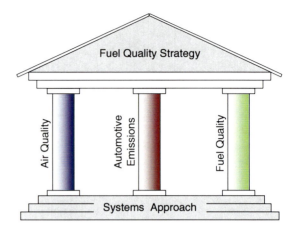

Fig. 5.2 The 'systems approach'[3]

and after treatment technology on that vehicle and the quality of the fuel used in the engine. This holistic approach has been integrated into a coherent body of legislation across the globe under the leadership of similar auto oil programs developed in the US, the EU and Japan.

Once air quality objectives are defined and source apportionment has occurred, e.g. the actual impact of transport on urban air pollution has been calculated, determination can be made of which automotive emissions must be reduced and by how much. This in turn will determine engine technology needs and the quality of fuels necessary to enable the engine and/or after treatment technology to meet the emission requirements. The intimate link between setting air quality objectives and the necessary changes in vehicle technology and fuel formulation to achieve those objectives has driven phenomenal development in vehicles and fuels since the 1990s.

The synergy between cleaner fuels and new vehicle technology has achieved dramatic reductions in all of the major air pollutants across European countries except for CO_2. Such an impact was already measured at the start of the Auto Oil program as can be seen in Fig. 5.3. Although the reductions in air pollutants have not amounted to the extreme reductions predicted, the point is that large reductions were still achieved and continue to go down as the new Euro V and Euro VI automotive emission and fuel requirements are implemented. Most importantly what the figure shows is that the body of clean fuels and vehicle emissions legislation under the systems approach would have a direct effect on air pollutants but would not reduce CO_2 emissions. This would require a new set of requirements.

The question now is how much further can we go with existing vehicle technology and conventional fuels to meet continued air quality and growing CO_2 concerns. Does it make sense to continue to make changes to existing conventional fuels? If we

---

[3] Figure designed by Sandrine Dixson-Declève whilst working for Hart Energy Consulting (2003) http://www.hartenergy.com

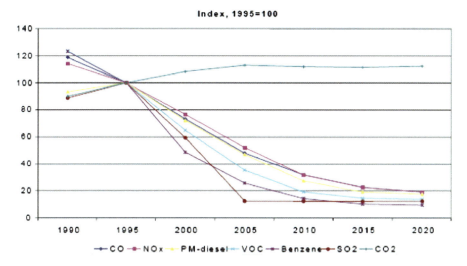

Fig. 5.3 Predicted air pollution reductions from auto oil I and II[4]

take into consideration the most recent and projected technological innovations for gasoline and diesel vehicles (Euro V and VI) and fuels (sulfur content of 10 parts per million or ppm), we are arriving at the point of diminishing returns for conventional fuel sources on the basis of cost and complexity with little to no health, environmental and climate benefit. In addition, one should not forget that the body of fuels and automotive emissions legislation implemented in Europe is relatively recent and that the new EU Member States in particular from Central and Eastern Europe have only started implementing the full package of measures including biofuel blending requirements since 2004 and later thus putting a great deal of economic pressure in particular on their refineries to comply.

Finding the right path forward requires a bit of back casting to the development of our clean fuels legislation and possible lessons learned.

5.2.2 Clean Fuels Policy: Lessons Learned

In the mid 1980s, transport fuel quality specifications first appeared in Europe in certain Member States such as the UK and Sweden with a focus on phasing out lead from petrol predominantly due to health concerns. During the early 1990s Finland and Sweden started bringing down sulfur levels, and limited fuel quality specifications were set by the European Committee for Standardization – Europe's standards body. Fuel product standards EN 228-1993 and EN 590-1993 for petrol and diesel

---

[4] European Commission, DG Environment, personal communication with Peter Gammeltoft, Head of Unit (2004).

respectively were adopted in 1993 in an attempt to standardize fuel products across Europe. These standards were product quality related but did not fulfil environmental criteria nor were they mandatory under EU law as no EU wide framework legislation existed at the time.

The full gamut of fuel quality and environmentally related specifications for both diesel and gasoline was only regulated in the late 1990s under Directive 98/70/EEC, which set gasoline and diesel specifications for 2000 and 2005 and its amending Directive 2003/17/EC. These fuel quality specifications were set to meet EU air quality targets and from the start were intimately linked to automobile emissions targets. Since the inception of the fuel quality directives, the CEN was tasked to continue to set technical specifications to complement existing EU legislation, or where legislation was absent, to fulfil industry demands.

Fuel Quality Directives 98/70/EEC and 2003/17/EC emanated from the EU Auto Oil Program, which was split into two key phases and time periods:

- Auto Oil I: 1991–1998
- Auto Oil II: 1997–2003

The Auto Oil Program was the first of its kind in Europe to bring together the resources and expertise of the automobile sector through the European Association of Automobile Manufacturers (ACEA) and the oil refiners through the European Petroleum Industry Association (EUROPIA) in collaboration with the services of the European Commission. An investigation of the Auto Oil Program setup and objectives is essential to understand the differences in policy approach and stakeholder dialogue between the early years of setting fuel quality legislation and the most recent decade of focusing on new fuel products such as biofuels and efforts to de-carbonize fuel.

The EU Auto Oil Program was based on the 'systems approach' bringing together the auto and oil industry under one umbrella program to work with the European Commission to find the most cost-effective solutions to air quality abatement from transport. During the period between November 1991 and July 1993, many meetings took place between the two industries and between industry and the Commission Services. In July 1993 a formal agreement between EUROPIA and ACEA was signed to commit funds to a joint research program aimed at furthering the understanding of the complex relationships between fuels, engines and emissions.

One of the major inputs into the final development of the Auto Oil I proposal was the European Program on Engine Fuels and Emissions (EPEFE), which analyzed the technical and economic feasibility of changes in engine technology and fuel quality to decrease vehicle emissions. The EPEFE program ran from 1992 to 1995. From the outset, EPEFE was only a small yet important part in the process of determining the measures necessary to protect European air quality.

The Auto Oil Program was comprehensive in scope, and EPEFE played an important role in ensuring scientific rigor and thorough data analysis. In addition, due to the large number of stakeholders in the process (industry, government and non-governmental organizations) it was necessary to ensure absolute transparency so that all stakeholders could check the process and recommendations. Progress and results from EPEFE were therefore publicly reported through quarterly bulletins and through meetings with experts from the Commission Services and Member States. Detailed information was made available for all interested experts through interim reports and a final report.

At that time the World Health Organization (WHO) was in the process of revising its air quality guidelines with the assistance of the Environment Directorate of the European Commission. These proposals were stricter than existing air quality targets set in Member States and the EU, but were seen as a valid benchmark for future legislation. The Auto Oil partners agreed to address several European air quality objectives following the WHO proposals. In order to set valid targets, nitrogen dioxide (NO_2), carbon monoxide (CO), benzene, particulate matter (PM) and tropospheric ozone (O_3) were measured in seven major urban areas: Athens, Cologne, The Hague, London, Lyon, Madrid and Milan.

Based on these results, Auto Oil I developed cost-effective combinations of measures sufficient to achieve, by 2010, stringent air quality standards with respect to NOx, benzene, CO, PM and tropospheric ozone including total NOx and volatile organic compounds (VOCs). These combinations demonstrated that the greatest effort and cost would be linked to the decrease of NOx, PM and tropospheric ozone.

Four specific measures were included in the cost-effectiveness assessment:

- Technical measures covering improved vehicle emission standards and improved fuels;
- Improved inspection and maintenance of vehicles;
- Local measures including public transport, alternative fuels in captive or city fleets, selective traffic bans;
- The possibility of using national fiscal measures.

One of the major results of this analysis was that non-technical measures, measures undertaken to improve vehicle technology and inspection and maintenance schemes were potentially the most cost-effective. In the area of inspection and maintenance this was disputed by the auto industry especially since there was less confident data on the reduction of vehicle emissions from better inspection and maintenance. In order to get a non-biased assessment of the most optimum combination of measures and true costs, the Commission appointed two separate consultancy firms to review the vehicle technology packages and the fuel quality packages. An analysis of the impact of fuel quality changes on emissions and related costs was based on the original EPEFE equations and the costs related to changes in each fuel parameter were calculated.

Non-technical measures were not mentioned nor further investigated at all which also angered the car industry alongside many environment and health non-governmental organizations. Note that non-technical measures and demand management have received little attention to date by policy makers. In addition, if one looks at the full body of transport legislation little effort has gone into the impact vehicle efficiency measures could have on air quality, energy security or de-carbonization. The results from the cost-effectiveness analysis carried out by the appointed consultancies concluded that cleaner fuels were not cost-effective in comparison with changes to vehicle technology and improved inspection and maintenance schemes. This was a somewhat logical result since the cost calculations were based on investment costs, and oil refining is heavily capital intensive whereas vehicle technology changes are moderately capital and labor intensive and therefore were evaluated as moderately cost-effective. Inspection and maintenance schemes which are very labor intensive and not capital intensive were therefore seen as extremely cost-effective.

Based on the above results, the Commission decided upon a package of legislative proposals targeting the transport sector yet from a cost-effectiveness point of view focusing on vehicle technology changes and inspection and maintenance. The package of proposals would therefore foresee the implementation by 1 January 2000 of petrol and diesel fuel reformulation, emission limits for passenger cars, emission limits for light duty vehicles, and a proposal strengthening vehicle inspection and maintenance rules.

After three years of analysis and debate under the Commission-led Auto Oil I Program and the industry-led EPEFE program (the technical portion of the Auto Oil I program) the European Parliament received proposals from the Commission for a new Directive on fuel quality and for two new Directives on car emissions amending earlier Directives.

At this time, it became quite clear that the automobile and oil industries were no longer in agreement over the Commission's proposals. The automobile sector felt that they were paying an unnecessary amount of the clean air bill while petrol and diesel quality had barely been touched. Their position was supported by a variety of stakeholders including EU Member States, civil society groups, many Members of the European Parliament (MEPs) most importantly the Finnish Rapporteur from the Green group Heidi Hautala, and a report sponsored by the governments of Sweden and Finland on tax incentives and the cost-effectiveness of phasing-in cleaner fuels. The criticisms focused on the fact that the oil industry was faced with little to no changes to fuel quality by the year 2000, and fuel quality changes, whether indicative or mandatory, had not been set for 2005. It is important to note that although the results of the Program were not perfect and claimed by some to be flawed, no one faulted the process and the importance of undertaking such a comprehensive stakeholder program.

On 29 July 1998, the technical and legal details for the proposed Directives on fuels and automotive emissions were unanimously agreed upon by the European Parliament and the Council of Ministers. Politically this was a huge institutional coup as this was one of the first environmental dossiers where the European Parliament showed its strength and was able to substantially tighten a proposal from the European Commission based on reports coming from outside. The joint decision by the Council and the Parliament managed to ensure the introduction of stricter mandatory fuel quality parameters by 2000, with even stricter mandatory specifications required as of 2005 whilst ensuring that those countries who would have economic difficulty in implementing the measures could apply for derogations. It also promoted the use of tax incentives and the concept of phasing-in cleaner fuels for the first time in EU history.

The main focus of the fuel product changes and reductions were on those fuel parameters having a direct impact on human health and the environment. This included the phasing-out of lead as well as reductions in aromatics, benzene, olefins and sulfur.

The second Auto Oil Program was established in 1997 just before Directive 98/70/EC was adopted to provide policymakers with an objective assessment of the most cost-effective package of measures necessary to reduce road transport emissions to a level consistent with the attainment of the new air quality standards being developed for adoption across Europe. As part of the extensive Auto Oil Program, seven Working Groups (WGs) were established. However, whereas Auto Oil II was to focus on those parameters not regulated post-2000 or potentially needing to be further regulated, its focus changed to reducing sulfur levels in diesel and gasoline.

This change in focus was due to increased pressure from car manufacturers claiming to need lower levels of sulfur to enable new automotive after treatment technologies supported by several strong Member States, in particular by Germany. As a result, in 2003 Directive 98/70/EC was amended by Directive 2003/17/EC to include the appropriate geographical distribution of 10 ppm sulfur fuels in 2005 and the full market penetration of 10 ppm gasoline in 2009. Further changes were made to the directives in 2009 under Directive 2009/30/EC as regards the specification of petrol, diesel and gas-oil, fuel used by inland waterway vessels and introducing a mechanism to monitor and reduce GHG emissions. This was the first regulatory link between fuels and GHG emissions.

With the adoption of Directive 2003/17/EC Europe leapfrogged in the short period of 12 years from leaded, high sulfur and high aromatics gasoline and very high sulfur diesel fuels to almost zero sulfur fuels and the banning of lead. From the very beginning the fuel quality Directives were introduced by all Member States due to a series of creative policy mechanisms, which allowed the leading countries to move forward faster and the laggards a bit more time through derogations. Such mechanisms were:

- Availability of two types of diesel and petrol fuel on the market e.g. a higher quality and lower quality fuel;
- Phase-in option: As of year 2000 Member States could permit the marketing of gasoline with year 2005 specifications (35% aromatics and 50 ppm sulfur) and diesel with 2005 specifications of 50 ppm sulfur;
- Derogations (exemptions) were only to be given on lead (from 2000) and on sulfur content in gasoline (from 2000 to 2003), and on sulfur content in diesel if adequate quantities of good quality fuel, e.g. cleaner fuel, existed on the market;
- Fiscal incentives were clearly seen as the facilitator for the early phase-in option. The Commission even went so far as to promise the expeditious implementation

of fiscal incentive demands by Member States to promote the market penetration of cleaner fuels.

In addition, a review process was developed to address remaining issues such as the introduction of 10 ppm diesel, the outstanding fuel quality parameters e.g. olefins, density, cetane number etc., alternative fuels including biofuels and issues related to volatility, total aromatics, lubricity, phosphorus, silicon and metallic additives. The European Commission would have to forward a proposal to the European Parliament and Council by December 31, 2005.

5.3 The Move from Clean Conventional Fuels to Biofuels

As seen in the previous section, until 2003 fuel legislation was focused on clean conventional fuels. Although the European Commission has discussed energy security concerns and the need for energy diversification, to date no mandatory legal text has been enacted covering the full gamut of alternative fuels. In fact, it is not until recently that the European Commission has changed its focus from conventional fuels to biofuels and now to low carbon fuels.

The same year the fuel quality directive was amended, a voluntary biofuels framework legislation (Directive 2003/30/EC) on the promotion of the use of biofuels or other renewable fuels for transport, was adopted to stimulate the uptake of biofuels across the EU.

The Directive required EU Member States to transpose the legislation into national legislation and action plans. Member States were asked to voluntarily ensure a minimum share of biofuels sold on their markets of 2% (by energy content) by 31 December 2005 at the latest, and 5.75% by December 2010. Biofuels were defined as liquid or gaseous fuels used for transport and produced from biomass, i.e. biodegradable waste and residue from, for example, agriculture and forestry. Member States setting lower targets would have to justify this on the basis of objective criteria.

The different types of biofuels referenced in the directive are bioethanol, biodiesel, ETBE, biogas, biomethanol and bio-oil. The biofuels could be made available as pure biofuels, blended biofuels or liquids derived from biofuels.

The European Commission and in particular the Directorate General (DG) in charge at the time – DG Transport and Energy – stated that the Directive was adopted 'in order to reduce greenhouse gas emissions and the environmental impact of transport, and to increase security of supply'. Rural development was mentioned as a third reason for action. The Directive was promoted as an opportunity for Europe to solve several problems with one solution. However, not all stakeholders and Commission DG's supported this perspective nor did they believe in either the air pollution or GHG benefits of biofuels. For them, the push for biofuel based agricultural feedstocks traditionally grown in Europe such as sugar beet, wheat and rapeseed was purely another way to assist farmers hit by Common Agricultural Policy (CAP) reform. Therefore a certain amount of distrust reigned from the very beginning between the different actors.

The logic was that the biofuels Directive would focus on promoting biofuels use and would be under the jurisdiction of DG Transport and Energy, while the fuel quality Directive would ensure that the final fuel with a percentage of biofuel or pure biofuels – 100% biodiesel and 85% ethanol and 15% gasoline blends (E85) – only allowed in designated vehicles met both vehicle emission requirements and air quality objectives. The latter objective would be regulated by DG Environment.

However, fuel quality issues related to biofuels soon surfaced. After 10 years of working together on fit for purpose fuels, the auto and oil industry were thrown into working with a nascent biofuels industry not at all used to the complexities of fuel quality and engine requirements. Many representatives of the biofuels industry who had responded to the call for greater biofuels use under Directive 2003/30/EC and new national subsidies did not fully understand the extent to which their products actually needed to function in an engine and meet fuel quality specifications. Note that such issues were also prevalent in the US and other countries across the globe promoting the use of biodiesel and biogasoline.

This distrust and frustration with biofuel producers was compounded when the ethanol industry in particular started pushing for the elimination of certain environmental criteria in order to ensure the uptake of their fuel. For example, by increasing vapor pressure requirements to ensure 10% (by volume) blending of pure ethanol. Higher vapor pressure limits result in the release of more VOCs, which are major contributors to ground level ozone formation. This created a great deal of frustration within the auto and oil industry as it was their job at the end of the day to ensure that the final product met fuel quality specifications and enabled vehicles to meet emission standards and air quality requirements under the systems approach. This dilemma would get even more complicated when the links between the Fuel Quality Directive and the Biofuels Directive became increasingly intertwined and discussions would seriously start on changing the CEN fuel quality standards to reflect the new biofuel blends as well as their sustainability requirements.

Already in Spring 2007, the European Council started to see that promoting biofuels was more complex than expected. Whilst they continued to push for a 10% (by volume) biofuels target they insisted that biofuels needed to be sustainable. In January 2008 the European Commission, guided by the EU Environment Commissioner at the time echoed this call and announced that the EU was rethinking its biofuel program due to environmental and social concerns.[5] It was announced that new criteria and guidelines would have to be adopted to ensure that EU

---

[5] 'EU rethinks biofuels guidelines', Roger Harrabin, BBC News http://news.bbc.co.uk/2/hi/europe/7186380.stm [last accessed March 2011].

biofuel requirements were not an environmental hazard. In particular, biofuels use needed to be looked at in light of issues related to increases in food prices, rainforest destruction (due to palm oil and soy production) and concern that wealthy landowners or multinationals would drive poor people off their land to convert the land to fuel crops. The UK House of Commons Environmental Audit Committee raised similar concerns, and called for a moratorium on biofuel targets (UK Parliament 2008).

5.4 Low Carbon Fuels Policy: De-Coupling CO_2 Growth from Transport Fuels

5.4.1 Low Carbon Fuels Policy: A New Legislative Package

On April 6, 2009, the European Council formally adopted the Energy and Climate Change Package, which attempted to partly answer the above concerns and further promote clean energy as an abatement solution to GHG emissions. The package is split between six separate legislative acts (including Directive 2009/30/EC amending Directive 98/70/EC as regards the specification of petrol, diesel and gas-oil, amending Directive 99/32/EC as regards the specification of fuel used by inland waterway vessels and repealing Directive 93/12/EEC (here forth the 'Fuel Quality Directive'), and the Directive 2009/28/EC on the promotion of the use of energy from renewable sources (here forth the 'Renewable Energy Directive').

The Fuel Quality and Renewable Energy Directives contribute to attaining an over arching GHG reduction target of 20% by 2020. This complements a 20% voluntary target in energy efficiency and a 20% mandatory share of renewable energy in the EU's total energy consumption by 2020. The three targets combined are known as the '20/20/20 target'.

With regard to the 20% renewable energy objectives, the Renewable Energy Directive would amend the existing voluntary target scheme and would instead set mandatory biofuel targets for 2020 as discussed below. It would also set a 10% (by volume) target for renewable energy use in transport, which according to recently submitted national renewable energy action plans will actually be closer to 11% by 2020.

After much debate in the European Parliament and Council, GHG emissions requirements and biofuel sustainability criteria were introduced under Article 7a and 7b respectively of both the Renewable Energy Directive and the new Fuel Quality Directive. Article 7a requires fuel suppliers to reduce the GHG intensity of energy supplied for road transport (Low Carbon Fuel Standard) by up to 10% by 31 December 2020, compared with the fuel baseline standard (to be determined). The main objectives of the Directive and article 7a as agreed by the European Parliament and Council are the following:

5 Fuel Policies in the EU: Lessons Learned from the Past and Outlook ...   111

- 6% mandatory lifecycle GHG reduction target for fuel suppliers by 2020[6];
- 4% non-binding lifecycle GHG target for fuel suppliers by 2020 split between 2% from electric vehicles and Carbon Capture and Sequestration (CCS) and 2% from the UN-led Clean Development Mechanism (CDM) credits in the fuel supply sector, which would in effect relate to global flaring and venting reductions[7];
- Intermediate non-binding targets of 2% by 31 December 2014 and 4% by 31 December 2017, which could be required by Member States.

Member States are required to designate the supplier or suppliers responsible for monitoring and reporting life cycle GHG emissions per unit of energy from fuel and energy supplied. In the case of providers of electricity for use in road vehicles, Member States have to ensure that each electricity provider when choosing to become a contributor to the reduction obligation can demonstrate that they can adequately measure and monitor electricity supplied for use in those vehicles.

With effect from 1 January 2011, suppliers (including electricity suppliers) are obligated to report annually, to the authority designated by the Member State, on the greenhouse gas intensity of fuel and energy supplied within each Member State by providing, as a minimum, the following information:

- the total volume of each type of fuel or energy supplied, indicating where purchased and its origin; and
- life cycle GHG emissions per unit of energy. Member States are required to ensure that reports are subject to verification.

In order to facilitate implementation, the European Commission would establish implementing measures, which would clarify the data gathering and reporting requirements for compliance.

It is the first time that a package of legislation focuses on the joint goals of reducing GHG emissions whilst meeting energy needs. In addition, for the first time sustainability criteria are integrated into the body of two legislative acts, a Fuel Quality Directive and a Renewable Energy Directive and will have to be met by all fuels (including non-road fuels) if they are to count toward the greenhouse gas intensity reduction obligation. The only difference between the two directives

---

[6] Due to direct links with the 10% transport target in the Renewable Energy Directive and uncertainties in the role of biofuels as a result of indirect land use concerns, the 6% target will therefore be reviewed in 2014.

[7] This chapter will not focus on the remaining 4% non-binding target as Chapter 4 discusses vehicle electrification. Needless to say the methodology necessary for assessing low carbon electrification must still be defined and linked to the Renewables Directive. In addition, it is not clear when CCS will truly be a cost effective and viable solution although it was agreed that CCS could be included in the Clean Development Mechanism (CDM). That said the CDM is currently under scrutiny and many countries believe that reform is necessary as biofuel and biomass projects are not readily accepted whilst loopholes allow other projects in the area of HCFC's for example to be accepted.

is their core objectives and structure. For example, the focus of the Fuel Quality Directive is on reducing all emissions from transport fuels and the lower the GHG emissions the more competitive the fuel is; whereas the focus of the Renewable Energy Directive is to increase the uptake of biofuels, which have emissions below a certain level e.g. 35% lower than the reference fuel (this raises to 50% in 2020) in order to count toward the target and be eligible for financial support.

Regarding the further definition of sustainability criteria in particular the integration of indirect land use change (ILUC) and the GHG default values for individual fossil fuels both directives called for more analysis and a report back to the European Parliament and Council by December 2010.

5.4.2 New Low Carbon Fuels Legislation: Analysis

The new Fuel Quality Directive applies to all energy supplied to road transport, inland waterway transport, non-road mobile machinery, and diesel for trains. Energy for electricity used by trains is excluded. Across Europe fuel suppliers will be obligated to report on an annual basis on the carbon intensity of their fuels and the energy supplied for road transport. A base year of 2010 has been set. From the outset, setting a future baseline such as 2010 was criticized by different stakeholders. In particular, refiners who had already made improvements to their refineries prior to 2010 complained that these investments would not be taken into account and ironically that they would be penalized rather than rewarded for their leadership position.

Under the new Fuel Quality Directive, fuel emissions are calculated through the full lifecycle of greenhouse gases. This includes gases emitted during exploration, refining, distribution and the combustion of the fuels. Although regarding the latter, as CO_2 emissions from combustion (about 85% of current lifecycle emissions) cannot be influenced by the supplier the main channel of influence here is through the vehicle engine (see Chapter 4 of this book). The Directive applies to all fuels imported into the EU; therefore, fossil or alternative fuels will have to comply with the legislation and in particular biofuels, will have to comply with the agreed upon sustainability criteria.

Based on the lifecycle approach, it is up to fuel suppliers to decide where they can achieve the most cost-effective cuts in emissions. The options are the following:

- Exploration: by opting for cleaner exploration practices entailing less flaring and venting
- Crude selection: by using different crudes with a lower carbon content (this eliminates high carbon options such as tar sands or coal for the production of coal to liquid fuels (CTL)
- Refinery Process: by improving refinery practices and increasing efficiency or using biomass for gas co firing
- Product Selection: by producing and supplying alternative fuels such as biofuels, LPG (Liquefied Petroleum Gas) or CNG (Compressed Natural Gas).

In terms of implementation, the European Commission was in the process of finalizing the implementing measure necessary for setting guidelines for refiners in spring 2011. The draft position available in March 2011 did not place as much emphasis on the energy efficiency options at the refinery but rather focused on the creation of a single value for oil and product changes.

In the short term, it is feasible that reducing flaring/venting could be accounted for but the best way to do this is under discussion. Today the gas flared annually is equivalent to 30% of the European Union's gas consumption. The latest World Bank Global Gas Flaring Reduction (GGFR 2008) data shows that based on satellite monitoring approximately 80 million tonnes of CO_2 equivalent (Mt $CO_{2\,eq}$) GHG emissions are associated with EU-bound global oil production from both flaring and venting. One option could be to only count savings calculated using 'co-product allocation', under which the benefits of reductions in flaring and venting could be divided proportionately between all the different oil products including those that are not subject to the regulated GHG reduction target. Under this option the annual potential for savings from flaring and venting under the Fuel Quality Directive would be about 10 Mt $CO_{2\,eq}$. A tonne of CO_2 equivalent avoided through a flaring and venting reduction project is a million grams of CO_2 equivalent. This means it can provide a supplier with a 1gCO_2 reduction for every Terajoule of energy they supply. If the savings were not linked to 'co-product allocation', then this would imply that 20 Mt $CO_{2\,eq}$ emissions could theoretically count as the reduction potential (European Commission 2011a).

Not only would the reduction of gas flaring have a direct impact on the carbon content calculations of fuels but reductions are also seen by regulators as potentially assisting companies and countries in meeting Emission Trading Scheme (ETS) requirements. However, in Europe most refiners have already reduced their flaring and it is still unclear how to include flaring from those operations outside Europe in full life cycle analysis (LCA) calculations. In addition, most countries where flaring still occurs e.g. the Middle East, Former Soviet Union and Africa are not part of the ETS and are already slowly phasing out flaring as they move toward gasification. Mechanisms to reduce global flaring and venting of natural gas are also under way, involving host governments, industry, the EU and in particular initiatives such as the Global Gas Flaring Reduction and Methane to Markets partnership. Therefore, it was expected that by 2015 little flaring will exist.[8] It is therefore questionable how much flaring and venting reductions can assist the EU in meeting its de-carbonization challenge in the long term. Refiners will have to choose if it makes sense to use this as an option to meet the 6% target or focus more on crude selection if allowed (this is dependent on the final accounting system adopted by the European Commission), and/or on refinery product and process changes.

---

[8] Personal communication with Dr. Petr Steiner, World Refining and Fuels Service, Hart Energy Consulting (2010) www.hartenergy.com

The latter will also depend on deliberations on whether or not to include refinery efficiency as a feasible emissions reduction measure. Non-refining stakeholders claim that as refineries are already part of the EU-ETS, improving the efficiency of their production could bring double benefits to companies that go forward with the investment. Several studies suggest that increasing efficiency decreases production costs and thus adds to the competitiveness of the refinery. Refineries on the other hand argue that most of the energy efficiency benefits and investments have been realized and that the implementation of the Fuel Quality Directive alongside the ETS negatively impacts refineries twice.

Regarding crude selection, most European refiners prefer to purchase light sweet crudes rather than heavier sour crudes due to the lower amount of processing needed to reach European fuel quality requirements. In addition to indigenous production (the North Sea), sweet crude is imported mainly from Africa (North Africa and Nigeria) and the Caspian region with a small amount of sweet crude also coming from the Middle East. While the share of Caspian oil will grow in the future, currently these regions account for approximately 10% of imports. With regard to heavier crudes such as unconventional (shale) oil or tar sands, in early 2011 only one refiner in Estonia was investigating the use of local shale oil for domestic use.

The setting of carbon default values for tar sands has been particularly controversial. The Commission had originally proposed that fuel from tar sands should be given a value of 107 g CO_2 equivalent per MJ to take into consideration its more energy intensive production process. However, Canada, the main producing country of tar sands, feared that by treating tar sands as a dirty fuel the EU would set a precedent and other countries would follow the EU's lead. Therefore, due to strong political pressure from Canada, the European Commission scrapped the specific value for tar sands until the end of 2011. Environmental groups have complained that such a move defeats the purpose of the Fuel Quality Directive by failing to account for the real carbon footprint caused by the extraction and processing of tar sands. The lack of transparency related to access to documents on the Canadian tar sand case[9] has created another level of distrust regarding the decision making process and between different stakeholders, which could have been avoided through a more comprehensive and open stakeholder process. That said it is unclear how many oil companies actually use and will continue to use the feedstock for European product rather than invest in refining technology to upgrade heavier fuel oil produced in Europe or Russia.

Fuel de-carbonization efforts therefore are most likely to focus on product changes by promoting biofuel blends or natural gas. In this regard, biofuels are expected to deliver the largest portion of the 6% mandatory lifecycle GHG reduction target for fuel suppliers by 2020.

---

[9] 'EU yields to Canada over oil trade "barriers": sources'. Pete Harrison, Reuters. http://ca.reuters.com/article/domesticNews/idCATRE62N3T920100324 [last accessed March 2011].

However, the total amount of biofuels used and accounted for will very much depend on the availability of 'sustainable' first, second or third generation biofuels. The exact amount of biofuels to be blended is therefore not clear in particular because GHG savings vary enormously depending on the commercial process used and whether ILUC will be included in the feedstock calculations. The European Commission's Joint Research Centre (JRC) claims that even when using their own rigorous GHG methodology most EU commercial processes vary between 18 and 50% (excluding ILUC), which is already a huge variation in GHG reduction capabilities (JRC 2008).

Environmental NGO Transport & Environment (T&E) (2010) (the NGO watch dog on the development and adoption of article 7a and EU biofuels legislation) has made some rough calculations on the total portion of biofuels realistically expected to fulfil the GHG reduction target. They have calculated that only 1% 'sustainable' biofuels were available in 2010, whereas 10% 'sustainable' biofuels would be available in 2020 with an average GHG saving of 50%. This is based on the following assumptions:

- 2010 base year
- average climate performance of these biofuels in 2020,
- uncertainty about the inclusion of ILUC impacts on biofuels GHG emissions performance.

If the above reasoning is used, biofuels could be expected to offer a 4.5% savings but not the full 6%. With regard to other alternative fuels, T&E predicts that natural gas derived fuels such as CNG and LPG are not expected to contribute much to the 6% target. However, taking into consideration cheap gas and continuous unconventional gas discoveries, it could be that gas does pick up more of the 6% than expected (European Commission 2011a). In the meantime the remaining 1.5% would need to be met by the oil industry via refining efficiency if allowed or other products.

5.4.3 Article 7a Modelling Methodology: Creating Effective Default Values

There is unified agreement amongst all non-biomass stakeholders that Article 7a of the Fuel Quality Directive provides an important new tool for reducing GHG emissions from transport fuels. Environmental NGOs, the oil industry, the auto industry and other related industries, Government Ministries and MEPs alike have even indicated that they prefer the approach undertaken under 7a to a 2020 10% target for renewable fuels, which in essence was pushing more biofuel use and hence one technological solution without full accountability for real GHG emission reductions or land use change impacts. They maintain that a renewable fuels target did not necessarily incentivize the right renewables in transport e.g. second and third generation biofuels or other low carbon options which may be discovered in the future. Whereas under Article 7a if completed and implemented properly, fuel suppliers should be pushed toward all lower carbon options across the well to tank cycle from exploration, the choice of crude, the production process or the final product choice.

The main area of disagreement is how far Article 7a should go in assessing the carbon content of fossil and non-fossil fuels. Should it take into consideration ILUC factors, which are incredibly complex to measure or evaluate full exploration and production processes for fossil fuels, which could eliminate high carbon crudes such as tar sands or coal derived products such as CTL (coal-to-liquids)?

Figure 5.4 shows the carbon intensity of diesel. As can be seen from the chart, 84% of emissions are created during the combustion or use phase. Again, fuel suppliers can only reduce emissions from the combustion phase by changing the carbon content of their product. The other 15% of emissions are created during extraction and refining and can be reduced with efficiency measures. The graph only shows average values for refining and extraction. Note, however, that the emissions from extraction and refining vary depending on the type of crude used.

At the end of the day it all comes down to the modelling methodology and approach chosen to calculate emissions from the production of fossil fuels such as petrol, diesel, natural gas and biofuels. Depending on the methodology adopted, different actions will be rewarded or not. The European Commission from the outset understood the complexity of finding the right modelling methodology for evaluating the carbon content of all fuels. In its own consultation document, the European Commission (2011a) laid out the advantages and disadvantages with the different approaches proposed.

However in the end, with regard to fossil fuels, the original analysis undertaken by the European Commission was focused on the setting of a single GHG default value for each fossil fuel derived from oil. This would mean that irrespective of the feedstock a singular value would be set for all fossil oil based diesel, petrol, hydrogen and coal-to-liquid. Since the overall objective of the proposal is to address GHG intensity of fuel in terms of its GHG emissions per unit of energy, the

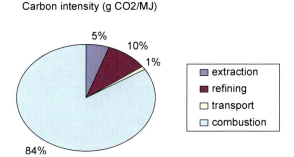

Fig. 5.4 Life cycle carbon intensity of average EU diesel
Source: T&E (2010)

European Commission established a baseline for GHG intensity per fossil fuel that is calculated as the following:

$$\frac{\sum_{a \text{ to } z} (GHGi_{\text{fuel x}} \cdot MJ_{\text{fuel x}})}{\sum MJ_{a \text{ to } z}} \tag{5.1}$$

Under the above baseline calculation *fuel x* refers to the different fossil fuels (diesel, petrol, diesel off-road gas-oil, liquid petroleum gas (LPG) and compressed natural gas (NG)) placed on the EU market in 2010 for use in any type of road vehicle and non-road machinery as defined in the Directive. Any other types of fossil based fuel that may be used are believed to represent less than 1% of the overall EU market and it is proposed that these should not be taken into account for the purposes of setting the baseline.

At issue with this approach is that several stakeholders have commented that not all fossil fuels are created equal nor for that matter are all biofuels created equal. In the case of fossil fuels, diesel produced from tar sands would have the same default value as diesel produced from sweet crudes, which are not as carbon intensive. Although it is clear that the main focus of the directive is to move away from fossil fuels in the first place and transition toward lower carbon alternatives, in the meantime as the full de-carbonization challenge is not expected to be met until 2050, and fossil fuels will remain a large part of the fuel mix until 2030 (IEA 2008), it is important to prevent the uptake of the most carbon intensive fossil fuels as well and encourage fuel suppliers to optimize their production processes.

With the current proposed system, some would claim that fuel suppliers are not incentivized to use cleaner crudes or invest in extraction efficiency, and most importantly oil companies that invest in cleaning up their production chain will not be recognized for doing so as they will be penalized in the same manner as those companies investing heavily in more carbon intensive crudes or products derived from such crude oil. The Commission originally favored separate default values for different oil-derived products including tar sands (refer to Table 5.1 below) as indicated before.

Whether or not these default values are accurate is important as is the extent to which Article 7a is extensive and effective in approach and fulfils full life cycle criteria from well to tank. If the default values set under Article 7a are perceived to be flawed from the outset, non-transparent regarding carbon intensity and not fully technology-neutral, the system will not work as no stakeholder will take it seriously. In addition, a system which does not apply the same rigor in well to wheel modelling across all fuels will fail from the outset.

The crux of this issue is therefore the accuracy of reporting and the development of a harmonized methodology for all fuels. Separate default values for high carbon oil could be important in order to ensure that reporting is as complete as possible and a comparative assessment is undertaken between different fossil fuel options. Some stakeholders believe that if oil companies are to be held accountable, their reporting should include the carbon intensity of all of their products, including the source of fuel and extraction method. Refiners would argue that the differential between

Table 5.1 Proposed default values (European Commission 2011a)

Proposed default values GHG (grams of CO_2 equivalent per MJ)	
Petrol	85.8
Diesel	87.4
LPG	73.6
CNG	76.7
Tar Sand	107
CTL	172
CTL with CCS	81
GTL	97
Hydrogen – wind based electrolysis	9
Hydrogen – steam reformed NG	82
Hydrogen from coal	190
Hydrogen from coal with CCS	6
Electricity (EU average)	48
Plastic based fuel	86

the different fossil fuels is not large enough to merit a default value per fossil fuel feedstock and that setting fixed industry average default values reflects the reality of producing diesel and petrol fuels. In comparison with biofuels, they see little potential for reducing GHG emissions from outside refineries given that the reported savings are of the order of 1.8% and, as discussed in the next section, these actual savings are difficult to calculate in comparison with biofuels which can vary up to 90%. Even if all emissions could be eliminated at the refinery this would only amount to around 10% of life cycle emissions.

In the area of fossil fuels, the Consultation paper supports the fuel industry view that fossil fuel pathways are different from biofuel pathways. The Commission indicates in its consultation paper that in the case of fossil fuels, it is the fuel pathway and combustion of the fuel which has the largest emissions impact whereas the way the fuel is processed varies only slightly. This is why the European Commission proposes little mark up between typical and default values for fossil fuels. The European Commission also claims in its Consultation document that the administrative burden of conducting actual GHG emission calculations rather than using default values has to be taken into account since in comparison with biofuels there is little variation in GHG emissions from fossil fuels. However, the Commission is considering the use of the same approach with a lower differential between typical and default values. This position is supported by the oil and gas industry.

In addition the oil and gas sector maintains that having specific default values for GHG emissions from fossil fuel feedstocks may result in a shift of product supply toward imports thus reducing supply flexibility and security, damaging EU refining competitiveness and potentially leading to carbon leakage (that is the loss of business in Europe to competition outside EU borders due to stricter carbon regulation). The issue of carbon leakage has proponents for both sides. Over the course of the year 2010, several reports have been published including one by the OECD (Reinaud

2008) indicating very little evidence of carbon leakage. However, the European Commission is concerned about the potential impact that GHG calculation methodology and the setting of default values could have on energy security and carbon leakage and points to data released by Wood MacKenzie supporting fears of carbon leakage. The European Commission has therefore listened to oil and gas industry complaints that an increase in supply of diesel from the FSU countries, for example, could flood the market with unutilized inexpensive residual fuel which could replace more efficient natural gas power and heat generation elsewhere. Cheap diesel product from Russia is already entering the EU market to be further refined for cleaner diesel to meet EU requirements. This scenario, they claim, could lead to carbon leakage.

As of this writing in March 2011, stakeholders were waiting on the European Commission's final proposed implementing measure, which would give all of the necessary criteria for proper implementation of Article 7a and the default values for fossil fuels.

5.5 Low Carbon Fuels Policy: Biofuels Use

The Renewable Energy Directive is very clear under Article 17 that the GHG emission savings from the use of biofuels and bioliquids need to be at least:

- 35% with effect from adoption of the directive,
- 50% by January 2017,
- 60% from 1 January 2018 for biofuels and bioliquids produced in installations in which production started on or after 1 January 2017.

The GHG emissions saving from the use of biofuels and bioliquids must be calculated in accordance with Article 19(1) of the Renewable Energy Directive:

$$\text{GHG SAVING} = (EF - EB)/EF \tag{5.2}$$

where

EB = total emissions from the biofuel or bioliquid; and
EF = total emissions from the fossil fuel comparator.

The fossil fuel comparator as reported under the Fuel Quality Directive has a starting value of 83.8 g CO_{2eq}/MJ. At present, this is the figure against which all biofuels are compared to determine GHG savings. This value will be superseded by the 'latest actual average emissions from the fossil part of petrol and diesel in the Community' when that information becomes available from national annual reports submitted under the Fuel Quality Directive – the first reporting taking place in 2011.

Under the starting value for the fossil fuel comparator of 83.8 g CO_{2eq}/MJ, in order to meet the GHG-saving threshold of 35%, a biofuel would have to emit 54.47 g CO_{2eq}/MJ or less, calculated as follows:

$$\text{GHG SAVING} = (83.8 - 54.47)/83.8 = 35\% \qquad (5.3)$$

The European Commission believes that this overall methodological framework has several advantages. It provides flexibility when calculating GHG savings, allowing the economic operators to use either the default GHG savings or to calculate the actual GHG savings themselves. The European Commission is also of the opinion that the Renewable Energy Directive in combination with the Fuel Quality Directive will through this framework incentivize higher GHG savings and will inherently promote more performing clean technology. It is also flexible enough to continuously allow for new entrants on the market. That said some will claim that any methodological framework is only as strong as the data it is based on. If the data inputs have been manipulated from the outset in the interest of certain feedstock choices (as claimed by some stakeholders) then the framework will have little credibility.

Although a methodological framework is in place to calculate savings (excluding ILUC), due to the complexity of assessing the exact nature of these high carbon stocks, the European Commission by the beginning of 2011 had already issued four draft communications for discussion focusing on the following two key topics:

- The practical implementation of the EU biofuels and bioliquids sustainability scheme;
- The type of information about biofuels and bioliquids to be submitted by economic operators to Member States; on voluntary schemes and default values and on guidelines for the calculation of land carbon stocks.

5.5.1 Consultation on Indirect Land Use Change (ILUC)

The draft communications were issued in particular due to the second issue above, as there continued to be a clean split between those government and environmental NGO stakeholders who believe that if biofuels are going to be evaluated by their GHG emissions potential they must be assessed under a full LCA including ILUC impacts and that an ILUC factor is necessary.

The above split on opinion is only compounded by the complexity and difficulty in reaching agreement over the LCA methodology regarding biofuels, and the global controversy surrounding actual ILUC impacts and LCA modelling of these impacts. Although the modelling process led by the European Commission's Joint Research Centre (JRC) was commended by NGOs for being quite complete, they complained that the discussions and data were not accessible enough during the final decision making process.

Instead of launching a full scale stakeholder process in the same manner as the Auto Oil programs, the European Commission launched a restricted internet consultation on different aspects of the directive still open for resolution and upon which a report was required by end 2010. The Directorate General for Energy (previously the Directorate General for Energy and Transport) in its consultation only sought stakeholder views in two separate face to face meetings.

From a procedural perspective, it is questionable whether the internet consultation and two meetings were sufficient for getting detailed feedback from all stakeholders, and some environmental groups questioned whether it was fully transparent and equitable. Criticism was most forthcoming once again from T&E. As a result of what they felt was an extreme lack of transparency, T&E and several NGOs launched an access-to-documents request asking for access to studies on ILUC that the European Commission was undertaking under its legislative mandate.[10] As of this writing they claim they have not been granted full access to these documents and have therefore taken the European Commission to court.

In comparison other stakeholders, predominantly third country governments exporting biofuels to Europe and biofuel producers, argue that a full ILUC is impossible to assess since no internationally accepted model exists and a decision today would not be based on sound scientific evidence. In fact a group of protesting countries (Argentina, Brazil, Colombia, Indonesia, Malaysia, Mauritius, Mozambique, Sierra Leone and Sudan) sent a letter already in December 2009 to the EU Energy Commissioner at the time indicating that 'the lack of appropriate data at a global level raises important concerns about the possibility of building a well-designed and comprehensive methodology in a very short time.' These countries accurately point to the lack of internationally agreed methodology and the need for a global consensus on sustainability criteria and proper GHG emission calculations and certification for all fuels. In addition, countries from this camp claim that setting an ILUC factor has the greatest impact on imports and thus they ask if this is not just a clever way for European producing countries and companies to protect their markets.

The above countries would argue that, due to its impact on trade and international GHG emission policies, this issue should not be settled by national jurisdictions but at the international level through an appropriate international framework, such as the United Nations Framework Convention on Climate Change (UNFCCC). However, those European countries wanting to resolve the ILUC issue rapidly do not think that the cumbersome UN process will bring resolution fast enough.

Evidence from across the Atlantic has further shown how complex the ILUC issue is, in particular with regard to verification. When delving further into the ILUC problem, the US Environmental Protection Agency (EPA) undertook a full analysis of existing verification programs. The extensive review was undertaken between the

---

[10] The lawsuit, brought by ClientEarth, Transport & Environment, the European Environmental Bureau, and BirdLife International, challenges the Commission's failure to release documents containing previously undisclosed information on the negative climate impacts of widespread biofuels use in the European Union. It is the second time the Commission has been sued for lack of transparency on EU biofuels policy.

proposed rule released in May 2009 and the final rule released in February 2010. The EPA review investigated non-governmental, third-party verification programs used for certifying and tracking agricultural and forest products from point of origin to point of use both within and outside the US. This included an analysis of the Roundtable on Sustainable Palm Oil, the Basel Criteria for Responsible Soy Production, and pending criteria from the Roundtable on Sustainable Biofuels, Soy Working Group, Better Sugarcane Initiative, Sustainable Agriculture Network and Forest Stewardship. The intention of this review was to assess whether a system existed for the tracking of biomass. Interestingly, the EPA was confronted with three challenges:

1. Most of the third-party certification systems listed above do not offer certification for the full scope of products needed.
2. A US-specific issue was that the acreage of agricultural land or actively managed tree plantations currently certified through third parties in the US covers only a small portion of the total available land and forests estimated to qualify for renewable biomass production under the EISA definition.
3. None of the existing third-party systems had definitions or criteria that perfectly match the land use definitions and restrictions contained in the EISA definition of renewable biomass.

Each of these three challenges could be applied to the EU; this is why it has been so difficult at the EU level to agree on a set of sustainability and ILUC criteria relevant to the European situation. Due to this review, the EPA concluded that it could not, at that time, solely rely on any existing third-party verification program to implement the land restrictions on renewable biomass under its Renewable Fuels Standard 2 legislation. Although the EU came to this conclusion as well, both the European Commission and US EPA have agreed that if the above issues could be solved third-party verification programs as listed above could potentially be used.

In addition, the European Commission has agreed to conclude bilateral or multilateral agreements with third countries that contain sustainability criteria for biofuels in line with the criteria in the Renewable Energy Directive. Voluntary national or international schemes or standards for the production of biomass products and for measuring GHG emission may also be accepted by the European Commission.

In December 2010, the Commission submitted the ILUC report that was promised to the European Parliament and the Council of Ministers under the Renewable Energy and Fuel Quality Directives. For some, the Report falls well short of its intended goal to give further direction regarding ILUC criteria.

The report includes a proposed package of non-binding proposals including two Communications and a Decision, which should help businesses and Member States to implement the Renewable Energy Directive. They focus especially on the sustainability criteria for biofuels and what must be done in order to control that only sustainable biofuels are used. In general the package covers concepts which had already been discussed under the Renewable Energy Directive yet needed further clarification, but does not go as far as giving the detailed clarification expected by national governments and businesses related to calculating land use change in GHG reduction methodology.

After a two-year investigation, the Commission report has accepted that ILUC will reduce carbon savings from biofuels, but it has stopped short of immediately recommending new barriers against unsustainable biofuels or giving a set ILUC threshold factor. Instead, it delayed the announcement of its ILUC strategy to July 2011.

The additional studies scheduled to be undertaken in 2011 to feed into a final strategy would focus on four key approaches regarding ILUC:

1. take no action for the time being, while continuing to monitor,
2. increase the minimum GHG saving threshold for biofuels,
3. introduce additional sustainability requirements on certain categories of biofuels,
4. attribute a quantity of GHG emissions to biofuels reflecting the estimated ILUC.

If point four is chosen as the European Parliament, certain Member States and NGOs have called for, biofuels will most likely be attributed an ILUC factor. If that ILUC factor is high it is assumed that most biofuels will not meet the 35% GHG savings required relative to fossil fuels. Such a scenario would in effect force the European Commission to re-evaluate the current 10% renewables target in transport fuels until second and third generation biofuels are available in large quantities, as no first generation biofuel made from existing biofuel feedstocks would comply. This makes fuel de-carbonization extremely difficult as few alternative options are available in the short to medium term.

5.6 Conclusions: The Road to 2050

This chapter has only addressed the complexity of finding low carbon fuel options to reducing GHG emissions from transport but, as indicated in the opening remarks, a series of other tools such as vehicle electrification, energy efficiency, new vehicle technology, carbon capture and storage (CCS), traffic management, greater public transport use etc. must be used to complement low carbon fuels and curb transports impact on the environment. These tools are addressed in other chapters of this book.

To meet today's EU objectives of energy security and de-carbonization, alternative fuels only offer part of the solution and as shown even this partial solution is not clear or fully accessible at present due to complexities surrounding carbon modelling and ILUC. There is therefore still a great deal of work to be done in finding alternatives to fossil fuels that truly offer a lower carbon footprint.

Some of this more holistic thinking in the area of fuels has already been undertaken by the European Commission and should be further built upon in light of the continued difficulties in setting sustainability criteria for fossil fuels and biofuels. The European Commission's Clean Air and Transport Unit of the Environmental

Directorate (now part of DG Climate Action) commissioned AEA Technology to undertake a series of studies which looked at possible pathways for the reduction of GHG emissions from the transport sector to 2050. The studies and stakeholder discussions were similar to the Auto Oil Program approach in the sense that they took on board a broader 'systems approach' when looking at the total GHG emissions from transport and different possible pathways (AEA 2010).

This was an extensive analysis which addressed a series of solutions but due to the European Commission's focus on Article 7a and sustainable biofuels this low carbon fuels work has not received the attention it deserved. The study is interesting in that it looks at the need for a more comprehensive approach that can take into account the full range of parameters that play a role in determining energy security in the transport sector and hence energy diversification and switching to low carbon fuels. In particular, it looks at the:

- linkage between price of new energy sources and the oil price
- proportion of vehicle fleet able to use the new energy source
- cost of new energy sources compared to oil
- surplus of supply capacity over demand
- susceptibility of new energy source to disruptions (extreme events and inadequate market structures)
- resource concentration for the supply of the new energy source.

The study claims that in order to be able to develop a full quantitative approach for assessing the energy security impacts of possible GHG reduction policies for the transport sector, it would be necessary to have access to quantitative data for all of the above parameters for each potential new energy source. Only in this way could all the energy security impacts and benefits be fully evaluated. A new study was commissioned to follow up on these results, which should address co-benefits and a quantitative approach for energy security.

The release of the 'Roadmap for moving to a competitive low carbon economy in 2050' in February 2011 (European Commission 2011b) stimulates further studies and policy proposals in transport de-carbonization. The Roadmap directly mentions the setting of sectoral GHG reduction targets for 2050 including a proposal to reduce GHG emissions from the transport sector by 54–67%. The challenge will be to fully understand the role biofuels will actually be able to play in helping meet the transport target, taking into consideration land use impacts as well as finding the right mix of low carbon electrification and other non-biomass alternative fuels such as natural gas and hydrogen.

Clearly due to the complexity of the issues related to fossil fuel and biofuel sustainability as well as particular biofuel modelling concerns regarding ILUC, it is hoped that a comprehensive stakeholder program similar to the Auto Oil Program may be put in place to ensure the adoption of the next phase of low carbon fuel legislation. NGOs have already asked the European Commission to create a stakeholder group to follow up on these issues and implementation in particular with regard to biofuels sustainability criteria.

In this respect, it is important not to forget the quantum leap industry and governments have made over the last decade by working together to develop new more environmentally friendly vehicles and fuels as a result of the Auto Oil Programs. In 2009 the European Union reported 100% sulfur-free petrol and diesel fuel market penetration as a result of Fuel Quality Directives 98/70/EC and 2003/30/EC (AEA 2011). It is because of the benefits from these cleaner fuels and vehicles and the road they have paved toward sustainability that we can worry less about our air quality concerns and focus more on CO_2 reduction and minimizing energy consumption overall.

At the end of the day industry and governments have a proven track record of working together to produce a more sustainable transport system and they must continue in this vain. The challenge will be to establish a transparent stakeholder process that can bring all actors to the table on equal footing to willingly address the shift from fossil fuels and the complex issues of LCA modelling and product sustainability without too much focus on special interests. Only through bold political and industry action to address real low carbon options and innovative solutions will Europe enable transformational change in the fuels industry and meet its 2050 GHG emissions targets of 80–95%.

References

AEA (2010) Towards the decarbonization of the EU's transport sector by 2050', Final Report for the EU Transport GHG: Routes to 2050? Project for DG Climate Action (Contract ENV.C.3/SER/2008/0053). 2010. http://www.eutransportghg2050.eu. Last accessed Mar 2011

AEA (2011) EU fuel quality monitoring – 2009 summary report. Report to the European Commission, DG Climate Action. Report ED05471 Final. January. https://circabc.europa.eu/d/d/workspace/SpacesStore/1266fd51-938c-4cf4-93b9-1225ea4bb018/FQD_Summary_2009(0).doc. Last accessed Mar 2011

European Commission (2010) Unlocking Europe's potential in clean innovation and growth: analysis of options to move beyond 20%. Communication SEC(2010)505 final. Brussels, May. http://ec.europa.eu/energy/infrastructure/studies/doc/2010_0505_annex_en.pdf. Last accessed Mar 2011

European Commission (2011a) Directive 2009/30/EC amending Directive 98/70/EC on fuel quality – consultation paper on the measures necessary for the implementation of Article 7a(5). https://circabc.europa.eu/d/d/workspace/SpacesStore/0a750c71-26c7-4959-b61c-553779fc13ff/art7aconsultation.pdf. Last accessed Mar 2011

European Commission (2011b) A roadmap for moving to a competitive low carbon economy in 2050. COM (2011) 112 final, March. http://ec.europa.eu/clima/documentation/roadmap/docs/com_2011_112_en.pdf. Last accessed Mar 2011

GGFR (Global Gas Flaring Reduction Partnership) (2008) Data available at http://go.worldbank.org/NEBP6PEHS0. Last accessed Mar 2011

Hill N, Morris M, Skinner I (2010) SULTAN: development of an illustrative scenarios tool for assessing potential impacts of measures on EU transport GHG. Task 9 Report VII produced as part of contract ENV.C.3/SER/2008/0053 between European Commission Directorate-General Environment/Climate Action and AEA Technology plc. See website: www.eutransportghg2050.eu

IEA (International Energy Agency) (2008) *World energy outlook*. Paris, France

JRC (EC Joint Research Centre) (2008) Biofuels in the European context: facts and uncertainties. http://ec.europa.eu/dgs/jrc/downloads/jrc_biofuels_report.pdf. Last accessed Mar 2011

Reinaud J (2008) Issues behind competitiveness and carbon leakage. IEA Information Paper, International Energy Agency, Paris, France, October. http://www.iea.org/papers/2008/Competitiveness_and_Carbon_Leakage.pdf. Last accessed Mar 2011

Transport & Environment (2010) T&E briefing: ensuring carbon reductions from fuel production. http://www.transportenvironment.org/Publications/prep_hand_out/lid:580. Last accessed Mar 2011

UK Parliament (2008) Announcement of publication: environmental audit committee calls for moratorium on biofuels, 21 Jan 2008. http://www.parliament.uk/business/committees/committees-archive/environmental-audit-committee/eac-210108/. Last accessed Mar 2011

Chapter 6
Fuel Taxation, Regulations and Selective Incentives: Striking the Balance

Per Kågeson

Abstract This chapter discusses to what extent fuel taxes may need to be complemented by other government-induced measures for achieving a cost-effective reduction of carbon emissions from cars. The conclusion is that market failures and the need to prepare for longer-term climate objectives make it essential to regulate the specific fuel consumption of new cars. The current European regulation, being full of derogations and other loopholes, makes support from national incentive schemes important. However, for efficiency, and in order to prevent excessive taxation, common guidelines are needed. They should prescribe that all incentives must comply with certain basic principles, the three most important being technical neutrality and equal treatment of all cars; continuous incentive (rather than a number of CO_2 thresholds); and that the size of fees and bonuses should not depart substantially from the marginal abatement cost in other sectors. However, selective incentives to emerging technologies by definition cannot be technologically neutral. Such incentives, thus, should only be allowed if the member state can show that there are good reasons to expect that the learning curve and economies of scale will make production cost decline considerably. This requirement will prevent them from subsidizing mature technologies and minimize the risk for lock-in effects and the distortion of competition.

6.1 Background

Road transport accounted for 22% of Europe's carbon emissions in 2008, and for 18% of its overall greenhouse gas (GHG) emissions. Cars and light duty vehicles are responsible for approximately two thirds of these emissions. The emissions from road transport grew by 28% between 1990 and 2008 despite an overall decline of European GHG emissions.

The European Union has decided to introduce a cap and trade system for carbon emissions from power plants, large boilers and a number of energy-intensive industries, the EU Emissions Trading System (EU ETS). Domestic transport emissions are not subject to any cap. To cut them remains the responsibility of the individual member states, which on average must reduce emissions from their non-emissions-trading sectors in 2020 by 10%[1] from 2005 levels.

When trying to reduce road transport emissions, the member states have a large menu of potential policy instruments and measures to choose from. Economic theory suggests that the most efficient and most cost-effective instrument would be to enforce a uniform tax on carbon emitted from sources in the non-trading sectors. However, experience shows that a number of market barriers prevent fuel taxation from becoming fully effective. In addition, the current European target is short term and will be followed by more stringent limits post 2020. This may call for the use of supplementary policy instruments to help Europe prepare for future commitments.

The aim of this chapter[2] is to discuss to what degree carbon abatement in the road transport sector can be achieved by general policy instruments such as fuel taxes, and to what extent these should be complemented by other government-induced measures. A second objective is to analyze the need for guidelines for incentives that are introduced in combination with fuel taxation. Although the focus of this chapter is on Europe, striking a balance between fuel taxation, regulatory measures and complementary economic incentives is a universal challenge.

6.2 European Taxation of Road Fuels

As the negative effect of carbon emissions is global, the location of the emitter does not matter. This makes CO_2 an ideal case for general policy instruments such as carbon taxes and emissions trading. However, the taxation of road fuels, recalculated into Euros (€) per tonne emitted, is very high compared to the implicit taxation of carbon in other sectors of society, the reason being that fuels have been the tax base used by governments to recover some of the direct and indirect costs caused by traffic. According to current legislation, taxation of petrol and diesel in the EU must not fall below respectively €359 and €330 per 1,000 l, while there is no upper limit. For example, the United Kingdom charges £572 per 1,000 l diesel fuel, equal to €250 per ton CO_2, and the Netherlands enforce an excise duty of €714 on unleaded petrol, equal to €302 per ton CO_2 emitted. A detailed table with automotive fuel taxes and final fuel prices around the EU in year 2010 is provided in Table 9.1, in Chapter 9 of this book.

By comparison, the price of emission allowances within the EU ETS has recently been about €15 per ton CO_2. However, the price is expected to rise as the cap is gradually lowered between 2012 and 2020 by a total of 21%. The price would have been substantially higher had the trading scheme been allowed to cover transport emissions. The EU's main reason for not suggesting a wider coverage was fear that the participation of the transport sector would drive the price to a level

---

[1] http://ec.europa.eu/energy/climate_actions/doc/2008_res_ia_en.pdf [last accessed March 2011].
[2] This chapter partially draws on Kågeson (2010).

that might cause 'carbon leakage', i.e. it might force energy-intensive production to move to other parts of the world. However, the introduction of road vehicles that use grid electricity (plug-in hybrids, battery cars and electric buses) implies a shift of emissions to power production, which is subject to the EU ETS.

There may also be cause to remember that the current way of taxing road fuels and road transport is far from perfect. Using the TREMOVE and MARKAL-TIMES models for Belgium, Proost et al. (2009) have shown that the introduction of a common tax on carbon (same rate in all sectors), combined with kilometer taxation (for internalizing the remaining externalities of road transport, including congestion), would achieve the climate change target at a lower cost and make GDP increase by an additional 1.2% in 2020. The tax revenue would increase by 3.1% of GDP. Their results show that in 2020 conventional fuels and vehicles are still cost-effective but in the longer term and with more ambitious targets, the marginal compliance cost increases and gradually allows a shift to new engines and fuels.

However, an advantage of fuel taxation is that it affects both the choice of car and the use of the vehicle. Several studies indicate that the long-term fuel price elasticity may fall in the interval between –0.6 and –0.8 and that reduced annual mileage accounts for about half of the consumers' adjustment, while the other half mainly consists of improved vehicle fuel efficiency (Goodwin 1992, Jansson and Wall 1994, Johansson and Schipper 1997, Kemel et al. 2009). Driving behavior may also be affected, provided that there is a potential for individual improvement that the driver is aware of.

As explained in detail in Chapter 9 of this book, high taxation has kept demand for road fuels in Europe at a relatively modest level by international comparison. However, there has been considerable resistance among motorists and hauliers to any further increase in tax rates. The increase in duties on petrol and diesel oil between 1 January 2002 and 1 July 2010 in the member states belonging to the euro zone did not even match inflation (17%). A few countries, notably Greece, Ireland and Portugal, that at the beginning of the period had rates well below average, have raised them by more, while the increase in Germany, Finland and Italy falls well below inflation. Among the old member states that did not join the euro, Sweden and the UK have raised the rates somewhat faster than inflation, while Denmark has not.

6.3 Should Fuel Taxation be Supplemented by Other Policy Instruments?

Against a background of high implicit taxation of carbon emitted from road fuels, there may be cause to question whether there is any need for supplementary policy measures in order to reduce emissions in a cost-effective way. Ideally, each tonne of CO_2 should be equally taxed regardless of where it is emitted. Adopting special measures in the transport sector, thus, could be seen as a deviation from the principle of equal treatment.

However, the European Union does already enforce several parallel policy instruments aimed at CO_2 abatement in the transport sector, among them minimum levels for taxation of fuels, a minimum level for the use of renewable energy in transport (10% in 2020), a directive on energy efficiency improvement (20%), special rules on fuel efficiency applying to new cars, and the proposed introduction of speed-limiters on vans and light trucks. Neither the European Commission (the EU's executive body) nor the member states appear to trust taxation as the sole instrument for curbing emissions. Many of the latter have, in addition to the implementation of common rules and regulations, chosen to use different types of incentives for the promotion of low-emitting cars and a speedy introduction of new technologies.

6.3.1 Arguments Pro and Against Using Double Instruments

Economic theory has presented arguments both against and in favor of using two or more policy instruments for achieving one objective. One counter-argument is that duplication wastes the time of both authorities and companies/citizens. Transaction costs increase without providing additional value, and there is a risk that the different measures disturb each other and give rise to higher overall costs than would have resulted from using just one instrument. The latter is particularly obvious when emissions trading is supplemented by carbon or energy taxes (Smith 2008).

When emissions trading is the general policy instrument, achieving the target is guaranteed by the cap. However, there is no guarantee that the respondents will always choose the least expensive and most efficient measures in response to the cap. Information, including mandatory energy labelling, and education, therefore, are natural complementary measures to cap and trade. Alternative forms of financing may also need to be contemplated.

Using taxes instead of emissions trading brings uncertainty with regard to the level needed for achieving the objective, and tax resistance among voters may cause politicians to set the level too low. The need for complementary policy levers may thus be greater than in a case where emissions are capped. JTRC (2008) says the introduction of mandatory standards for fuel efficiency as a means of avoiding an increase in fuel taxation may be regarded 'a trade-off between lower political costs and higher economic costs'.

Moreover, Clerides and Zachariadis (2008) found when comparing variations in crude oil prices, fuel taxation and the use of mandatory and voluntary standards in 18 countries over a period of 30 years, that standards affected specific fuel consumption significantly more than fuel taxes. Acemoglu et al. (2009) arrived at a similar conclusion in a study based on modelling. They conclude that a combination of modest fuel taxation and support for research and market introduction of new technologies will achieve reductions that diminish the future need for raising the tax rates.

6.3.2 Compensating for Market Failures

The question of whether one should or should not supplement a general market-based instrument with more selective policy instruments is complicated. One reason for doing so is the perceived risk of market imperfections or market failures. These may result from ignorance or occur in situations where those who actually pay are not in a position to decide. An example of the latter is tenants that have to pay the bill regardless of which heating system the landlord has chosen.

Another problem is the fact that citizens often have pay-off requirements on investment that differ from those of the government policy makers. The reasons may be uncertainty concerning the second-hand value or that financing the investment implies rents higher than those used in a traditional cost-benefit analysis. However, this argument for policy intervention is controversial. Regulators do not generally interfere because private discount rates are believed to be higher than those used in public projects appraisal (OECD and ITF 2008).

A basic problem is that rational decision-making requires informed decision makers. In the case of new cars, the person making the decision must have a clear view on the anticipated annual mileage, the future price of fuel, and how the second-hand value of the car is affected by his choice. In addition he/she must master elementary math and preferably also know how to discount future costs. Deep interviews with American households show that many have difficulties understanding the basic elements of the issue (Turrentine and Kurani 2007).

According to Greene (2010), referring to Della Vigna (2009), behavioral economics show that consumers who must take decisions under uncertainty tend to put greater weight on potential loss than on potential gain, and that they exaggerate the risk of loss. This, and the difficulties involved in trying to calculate the long-term effects of improved fuel-efficiency, may explain why most consumers implicitly require additional costs of fuel efficient cars to pay-off within 3–4 years or less (Greene 2010). However, Sallee (2010) claims the results of studies looking into how consumers value fuel economy have been mixed.

Another type of market failure occurs when firms hesitate to involve in technological development and demonstration because of fear that only a part of the added value will fall on them. Patents do not always sufficiently capture all aspects, and technological development therefore sometimes creates a large spill-over. This is an example of a positive externality and evidence of differing private and social return on investment. Promoting technological change may thus be seen as a natural supplement to putting a price on emissions (Fischer 2008). Barla and Proost (2008) rate the spill-over effect from investment in more efficient vehicle technologies as substantial. Sallee (2010), on the other hand, finds little empirical evidence of technological spill-overs or network effects.

An aspect of the issue is whether there exists a potential for cost-effective efficiency improvement that is not utilized by the market. This matter is complicated by the fact that producers are in the position to take all decisions concerning choice of technology and design. In this respect they act as the consumers' deputies. A recent example of this problem was the hesitancy among car producers concerning market introduction of the stop-start function and regeneration of breaking energy, which, according to the French automakers' own estimates, was cost-effective based on reasonable assumptions concerning private rent and future fuel prices.[3]

The annual reduction of CO_2 from new cars rose from an average of 1.4% per annum in 2005 to 2008 to 5.1% in 2009. This was probably an effect of the European Commission's proposal in 2008 for a CO_2 regulation and of increasing use of economic incentives in the member states. This progress is in stark contrast to the gloomy prospects outlined recently by Cuenot (2009), and Fontaras and Samaras (2010) who believe that it will be very challenging, indeed, to reach the EU target for 2020 without strong additional incentives that make the market go for smaller cars. In Cuenot's most optimistic scenario, the average new large diesel car will emit 180 g CO_2 per kilometer (g/km) in 2015, which is a stunning 60 g/km more than achieved by the best 2010 models. Contrary to Cuenot's belief, it therefore seems likely that the average new car in 2015 will emit no more than the best of today's cars. That reduces the need for down-sizing. However, in order to make customers choose the best technology in each segment, economic incentives may still be needed.

New technologies often need to go through several stages of development before being ready for broad market introduction. The *learning curve* or *experience curve* makes the technology more and more complete and often also less expensive. The economy of scale may be crucial in this context. Mass production may be a prerequisite for bringing battery costs to a level where the incremental cost of electric propulsion compared to an internal combustion engine is balanced by the expected reduction in fuel cost over the life of the vehicle. High taxes on fuel may not alone make such production volumes possible.

Yet another reason for contemplating the use of complementary incentives in order to cut emissions is the fact that a majority of new vehicles are bought by companies and institutions, many of them to be used as private cars. Company cars are typically provided to middle and high income families, which are potentially less sensitive to fuel cost than people who buy their vehicles in the second-hand market. Consequently subsidies offered to users of such cars will benefit the rich. In this sense a tax regime that favors company cars is not only environmentally harmful but also likely to have adverse distributional consequences (Copenhagen Economics 2010). According to Van Dender and Crist (2009), 'the case for a standard is particularly strong when fuel taxes are low and incomes high, as both factors exacerbate the gap between consumers' aspirations, which drive supply decisions, and policy targets for fuel economy'.

Another factor of potential importance in the context of adding policy instruments to the use of carbon taxes or emissions trading is that Europe's current climate change commitment concerns a short-term target (2020) and that more far-reaching measures may be needed in order to reach future longer-term objectives. The Intergovernmental Panel on Climate Change (IPCC) shows in its most recent

---

[3] Ulf Perbo, deputy director of Bil Sweden, personal communication.

assessment that the atmospheric concentration of greenhouse gases might have to be lower than previously believed when the objective is to prevent the mean global temperature from exceeding its pre-industrial level by more than 2°C. The industrialized countries may have to be prepared to reduce their emissions by something like 80% already by 2030 or 2035 to prevent this from happening. Considering the lead-times for technological change and the time it takes to renew fleets of vessels and vehicles, the EU and its member states should probably start the process of preparing for post 2020 commitments now. Low-consuming vehicles improve the resilience of the transport sector.

When the OECD and International Transport Forum (ITF) arranged a Round Table on the possible need for complementary measures to improve the fuel efficiency of new cars, an over-whelming majority of around 40 international experts concluded that there are good reasons to supplement fuel taxes with binding requirements on specific fuel consumption (JTRC 2008). The Round Table did not discuss the issue of fiscal incentives. However, IEA (2010) thinks that a carbon-differentiated registration tax may be needed to make the market make full use of the technical potential for halving the specific fuel consumption of new cars by 2030.

In should also be kept in mind that most member states of the EU have already introduced financial incentives aimed at the fuel consumption of, and CO_2 emissions from, new cars. However, many of them are poorly designed, and a higher degree of European harmonization may be needed in order to improve the effectiveness of these instruments.

6.4 EU's Regulation of CO_2 Emissions from New Cars

As outlined in Chapter 4 of this book, the European Union has started regulating CO_2 emissions from new cars (Regulation 443/2009), which on average for each manufacturer must not emit more than 130 g/km by 2015 for cars of average weight. Large cars are compensated for 60% of the additional fuel consumption that theoretically comes with increasing weight. The baseline, adjusted for vehicle weight, is shown in Fig. 6.1. The preliminary target for 2020 is 95 g/km.

The regulation is burdened by a number of exemptions and special rules. *Super credits* means that new cars that emit less than 50 g/km CO_2 are allowed to be counted as:

- 3.5 cars in 2012 and 2013
- 2.5 cars in 2014
- 1.5 cars in 2015

Being scrapped at the end of 2015, this rule will be of limited importance as producers anyway do not have to comply fully with the 130 g/km limit until 2015.

Sweden managed to get a discount of 5% for cars that can run on E85 (a mixture of 85% ethanol and 15% petrol). This exemption expires at the end of 2015 and

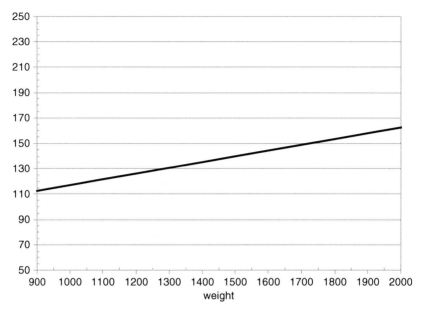

Fig. 6.1 The maximum average highest permissible CO_2 emission from new cars in the EU in 2015 for vehicles of varying weight, g/km

can only be applied to cars sold in countries where at least 30% of the fuel stations provide E85 (i.e., only Sweden). It could in theory be applied to about 1% of all new registrations in EU27 provided that all car producers are in need of the subsidy and that nearly all petrol cars sold in Sweden are equipped for E85. In practice this derogation is meaningless.

A special rule of greater potential importance is that for *Eco-innovations,* which allow the motor industry up to 7 g/km credits for innovations that are not captured by the official test cycle. The European Commission is expected to announce during 2011 what measures may be regarded *Eco-innovations.*

The regulation exempts carmakers with between 10,000 and 300,000 sales in the EU by allowing them to apply for a default target of a 25% reduction compared with 2007.

The Commission proposed that a penalty of 95 Euros per g/km and car should be enforced on sales that do not meet the average CO_2 requirement. The European Council and the European Parliament decided that the penalty should be gradually introduced and apply fully from 2018.

T&E (2010) thinks that all these loopholes together in practice mean that the target for 2015 is close to 140 g/km, rather than 130 g/km – see also Chapter 7 of this book. However, the outcome does not need to be this bad as the manufacturers have to prepare for years 2016–2018 when the derogations end. The risk remains that the low initial penalties will delay the achievement of the target by a year or two but is probably limited to a few g/km on average. One way of preventing this from happening is for the member states to introduce economic incentives that make the market fully take advantage of the technological achievements. Complementary incentives would have been less needed had the regulation been free from loopholes and temporary derogations.

6.5 European Vehicle Taxation and Incentives to Low-Emitting Cars

Most member states enforce both registration tax and annual vehicle tax, the first being a one-off sales tax and the latter a recurrent tax on cars in use. Seven countries do not use a registration tax, and eight refrain from taxing the use of cars. Three countries, Estonia, the Czech Republic and Slovakia, charge neither purchase nor ownership. Two major member states, Germany and the United Kingdom, do not employ a registration tax.

The trend in recent years has been to shift the taxation of cars away from vehicle weight and engine capacity. Today most member states use specific CO_2 emissions or fuel consumption and/or cylinder volume as parameters for the differentiation of their vehicle taxes. A few, notably Denmark, Finland, Malta, Portugal and Slovenia, relate the registration tax partially to the market price of the car (ACEA 2010). In some countries, taxation differs by region.

There is a large variation among member states in the rates enforced. The Danish registration tax starts at 105% of a car's price and rises progressively with the price of the vehicle. The Irish registration tax falls in the range of 14–36% of the retail price depending on the specific emission of CO_2 (divided into seven bands). Several other countries employ progressive taxes, among them France, Italy, Latvia, the Netherlands, Poland, Norway, Portugal, Rumania, and Spain.

Several European countries use *feebates* or bonus-malus systems to provide disincentives to large gas guzzlers and incentives to cars with low fuel consumption and emissions. The idea is to make the system more or less budget neutral by enforcing penalties on high-consuming vehicles and using the proceeds for financing grants to low-emitting cars. Such systems exist in Belgium (bonus below 115 g/km), Austria, Cyprus and Norway (bonus below 120 g/km), and France (bonus below 125 g).

The French model is worth a closer look. It went into force on 1 January 2008. The malus (i.e. fees) is shown in Table 6.1 and the bonuses (i.e. rebates) in Table 6.2.

The annual vehicle taxation is less interesting as most member states enforce rates that are too low to have a significant impact on market preferences. Ireland is an exception: The rate rises with the emission level and becomes rather high in CO_2 bands E (171–190 g/km), F (191–225 g/km) and G (>225 g/km) where it amounts to, respectively, €630, 1,050 and 2,100. The heaviest segments among diesel cars in the Netherlands are taxed equally high, and in the United Kingdom cars in the band representing emissions above 255 g/km have to pay £950 during the year of registration and £435 thereafter. The tax is progressive. Cars emitting 131–140 g/km

Table 6.1 Malus (€) on first registration of passenger cars in France

CO₂ g/km	Year of registration				
	2008	2009	2010	2011	2012
≤150	0	0	0	0	0
151–155	0	0	0	200	200
156–160	0	0	200	750	750
161–165	200	200	750	750	750
166–190	750	750	750	750	750
191–195	750	750	750	1,600	1,600
196–200	750	750	1,600	1,600	1,600
201–240	1,600	1,600	1,600	1,600	1,600
241–245	1,600	1,600	1,600	2,600	2,600
246–250	1,600	1,600	2,600	2,600	2,600
>250	2,600	2,600	2,600	2,600	2,600

Note: Flex-fuel cars (E85) that emit less than 250 g/km get a 40% discount on the malus

Table 6.2 Bonus (€) to first registration of passenger cars in France

CO₂ g/km	Electric hybrids and cars that can run on LPG or CNG				
	Year of registration				
	2008	2009	2010	2011	2012
≤130	2,000	2,000	2,000	2,000	2,000
131–135	2,000	2,000	2,000	2,000	0
136–140	2,000	2,000	0	0	0

	Other cars and the cars mentioned above if they emit less than 60 g/km				
	Year of registration				
	2008	2009	2010	2011	2012
≤60	5,000	5,000	5,000	5,000	5,000
61–90	1,000	1,000	1,000	1,000	1,000
91–95	1,000	1,000	1,000	1,000	500
96–100	1,000	1,000	500	500	500
101–105	700	700	500	500	500
106–110	700	700	500	500	500
111–115	700	700	500	500	100
116–120	700	700	100	100	100
121–125	200	200	100	100	0
126–130	200	200	0	0	0

pay only £110 per annum. Chapter 8 of this book provides detailed information on current European vehicle tax regimes.

Most European governments have either decided on incentives to electric cars and plug-in hybrids or are in the process of doing so. In addition to R&D programs and subsidies to demonstration fleets and charging infrastructure, they exempt electric vehicles from taxation and/or provide economic benefits. The value of these subsidies is particularly high in countries such as Denmark, France, Ireland, Portugal, Spain, the United Kingdom and the Netherlands, who in different forms offer €5,000 or more per electric car. By comparison, the US federal government offers a tax reduction worth between $2,500 and $7,500 (depending on battery capacity) to the 200,000 first buyers of electric cars.

Large variations in vehicle taxation among member states, no doubt cause a fragmentation of the internal market. The motor industry is forced to cope with differing national regulations and incentive systems, which add to the cost of developing and introducing the models. Notably, only Germany (annual taxation) and Finland (registration tax) enforce carbon taxes that increase linearly with the emission. In the case of Germany the penalty is €2 on each g/km above 120. The German limit value will be gradually reduced to 110 g/km in 2012 and 95 g/km in 2014. Most Member States use thresholds that provide a strong incentive to manufacturers to make cars that just barely pass the bottom-line of a given band. In total, the current national tax regulations of EU27 encompass 41 different thresholds for CO_2 and numerous different time-tables for the duration of them. However, a few are rather common, in particular the threshold of 120 g/km, which was in the autumn of 2010 used in 12 member states either for registration tax or annual circulation tax.

Portugal and the Netherlands make matters extra complicated by enforcing different limit values for diesel and petrol cars, expressed as g/km CO_2 (OECD 2009).

ACEA, the branch organization of the European motor industry, in April 2010 expressed the following principal view on the carbon taxation of new cars:

> Failure to harmonise tax systems weakens the environmental benefits that CO_2-based taxation and incentives can bring. European automakers have long called for the abolition of car registration taxes which are still widely applied in the EU. Generally, registration taxes threaten fleet renewal. A harmonised CO_2-based tax regime for cars should be a priority, applying a linear, technology-neutral system that is budget neutral in end effect. It would maximise emission reductions, support manufacturers and maintain the integrity of the single market.[4]

Bastard (2010), Renault's vice president for environment and taxation, concludes in an extensive analysis that the fragmentation of the European taxation of cars represents a significant cost to the industry and its customers and jeopardizes the concept of the single market. Furthermore, Bastard finds the tax environment not to be predictable. He highlights the problems connected to thresholds that may either give a manufacturer a disproportional benefit of an improvement by one g/km or no benefit at all. Bastard also notes the large differences in implicit valuation of CO_2 among the national schemes.

In summary, the automotive industry wants the incentives to be:

---

[4] ACEA, press release 21.4.2010: http://www.acea.be/index.php/news/news_detail/an_increasing_number_of_member_states_levy_co2_based_taxation_or_incentivis [last accessed March 2011].

- harmonized within the EU,
- technically neutral, and
- designed in a manner that provides a linear incentive, i.e., no thresholds.

In addition, the industry wants to scrap any conventional taxes on registration but appears to be open-minded about budget-neutral registration fees that do not (in themselves) make the average new car more expensive, e.g. a bonus-malus (feebate) system.

6.6 Guidelines for Car Incentives

The previous section of this chapter displayed the risk of losing efficiency as a result of fragmentation and poorly-designed policy instruments. This section will focus on parameters and principles that are essential in designing incentives that are robust and cost-effective. It will address the following issues:

- Coverage and CO_2 versus energy use?
- Technological neutrality versus selective treatment?
- Thresholds versus continuity?
- Apply the same model to all cars regardless of owner?
- Take account of vehicle utility?
- Registration tax or annual vehicle tax?

6.6.1 Coverage and the CO_2 Versus Energy Issue

The EU currently regulates tank-to-wheel emissions of CO_2 from all cars that have engines and fuel tanks that allow them to run on fossil fuels. In this context well-to-tank emissions are disregarded for good reasons. Separate minimum rules and economic incentives apply to the introduction of renewable road fuels, and the limited potential, world-wide, for sustainable production of biofuels makes it important to make sure that all new vehicles are fuel efficient. It is also essential to understand that there is only one type of CO_2 molecule and that CO_2 emitted from the combustion of biomass in the short to medium term adds to the atmosphere's concentration of CO_2. There is also cause to note that increasing demand for electricity will add to the difficulties of phasing out fossil power production.

However, the current regulation does not cover battery cars as they do not themselves emit CO_2. Moreover, cars that alternatively can use E85 or petrol are measured according to their emissions when running on petrol, while cars that can alternatively use methane or petrol are treated as running on fossil methane.

At the time of first registration of a flex-fuel car it is impossible to know what average mix of fuel it will use during its life. The outcome will depend on future supply and on future relative prices and taxes. As reported in Chapter 11, evidence from the Swedish market shows that owners of ethanol cars are very price-sensitive and prefer petrol to E85 even when the difference in price is minimal. Where methane is concerned, the ratio between biogas and natural gas will depend on regional and local circumstances and on price.

The ongoing introduction of plug-in hybrids adds a new dimension to the regulation as these vehicles can alternatively use grid electricity or road fuels. Again, it is difficult to forecast what the actual mix will be over the life of the car.

The conclusion is that it may be better to regulate energy use per vehicle kilometer than the specific CO_2 emission. This is already the case in the fuel economy regulations of the United States, China and Japan and is currently being discussed in a working group on Environmentally Friendly Vehicles of the United Nations Economic Commission for Europe (UNECE).[5]

6.6.2 Technological Neutrality Versus Selective Treatment

Policy instruments and subsidies that favor certain technical solutions run the risk of making it difficult for competing options to make it to the market, and it is difficult and often impossible for policy makers to know in advance which of the alternatives will turn out to be most cost-effective (OECD 2005). The internal combustion engine still has a large potential for further development, and new engine technologies such as HCCI,[6] OPOC[7] and FPEC[8] may provide additional opportunity. Light construction material may turn out to be less expensive per unit of energy-use avoided than are lithium-ion batteries. Lock-in effects can be avoided by giving equal treatment to all competing technologies. In the road sector this can be achieved by providing the same benefit to any reduction by one unit of energy (or one gram of CO_2).

However, creating technologically neutral incentives has a down-side. In order not to mis-use resources, the incentives provided must be of the same magnitude as the expected marginal abatement cost in other sectors of society. This is no problem for established technologies but may turn out to be a hindrance for technological renewal. New technologies are often very expensive at the out-set, but production costs may fall as a result of gradual improvement and economies of scale. A problem in this context, however, is that governments cannot afford to spend money on all new inventions. Therefore, the general incentives that apply equally to all options may need to be supplemented by subsidies that are directed to certain promising technologies.

The alternative of raising the general incentive to a level where it will make room for costly inventions in their first stage of production would mean providing too much money to mature technologies and distorting sectoral efforts to combat climate change. Therefore, it may be cause to avoid progressive rates in the taxation of fuel economy.

---

[5] Advocated by participating experts from Germany and the UK.

[6] Homogeneous Charge Compression Ignition.

[7] Opposed Piston Opposed Cylinder.

[8] Free Piston Energy Converter.

Selective treatment of promising new technologies should only be given in cases where on reasonably good ground the slope of the learning-curve can be expected to be steep. This may be the case with car batteries which, in order to make the electric car competitive, need to cut costs by at least 50% while, at the same time, their reliability and durability are improved. If every doubling of production volumes above, say, 50,000 full-size units makes costs fall by 15% from their previous level, it would take a little more than four doublings to reach the cost-reduction target. This is equal to an accumulated production volume of about one million units. However, the need for substantial subsidies will, of course, gradually diminish on that journey. One way of taking account of this and of avoiding wind-fall profits would be to gradually reduce the size of the subsidy. The American Hybrid Vehicle Tax Credit featured a phase-out provision, although not ideally designed (Sallee 2010).

The conclusion may be that general incentives at a modest level should be complemented by selective treatment of those emerging technologies that are regarded as so promising that learning can be expected to cut costs substantially within a few years. To avoid wasting money and creating a lock-in effect for competing options, the subsidy to a promising technology must be limited in time and volume. If the technology turns out to offer less than expected, the scheme should be discontinued.

6.6.3 Thresholds Versus Continuity?

As noted in a previous section, the member states currently together make use of more than 40 different CO_2 thresholds in the taxation of new cars. In some cases the economic impact of being on the wrong side of a threshold is significant. This forces the automotive industry to make costly adjustments. Currently only Germany and Finland use incentives that increase linearly with falling emissions. However, in Finland the CO_2 reward is tied to a tax that progressively increases with the list price of the car. For economic efficiency each reduction by one unit of energy or by one gram of CO_2 ought to be equally rewarded. A linear model is therefore better than one based on bands or classes.

An argument in favor of thresholds is that some member states have environmental labelling systems based on bands. The thresholds of these systems currently differ among the countries. The United Kingdom uses 13 bands while Ireland has a system based on seven. European harmonization in this respect would be a way of limiting the negative side-effects on the internal market. This is particularly important if the same bands are used for economic incentives. However, still better would be to abandon any thresholds in the incentive systems. Linear incentives can co-exist with a labelling system based on bands.

6.6.4 Apply the Same Model to All Cars Regardless of Owner?

Copenhagen Economics (2010) questions the idea of transforming the treatment of company cars beyond tax neutrality by way of building in a premium for energy-efficiency that does not apply to privately owned vehicles. The authors believe that such treatment may create incentives to move cars to and from corporate ownership rather than affecting overall fuel-consumption levels. They conclude that special rules for company cars are associated with a substantial risk of overkill, ineffectiveness and non-transparency. According to the report, the first best approach to reduce emissions is by implementing CO_2 taxes, and the second best is to include energy efficiency in the taxation of cars.

The extra administrative cost of using different incentive systems for private cars and vehicles owned by companies and institutions is a second argument for treating all cars in one system.

6.6.5 Take Account of Vehicle Utility?

Large cars have a higher utility than small by making room for more passengers and luggage. However, few households in Europe have more than four family members. There were diverging views among the member states on whether large vehicles should be compensated for higher emissions in the European regulation of CO_2 from new cars. The discussion resulted in a compromise whereby 60% of the extra fuel consumption that is caused by additional weight is compensated for.

So long as the EU regulation compensates vehicles for additional size and utility it makes sense to relate the CO_2 or energy incentive to the official baseline. This is the only way of handling the issue if the goal is to reduce distortion by harmonizing tax rules. However, there are good reasons to contemplate a change from vehicle weight to vehicle foot-print (bottom area) as this would remove the risk of providing an incentive to the manufacturers to refrain from using light materials in order to reduce weight.

6.6.6 Registration Tax or Annual Vehicle Tax?

Based on a report by TiS (2002), the European Commission (2005) concluded that car purchases are more affected by retail prices than by lifetime costs. Thus, incentives tied to the price of a vehicle are significantly more likely to influence purchase decisions than equally large incentives distributed over the life of the car. To provide an incentive equal to that of a registration tax of a given size, the annual vehicle tax would (accumulated over the years) have to be raised to a level well above that of the registration tax (Trafikbeskattningsutredningen 1999). This would make it more expensive to own a car and would have a negative distributional effect.

The conclusion is that for a given size a registration tax is a more effective base for carbon or fuel differentiation than is a circulation tax. However, a negative consequence of registration taxes is that they make the purchase of vehicles more expensive, which delays the renewal of the existing vehicle fleet.

6.7 A European Model for the Taxation of New Registrations

This section will demonstrate the design of a European model for how member states can offer an incentive to reduce fuel consumption and carbon emissions from new cars. A problem with most existing national subsidies is their uncertain future because of the scheme's burden on state budgets. This creates difficulties for the industry, which wants the rules to be foreseeable and reliable over at least a full product cycle, i.e., 6–8 years. A budget-neutral registration tax might be the solution to this problem as well as a way of avoiding making the average new car more expensive.

6.7.1 Bonus-Malus

Feebates or *bonus-malus* (respectively, American and French jargon) are models for making the market finance new car incentives without in themselves making the average new vehicle more expensive. Another possibility is to allow the manufacturing industry to trade specific emission credits tied to a common baseline (Department of Transport 1992).[9]

The first full-scale experience of a budget-neutral model came with the French bonus-malus scheme that started in January 2008. The French model is designed to operate over many years but is not completely technology neutral as it offers a higher bonus to electric hybrids and gas-fuelled cars. Another disadvantage is that the system is built on a number of thresholds. It may also be noted that the scheme does not take account of vehicle size. The French model thus is a good point of departure when designing a European model, but some improvement is needed.

In the short term, while waiting for the EU to shift its regulation from CO_2 to energy, a European model would have to focus on CO_2 emissions per kilometer. In a bonus-malus model that provides equal treatment to each gram of CO_2 it is necessary to identify a point where a car of average size is neither subject to a fee nor will receive a bonus, the so-called 'pivot point' (Greene et al. 2005). In the next few years this point could be 130 g/km. However, in order to contribute to early fulfilment of the target for 2012–2015, the zero point needs to be shifted downward relatively early. The difference between 95 g/km, the preliminary target for 2020, and 130 g/km is 35 g/km. Thus, one way of moving the market gradually toward the long-term objective could be to move the pivot point by 5 g/km per calendar year between 2014 and 2020. To allow the industry ample time to adjust it is essential to signal well in advance how the pivot point will move.

---

[9] The US has recently proposed that the International Maritime Organization (IMO) should develop a world-wide baseline and credit system for the specific emissions from maritime vessels.

6.7.2 The Level of the Incentive

The bonus should be set approximately level with the marginal abatement costs in other sectors of society. One point of departure could be the future price of CO_2 allowances in a theoretical case where cap and trade covers emissions from all sectors. In a case where the medium target is a reduction by 30% from the 1990 level, the allowance price in such a scheme would probably exceed €50 per ton CO_2. Without parallel taxation of road fuel, the CO_2 price would, of course, be significantly higher.

In a case where the expected average total mileage over a car's lifetime is 300,000 km[10] and the price on CO_2 is €50 per ton, the socio-economic value of reducing the CO_2 emission by one g/km would be €15. This means, for instance, that the difference in bonus-malus between two cars of the same weight, emitting respectively 120 and 140 g/km, would be €300. However, the difference in fuel consumption would add incentive even in a case when the buyer only considers the effect on fuel cost of the first 4–5 years of the life of the car (and disregards any impact on the second-hand value of the vehicle).

The impact on the price of electric cars depends on how they are treated. So long as the system rewards low direct emissions of CO_2, vehicles using grid electricity would be regarded as emitting zero. For a car of average European weight this translates into a bonus of €1,950 so long as the pivot point is 130 g/km and the value of each g/km is €15. When/if the base of the scheme changes from CO_2 to kWh (tank/battery-to-wheel), the bonus of an electric vehicle shrinks to something like €1,200. This is obviously not enough to bridge the difference between today's battery prices and a price that can be balanced by lower fuel cost.

6.8 A European Model for Selective Incentives

The subsidies required for allowing a new technology to take advantage of its learning curve depends on the size of the cost-gap and the anticipated volumes needed. Current European grants to electric vehicles indicate that governments think €5,000 per car to be appropriate. This is approximately €3,000 more than a zero-emission car would gain from the bonus-malus scheme discussed above.

6.8.1 Subsidies to Electric Vehicles and Plug-In Hybrids

It is not self-evident that all electric cars should receive an equally large subsidy. The United States relates its electric car grant to battery capacity. This makes sense as it is the battery technology that needs to take advantage of mass production. The

---

[10] This is higher than the current average and reflects a trend toward higher vehicle age.

learning curve for other parts of the electric propulsion system could be expected to be much flatter, and therefore does not merit the same government support. Focusing on battery capacity has a second advantage by providing equal treatment to all types of car battery and by allowing battery cars, plug-in electric hybrids and electric cars with range-extenders to be covered by the same incentive.

Portugal has declared that cars must have an all-electric range of at least 120 km to benefit from the country's subsidy to electric vehicles. This is not a good idea. Battery development can benefit as much per unit of battery that is installed in cars with a smaller range, including plug-in hybrids. For optimal social and private return on investment it is essential not to invest in larger battery capacity than the owner of the vehicle can utilize. This is particularly evident in the case of plug-in hybrids. To invest in a range of 40 km is a waste of money, weight, and energy if the normal daily use is only 20 km. In Britain, a battery capacity of 20 kWh would provide the average car owner with sufficient electric propulsion for nine out of 10 days. To cover 19 out of 20 days, the battery capacity would have to double, adding 100 kg to the weight of the vehicle and perhaps £10,000 in battery cost (Royal Academy of Engineering 2010).

Special incentives should not be provided to larger volumes of cars than needed and not by greater grants than required. Nissan-Renault expects global sales of electric vehicles to reach 0.5–1.0 million within 4 years, and CEO Carlos Ghosn says that at this volume government support will no longer be needed.[11] Other large manufacturers such as Toyota and Volkswagen believe that it will take longer. If subsidies to one million cars are needed for taking full advantage of the steep part of the learning curve, and Europe wants to take its share of a joint OECD responsibility, the member states need to support sales of a total volume of 400,000 electric cars (or the equivalent battery capacity of hybrids). This is (per million inhabitants) approximately what Portugal has promised to do by making grants available for 5,000 cars. By comparison, the United States has limited its federal grant to the first 200,000 electric vehicles.

6.8.2 Selective Support for Other Emerging Technologies

Electric batteries may not be the only equipment needed for the propulsion of cars that on environmental grounds merit selective government support. Some of the new engine concepts mentioned in Section 6.6.2 above may also turn out to be technologies that would benefit strongly from growing production volumes.

However, providing support to relatively mature technologies in excess of the bonuses of a bonus-malus scheme would distort competition without providing much added value. When economies of scale cannot be expected to cut costs considerably there is no cause for subsidies. Equipment for using methane in passenger

---

[11] Carlines No. 3, 2010. http://www.walshcarlines.com/pdf/nsl20103.pdf [last accessed March 2011].

cars is an example of a relatively mature technology where additional support would not result in a steep learning curve. The biogas potential is limited and better used in heavy-duty vehicles that fuel from their own depots, including vehicles mixing 80% methane with 20% diesel (dual-fuel). Fossil methane (natural gas) in the Otto engine of a passenger car results in a CO_2 emission per km of about the same size as that from a diesel engine fuelled with fossil diesel.

6.9 Short Impact Assessment

6.9.1 Effects on Energy Use and Emissions

Experience from the French bonus-malus may be used for assessing the potential effect on emissions from employing the incentive suggested above for a European scheme. However, the two models differ with respect to thresholds and continuity. In the French case a reduction in 2010 from 130 g/km to 100 g/km results in a bonus of €500. This is equal to €17 per gram, which comes close to the proposed €15. A reduction from 190 to 160 in the French system reduces the malus by €550 or by €18 per gram.

An official assessment of the French scheme shows that the average emission from new cars fell dramatically after the implementation of the reform (Fig. 6.2).

The share of cars that are entitled to a bonus increased from 30.4% in 2007 (prior to the reform) to 44.7% in 2008 and 55.5% in 2009. The share of vehicles subject to a malus declined from 24.4% in 2007 to 14% in 2008 and 8.9% in 2009. However, the figures for 2007 and 2008 are to some extent influenced by tactical behavior among consumers. Buyers of high-emitting vehicles rushed to purchase new cars in

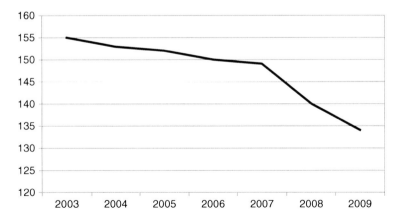

Fig. 6.2 Average CO_2 emission (g/km) from cars registered in France before and after the bonus-malus reform
Source: CGDD (2010)

December of 2007, while many of those planning to buy a new low-consuming car postponed the time of delivery until just after 1 January 2008 (CGDD 2010). Sallee (2010) provides similar evidence from the US CAFE program.

The number of car models emitting less than 110 g/km grew from 20 in 2007 to close to one hundred two years later. In 2009, cars emitting less than 120 g/km accounted for 48% of the French market.[12]

When analyzing the outcome it is essential to remember that cars emitting between 130 and 160 g/km were not subject to any incentive. Even so, the average emission shrank by 9 g/km during 2008, followed by an additional 6 g/km in 2009. This is equal to respectively 6 and 4.3% improvement on the previous year. Some of this would have happened in the absence of the bonus-malus, but the chart above indicates the strength of the model. After such a strong opening one cannot expect similar reductions every year, but the fact that the requirements change over time (see Tables 6.1 and 6.2 above) means that the scheme will continue to have a significant influence on market preferences.

It is also relevant to recognize that Norway's CO_2 differentiation of its registration tax resulted in a similar reduction. The average emission from new registrations fell from 177 g/km in 2007 to 159 g/km 1 year later and 151 g/km in 2009 (OFV 2008).

6.9.2 The Rebound Effect

Reducing fuel consumption per kilometer makes it cheaper to use the car and results in additional mileage. The magnitude of the *rebound effect* is not exactly known. However, the results from numerous studies of fuel-price sensitivity indicate that the fuel elasticity for mileage accounts for about half of the total elasticity, i.e., about –0.3. If the response to rising and shrinking fuel cost is symmetric, this figure indicates the upper limit of the rebound effect. In addition, time is a scarce resource. A large decrease in variable cost is not going to make us want to spend much more time at the wheel. As explained in Chapter 2 of this book, the time that humans spend on mobility per day is relatively constant over time and across cultures (Schäfer and Victor 1997). The best estimate of recent years, based on American data, suggests that the rebound effect erodes about 10% of the improved fuel efficiency (Small and Van Dender 2007). One should be aware of the rebound effect, but its existence is no valid argument against investing in improved fuel efficiency.

6.9.3 Costs and Other Economic Consequences

It is not possible to know in advance whether a bonus-malus scheme will turn out to be budget-neutral, but in a system based on a continuous incentive any surplus or

---

[12] Carlines No. 3 2010, based on figures from ADEME.

deficit is likely to be small. In the French case, the deficit grew big partly as a result of not charging vehicles that emit 130 to 160 g/km.

Whether a bonus-malus system will make the average car more expensive is an open question. This may be the case if most customers prefer high-performing cars to engine down-sizing. However, many are likely to adjust to the new situation by compromising a little. By choosing a car among the most low-emitting models of the preferred size customers have an opportunity to reduce emission at negative cost. Interesting in this context is that the average price of new cars sold in France fell by 8% between 2007 and 2009 (Bastard 2010). Other evidence of large reductions at low cost is the dramatic decrease in the fuel consumption of diesel cars, marketed by the industry under labels such as Efficient Dynamics (BMW), ECO2 (Renault), DRIVe (Volvo), BlueMotion (VW), BlueEfficiency (Daimler), Econetic (Ford) and Ecoflex (Opel). An assessment carried out on behalf of T&E (2010) shows that more than half of the reductions that took place in 2009 were a result of the implementation of better technology, while a minor part can be attributed to a shift to smaller vehicles or engines.

Using incentives to make the car market take advantage of technologies that reduce fuel consumption has a positive distributional effect as it puts most of the burden of adjusting to climate change on those that can afford to buy a new car, while at the same time making cars cheaper to drive. Roughly half of the original value of a new car is lost during its first 3 years.

6.10 Problems Connected to the Current European Test-Cycle

One argument against regulating fuel-efficiency and/or using complementary economic incentives tied to the fuel economy of new cars is that fuel economy ratings are imprecise and differ from the fuel consumed in real driving (Sallee 2010). This contra-argument carries some weight. However, the solution lies in allowing the test-cycle to reflect average driving and to incorporate the use of auxiliary systems such as air-conditioning.

6.11 Light Commercial Vehicles

Vans and light trucks represent about 12% of the European vehicle fleet. Their numbers grew by 50% between 1997 and 2007.[13] In early 2011, the Council and the European Parliament decided to limit the average CO_2 emission from vans to 175 g/km in 2017 and 147 g/km in 2020.[14] The average emission in 2007 was 203 g/km.

---

[13] T&E Briefing: *Vans and CO₂*, Updated September 2010. http://www.transportenvironment.org/tag/vans [last accessed March 2011].

[14] Based on COM(2009)593, Regulation of the European Parliament and of the Council, *Setting emission performance standards for new light commercial vehicles as part of the Community's integrated approach to reduce CO₂ emissions from light-duty vehicles*.

Experience from the United States and Europe shows that vans can, in many circumstances, substitute passenger cars and that favorable tax treatment and/or lower fuel economy standards will make part of the market shift from cars to vans. Light commercial vehicles are rarely subject to CO_2 taxes in the EU (Bastard 2010). Only a few member states have tax regimes that treat the two categories of vehicles in the same or in a similar way.

It should be possible to design a bonus-malus system that provides equal fiscal treatment to all light vehicles. This becomes particularly important when vans are allowed to emit more than cars. The difference between the existing requirement on cars and that for vans is 45 g/km in 2015–2017 and 42 g/km in 2020. For neutral tax treatment of vans and cars the pivotal point of the bonus-malus should be the same for both categories and the same compensation for higher weight should be applied. A problem in this context is that the regulation of cars allows compensation by 60% of the effect of weight on fuel consumption, while vans will be compensated by 100%.

6.12 Steps Toward European Harmonization

The apparent lack of European coordination of road fuel and vehicle taxation causes fragmentation of the internal market and distortion of competition. It also makes climate change mitigation unnecessarily complicated and expensive.

The current taxation of energy has to comply with the rules laid out in EU Directive 2003/96. The European Commission has tried several times to bring additional elements of harmonization into the rules governing taxation of fuels and vehicles. A reform of the current regulation should include equal tax treatment of diesel oil and petrol. Currently only the United Kingdom taxes diesel on par with petrol.[15] Some member states currently exempt large quantities of biofuels from fuel excise duty. This possibility was originally intended only for limited trials. For equal treatment, the discount on the taxes enjoyed by biofuels, but currently enforced on petrol and diesel, should be proportional to the value of lower emissions well-to-tank. The socio-economic value of CO_2 reductions could be determined by the price of EU ETS allowances, perhaps €25 per ton during the current decade. However, the fact that the price would have been a great deal higher had fossil fuels not been taxed also needs to be considered.[16] This may argue for a discount of €75 per ton in a case where an alternative fuel does not give rise to any emission of fossil carbon (well-to-tank).

€75 per ton CO_2 is, in the taxation of diesel fuel, equal to roughly €200 per 1,000 l. The larger part of the existing taxation of road fuels is thus unrelated to

---

[15] Based on each fuel's content of carbon and/or energy diesel should actually be taxed above petrol.
[16] Ten years ago the taxation of fossil fuels in EU15 on average corresponded to €45–50 per ton of CO_2 emitted (Kågeson 2001).

greenhouse gas emissions and is often seen as a way of making road transport pay for other externalities. If these externalities (exhaust emissions, noise, road tear, congestion and accident risk) are internalized by kilometer-based charging or road tolls, the reform of the minimum rates for fuel taxation should allow vehicles subject to the kilometer charge to be exempt from the part of fuel tax that does not relate to CO_2 emissions. In order to avoid large differences in tax levels across Europe, a member state that enforces road tolls that cover all externalities could avoid double taxation by returning part of the proceeds to owners of vehicles (regardless of nationality) that are subject to that toll.

Taxation of vehicles also needs a bit of harmonization. Current praxis reveals huge differences in the way CO_2 is valued. Portugal enforces a registration tax of €10,000 on diesel cars that emit more than 200 g/km, which makes such cars nearly unmarketable (Bastard 2010). The progressive nature of the Portuguese and Norwegian taxes makes high-emitting cars pay sums that correspond to €500 or more per ton of CO_2 when account is taken of the expected lifetime mileage of the vehicle. The annual vehicle taxes of Ireland and Denmark sometimes reach levels equivalent to more than, respectively, €700 and €500 per ton CO_2 (OECD 2009).

The EU needs guidelines or common rules that prevent excessive taxation and that prescribe that all incentives must comply with certain basic principles.[17] For general instruments such as registration and circulation taxes the most important principles are:

- Technical neutrality and equal treatment of all cars
- Continuous incentive (rather than a number of thresholds)
- Fees and bonuses that in size are proportional to the social cost or benefit of the measure

Selective incentives to emerging technologies by definition cannot be technically neutral. Such incentives should only be allowed if the member state can show that there are good reasons to expect that the learning curve and economies of scale will make production costs decline considerably. This requirement will prevent member states from subsidizing mature technologies and thus minimize the risk for lock-in effects and distortion of trade and competition. The subsidy should proportionally not exceed the incremental cost and it should be limited in time.

6.13 Conclusions

A main conclusion of this chapter is that the existence of market imperfections and the uncertainty about how fast industrial nations will have to curb their CO_2 emissions make it useful to regulate the fuel efficiency of new cars and to contemplate

---

[17] The EU Commission (2010) has declared that it is about to prepare 'guidelines on financial incentives to consumers to buy green vehicles'.

complementary use of economic incentives. The case for regulation is particularly strong when tax resistance makes it difficult or impossible to raise fuel taxes. The need for complementing regulation by incentives aimed at the first buyer is strong when the regulation, as happens to be the case in Europe, is full of loopholes.

The guidelines provided in this chapter underline the importance of making regulations and complementary incentives technology neutral, continuous (no notches) and in size proportional to the objective and to the marginal abatement cost in other sectors. To foster early introduction of new technologies selective support may be needed, provided that the learning curve of the technology is sufficiently promising. In order to avoid creating windfall profits the subsidy should be limited in time and be reduced or withdrawn as the technology matures. These guidelines are universal in character and should apply regardless of national circumstances, i.e., also to countries outside Europe.

References

ACEA (2010) ACEA tax guide 2010. Brussels
Acemoglu, D, Aghion, P, Burztyn, L, Hemous, D (2009) The environment and directed technical change. NBER Working Paper Series 15451. National Bureau of Economic Research, Cambridge, MA
Barla P, Proost S (2008) Automobile fuel efficiency policies with international knowledge spillovers. Discussion Paper 08.17. Center for Economic Studies, Catholic University of Leuven, Belgium
Bastard L (2010) The impact of economic instruments on the auto industry and the consequences of fragmenting markets – focus on the EU case. Discussion Paper No. 2010-8, Joint Transport Research Centre, OECD/ITF, Paris
CGDD (2010) Une évaluation du bonus malus automobile écologique [An ecological evaluation of the automobile bonus malus system]. Commissariat Général au Développement Durable, Le point sur No 53, Mai 2010, Paris (in French)
Clerides S, Zachariadis T (2008) The effect of standards and fuel prices on automobile fuel economy: an international analysis. Energy Econ 30:2657–2672
Copenhagen Economics (2010) Company car taxation. Working Paper No. 22 2010 (for the European Commission, DG Taxation and Customs)
Cuenot F (2009) CO_2 emissions from new cars and vehicle weight in Europe; How the EU regulation could have been avoided and how to reach it? Energy Policy 37:3832–3842
Della Vigna S (2009) Psychology and economics: evidence from the field. J Economic Literature 47:315–372
Department of Transport (1992) Tradable credits to reduce CO_2 emissions from motor cars, London, 11 February
European Commission (2005) Annex to the proposal for a council directive on passenger car related taxes, impact assessment. 261 final, SEC(2005) 809
European Commission (2010) A European strategy on clean and energy efficient vehicles. Communication from the Commission to the European Parliament, the Council and the European Economic and Social Committee COM (2010) 0186 final
Fischer C (2008) Emissions pricing, spillovers, and public investment in environmentally friendly technologies. Energy Econ 30:487–502
Fontaras G, Samaras Z (2010) On the way to 130 g CO_2/km – estimating the future characteristics of the average European passenger car. Energy Policy 38:1826–1833
Goodwin PB (1992) A review of new demand elasticities with special reference to short and long run effects on price charges. J Transport Econ Policy 1992:2

Greene D (2010) Why the market for new passenger cars generally undervalues fuel economy. Joint Transport Research Centre Round Table, 18–19 February 2010. OECD/International Transport Forum, Paris

Greene D, Patterson D, Singh M, Jay L (2005) Feebates, rebates and gas-guzzler taxes: a study of incentives for increased fuel economy. Energy Policy 33:757–775

IEA (2010) Energy technology perspectives 2010. Scenarios & Strategies to 2050. International Energy Agency, Paris

Jansson JO, Wall R (1994) Bensinskatteförändringars effekter [Effects of changing the tax rate on petrol]. Ds 1994:55, Ministry of Finance, Stockholm

Johansson O, Schipper L (1997) Measuring the long-run fuel demand of cars: separate estimations of vehicle stock, mean fuel intensity, and mean annual driving distance. J Transport Econ Policy 31(3):277–292

JTRC (2008) The cost and effectiveness of policies to reduce vehicle emissions. Discussion Paper No. 2008-9. Joint Transport Research Centre, Round Table, 31 January–1 February 2008, Paris, OECD/ITF

Kågeson P (2001) The impact of CO_2 emissions trading on the European transport sector. VINNOVA Report VR 2001:17, Stockholm. http://vinnova.se/upload/EPiStorePDF/vr-01-17.pdf. Last accessed Mar 2011

Kågeson P (2010) Med klimatet i tankarna. Styrmedel för energieffektiva bilar [Policy instruments for improving the fuel efficiency of cars], Ministry of Finance, Stockholm

Kemel E, Collet R, Hivert L (2009) How do French motorists react to a multi-annual fuel price increase? An econometric analysis based on 1999–2007 panel data. Presentation to the 12th IATBR Conference, December 2009, Jaipur, India

OECD (2005) Environmentally harmful subsidies: challenges for reform. Paris

OECD (2009) Incentives for CO_2 emission reductions in current motor vehicle taxes. OECD Environment Directorate, Environment Policy Committee, Paris

OECD and ITF (2008) The cost and effectiveness of policies to reduce vehicle emissions. Discussion Paper No.2008-9. Joint Research Centre, OECD and International Transport Forum

OFV (2008) Evaluering av avgiftsomleggingen i 2007 [Assessment of the tax shift in 2007]. Opplysningsrådet for Veitrafikken [The Information Council for Road Traffic], OFV-Rapport nr 1 – mars 2008, Oslo

Proost S, Delhaye E, Nijs W, Van Regemoter D (2009) Will a radical transport pricing reform jeopardize the ambitious EU climate change objectives? Energy Policy 37:3863–3871

Royal Academy of Engineering (2010) Electric vehicles: charged with potential. London

Sallee J (2010) The taxation of fuel economy, NBER Working Paper 16466. National Bureau of Economic Research, Cambridge, MA

Schäfer A, Victor DG (1997) The future mobility of the world population, IT/IIASA

Small K, Van Dender K (2007) Fuel efficiency and motor vehicle travel: the declining rebound effect. Energy J 28(1):25–51

Smith S (2008) Environmentally related taxes and tradable permit systems in practice, OECD

T&E (2010) How clean are Europe's cars? An analysis of carmaker progress towards EU CO_2 targets in 2009. European Federation for Transport and Environment, Bryssel. http://www.transportenvironment.org/how_clean_are_europe-s_cars/. Last accessed Mar 2011

TiS (2002) In cooperation with INFRAS, Erasmus University Rotterdam, and DIW. Study on Vehicle Taxation in the Member States of the European Union, Final Report prepared for the European Commission, January

Trafikbeskattningsutredningen (1999) Bilen, miljön och säkerheten [Cars, Environment and Safety]. SOU 1999:62, Ministry of Finance, Stockholm

Turrentine TS, Kurani, KS (2007) Car buyers and fuel economy. Energy Policy 35:1213–1223

Van Dender K, Crist P (2009) Policy instruments to limit negative environmental impacts from increased international transport. Discussion Paper No.2009-9. Joint Research Centre, OECD and International Transport Forum

Chapter 7
The Right EU Policy Framework for Reducing Car CO_2 Emissions

Jos Dings

Abstract This chapter analyzes how the EU should change its environmental policies on cars and fuels in order to help achieve the target for a 60% cut in greenhouse gas emissions from transport in Europe by 2050 announced in January 2011. In order to achieve this target, fuel economy standards for cars and low carbon standards for fuels should be tightened significantly, and minimum taxes on petrol and diesel should be raised. But the primary purpose of the paper is to make specific recommendations on how vehicle and fuel policy should be designed to facilitate such deep emissions cuts in the future. The CO_2 standards for cars the EU agreed in 2008 are a big step forward compared to the flawed voluntary commitment of the industry; it has taken CO_2 from a Corporate Social Responsibility issue to a bottom line issue and once again demonstrated that emissions can be cut more quickly and cheaply than previously thought possible. But apart from a still-too-low level of ambition, the law suffers from a couple of design flaws. First, the evidence is that the rather impressive recent cuts in official CO_2 emissions do not yet translate to equally impressive savings on the road. Carmakers have gained a lot of grams by testing their cars' consumption more cunningly, instead of making them really more efficient. Second, future standards should not be weight-based but rather size-based. Third, tailpipe CO_2 emissions as the regulated metric will have to be replaced by something more resembling energy efficiency. On taxation, Europe's major problem is too low a level of taxation of diesel. This is to a large extent explained by a fear member states have of losing revenue through diesel tourism by trucks if they raise diesel taxes. EU-level efforts to raise minimum levels of diesel taxation should be applauded. But North America's International Fuel Tax Agreement offers a structural way out – essentially it taxes truck diesel on the basis of fuel *used* in a state rather than *sold*, making it possible for states to raise diesel taxes without losing any of the extra revenue to other countries. But it is Europe's climate policy in transport fuels that is in most need of a clean-up. Current policy increases rather than reduces greenhouse gas emissions of a liter of fuel. Scrapping the 10% biofuels target, leaving only a low carbon fuel standard in place, and getting the carbon accounting right for both biofuels and fossil fuels are the two top priorities.

J. Dings (✉)
Transport & Environment (T&E), 1050 Brussels, Belgium
e-mail: jos.dings@transportenvironment.org

7.1 Views on Regulation – Standards Vs. Incentives

We start this chapter with a short overview of how different regulatory instruments on vehicles and fuels, and tax systems should be best used.

7.1.1 Can We Rely on a Carbon Price Such As the One Set by the European Union Emissions Trading System (EU ETS), Or Do We Need Additional Instruments?

First it is necessary to look at why we need different regulations. Isn't it enough to set one carbon price throughout the economy, let the market find the cheapest options and, if that turns out not to be enough, raise the carbon price? In short, why not include transport in the EU ETS?

For two economic reasons, and one very important political reason, such a single-instrument policy would be inadequate.

The first economic reason is that, as long as climate policy is more regional than global in character, as is clearly the case at the moment, political concerns of carbon leakage dramatically constrain carbon prices in the most vulnerable (exposed) sectors. In such an imperfect world of regional action, the true policy challenge is not just to minimize carbon abatement costs (which would indeed lead to equal carbon prices everywhere) but to minimize the total costs of carbon abatement, carbon leakage, and energy dependence. This is the central challenge of EU climate policy, although it is never explicitly stated as such. And inevitably if this policy objective is pursued, carbon prices in sheltered sectors (such as transport, buildings, households) will be higher than in exposed ones, instead of equal.

The second economic reason is that, even if climate policy were global, the lack of a price on carbon is not the only market imperfection to be corrected. Plenty of other imperfections exist that lead to higher-than-optimal carbon emissions, one of the most important ones being consumer short sightedness. People rarely buy products on the basis of lifecycle costs; upfront purchase costs dominate the decision. Gillette and HP know this; they keep their razors and printers cheap and earn money with blades and ink cartridges. Carmakers focus mostly on the retail price of the vehicle, for marketing purposes. For that reason, they do not include many fuel-saving technologies that are, on a lifecycle basis, attractive to consumers. While it is questionable whether regulators should act to change the way printers and razorblades are marketed, CO_2 emissions from cars are just too significant for society to ignore. Regulatory action on car fuel efficiency of cars is warranted, even if fuel taxes are already in place, because consumers themselves cannot be relied upon to pay upfront for something that would save them money over time.

Finally, the political reason that one instrument is not enough is that politics is primarily about fairness i.e. distribution of efforts or 'sharing the pain'. A carbon tax puts all the burden of climate policy on consumers and nothing on the car or fuel industries and hence violates that essential political principle of fairness.

There is also a further reason to use both economic and regulatory instruments: they are mutually reinforcing. Efficiency standards soften the impact of fuel taxation for consumers because the cars they buy need less fuel, and higher fuel taxes make efficient cars more attractive for manufacturers to produce.

For all these reasons we need a range of policies, not just one instrument. The next section sets out the essential elements of such a policy package.

7.1.2 Four Essential Elements of a Policy Package

7.1.2.1 Fuel Taxes

Fuel taxes are central – it has been said before and elsewhere in this book. Fuel taxes give every actor in the mobility chain the right incentives to cut fuel consumption, and therefore GHG emissions. The long-term price elasticity of road transport fuel is typically estimated at higher than –0.5 – so a 10% higher fuel prices cuts consumption in the long run by more than 5%. But one reason why fuel tax is so important is rarely, if ever, mentioned.

For climate policy to succeed, oil must be cheap from the producer's perspective, so that they are encouraged to leave it (and its carbon) in the ground. But it must be expensive for consumers – so they use it more efficiently and look for lower-carbon alternatives. The only way to solve this paradox is bridging the gap between producer and consumer prices with a fuel tax.

Despite how obvious it is that a low global oil price is an inevitable by-product of successful climate policy, this fact is almost never recognized. For maximum climate impact, fuel taxes should reflect as least the marginal carbon footprint of fuels.

7.1.2.2 Energy Efficiency Standards

As mentioned earlier, fuel taxes should be supplemented with technical standards to lower vehicle emissions. Such standards should be primarily based on the energy efficiency of vehicles. This is not exactly the same as lowering their CO_2 emissions, which is the basis of today's policy. Carmakers should not be held accountable for the carbon footprint of the fuels, but only for the amount they consume. Energy labels for white goods such as fridges do not depend on the greenness of the electricity they consume, cars should be no different. This will become much more important in the future, when ever more cars will not be powered by oil.

This also applies to vehicle taxation. This is not to say that all the hard-won consumer recognition of CO_2/km should be ditched in favor of kWh/km, but a future regulatory metric should as closely as possible represent energy efficiency, even if it is presented to consumers in a different way.

7.1.2.3 Fuel Regulation

Fuels should be regulated and taxed on the basis of their well-to-wheel carbon footprint, per unit of energy. In contrast to carmakers, fuel suppliers can control the carbon footprint of fuel and hence they should be given the incentive to supply the lowest carbon fuel possible.

By multiplying the amount of energy per km (regulated in cars legislation) by the carbon footprint of fuels per unit of energy (regulated in fuels legislation) a watertight system emerges that regulates the carbon footprint per car kilometer.

7.1.2.4 Vehicle Taxes

Vehicle tax systems should not directly or indirectly subsidize car, car trips, fuel purchases, or blunt incentives to choose efficient vehicles. As a rule subsidies are inefficient policy instruments, and subsidies that make the climate problem worse should be eliminated as a first priority. In the next section we will see that in particular most company car tax systems violate many of these principles.

7.1.3 How does Current Policy Relate to this Ideal Scenario?

7.1.3.1 Fuel Taxes

The key EU policy mistake in the area of fuel taxation is that the minimum level of diesel tax as enshrined in the Energy Tax Directive (Official Journal of the European Communities 2003) is €0.33 per liter, while for petrol it is 9% higher at €0.359 per liter. It should be the other way round because burning a liter of diesel leads to roughly 12% more CO_2 emissions than a liter of petrol. At the time of writing the European Commission is considering a proposal to review the energy tax directive to correct for this by increasing the minimum diesel tax to €0.39 per liter in 2018 prices while leaving the minimum petrol tax as it is.

A second problem is that quite a few member states offer generic tax exemptions for biofuels. This cannot be justified because, when indirect land use change effects of biofuels are taken into account, conventional biofuels often have a higher, not lower, carbon footprint than petrol and diesel (IEEP 2010), and their carbon footprint varies wildly by feedstock and process.

7.1.3.2 Energy/Fuel Efficiency Standards

Not a single region in the world has a pure energy efficiency standard. Currently there are fuel economy standards in the USA, China and Japan and CO_2 standards in the USA, the EU and South Korea. But electricity does not emit tailpipe CO_2 and electricity (as well as hydrogen) is not measured in gallons. Unsurprisingly all regions are grappling with how to treat plug in hybrid electric vehicles (PHEVs), electric vehicles (EVs) or fuel cell vehicles (FCVs) and currently tend to give them zero CO_2 ratings. Currently this does not yet pose major problems because the market share of such vehicles is close to zero. But it needs to be solved, and soon.

7.1.3.3 Fuel Regulation

While neither fuel taxation nor vehicle regulations are perfect, it is arguably the environmental regulation of transport fuel where European policy deviates most from the ideal scenario.

The main culprit is the 'Directive on the promotion of the use of renewable energy sources 2009/28' (the RES Directive). This law obliges all 27 EU Member States to source 10% of its transport energy from renewable sources by 2020. Recent national action plans demonstrate that the vast majority (8.8%) of this target will be met through the use of biofuels. Whilst formally the biofuels qualifying under this target have to reduce well-to-wheel (WTW) GHG emissions compared with their fossil equivalents by 35% initially, increasing to 50% by 2017, these reductions exclude the effects of indirect land use change (ILUC). Recent science shows that by including ILUC effects into the equation, the average extra biofuels that will come to the EU market between 2011 and 2020 do not reduce GHG emissions by 35–50%, but rather increase them by 80 to over 160% (IEEP 2010). In other words, they are worse from a climate change point-of-view than the fossil fuels they are designed to replace.

This serious carbon accounting error is the law's first flaw. The second is that the GHG standard is a pass/fail requirement, which means it lacks incentives to reduce GHG emissions beyond the standard.

Fortunately the EU has also adopted a much better, but not yet very effective, fuel law called the Fuel Quality Directive (2009/30) that does give such incentives. This law originally only regulated classical fuel quality issues such as sulfur and aromatics content. But since its revision in late 2008 it regulates well-to-wheel GHG emissions of petrol and diesel per unit of energy (Article 7a). It sets a 6% reduction target for 2020 and leaves fuel suppliers freedom of choice with respect to their compliance strategies. This law does give the incentives to slash WTW carbon emissions; for example, a biofuel that cuts GHG emissions by 40% is worth twice as much as one that reduces it only by 20%. California has a similar 'low carbon fuel standard' in place but is more advanced in its implementation. As a result, every percentage point of WTW carbon footprint reduction is now worth roughly a 0.05 Eurocent per liter. This modest incentive is already making fuel suppliers clean up their production processes.[1]

The EU law is not yet very effective because the 6% target is virtually met by default (10% of biofuels times the required at least 50% reduction by 2017 means already at least 5% reduction of WTW emissions – on paper). Another reason is that the law has not yet been fully implemented. For example, specific GHG values for high-carbon oil sources such as tar sands and oil shale have still not been introduced. This needs urgent attention; it would send a strong signal to oil companies that such fuels are not welcome on the world's premium motor fuel markets, seriously undermining their economic attractiveness.

---

[1] Personal communication with John Courtis, California Air Resources Board, January 2011.

In short, the EU has quite a 'to do list' on the fuels side: it needs to correct carbon accounting by including effects of indirect land use change and unconventional oil. And it needs to ditch the 10% quantity target for transport renewables so that the fuel quality Directive's GHG reduction target for fuels is the only game in town. Only then will we see a genuine market drive toward low carbon fuel.

7.1.3.4 Vehicle Taxes

In terms of vehicle and income tax systems, Chapter 8 of this book goes into vehicle purchase and circulation taxes, but it does not address company car taxation. Until recently this area was not very well studied. But a report by Copenhagen Economics for the European Commission demonstrates how environmentally and economically damaging company car tax schemes in most member states are (Copenhagen Economics 2010). The subsidies – primarily very mild, and often lump-sum, benefit-in-kind tax arrangements for driving a company car for private use – encourage employees to drive much bigger and more expensive cars for private purposes than they would otherwise be able to do. On top, private kilometers can often be driven at zero cost to the employee with company petrol cards. As 50% of European car sales are company cars, the numbers are vast. The fiscal subsidy amounts to 0.5% of GDP (€54bn a year), and it leads to 4–8% excess CO_2 emissions from cars.

The USA offers inspiration for a solution. Private miles driven with company cars are taxed on the basis of the full lease costs of these miles – including fuel. This largely eliminates the incentive for employers to offer employees company cars as a cheap fringe benefit, and it also largely eliminates the zero-cost private kilometers that many European employees can make.

In broad brush strokes we have outlined four major pillars of a policy package to reduce GHG emissions from cars. In the next section we will go into greater detail on how effective Europe's fuel tax policy and the cars and CO_2 regulation have been until now, how their effectiveness can be further improved, and what the estimated costs before and after regulation have been.

7.2 Fuel Tax and Europe

7.2.1 The Effects of European Policy on Road Fuel Taxes

Establishing a minimum fuel tax on petrol and diesel has arguably been the most effective climate measure in transport the EU has ever taken. Though it should be noted that the main motivation for the EU to intervene in fuel taxes has not so much been concerns about climate change but rather avoiding a 'race to the bottom' through fuel tourism, particularly with lorries. At the end of this section we will come back to this issue.

Chapter 9 of this book reviews fuel tax elasticities, and while the evidence is that elasticities have decreased somewhat with rising incomes, they are still significant enough to make a big difference. A quick glance at European and American

7 The Right EU Policy Framework for Reducing Car CO_2 Emissions

car markets, buyer preferences and the resulting fuel efficiencies should convince doubters of the importance of fuel taxes in influencing consumer choices. Simply put, driving American gas guzzlers is unaffordable with European fuel taxes.

So it is of utmost importance to keep fuel taxes up. Has that happened over the past decades? In order to answer that question we analyzed a database with fuel prices, fuel taxes, inflation and fuel consumption figures since 1980. Complete records for those 31 years are only available for the nine member states that made up the EU in 1980: Germany, France, Italy, UK, Netherlands, Belgium, Ireland, Denmark and Luxembourg. Together these nine countries account for two thirds of EU fuel consumption so they are reasonably representative for EU-wide developments. See Fig. 7.1.

The analysis demonstrates that while in the 1980–2000 period the average fuel tax paid on a liter of fuel went up by 49% (from €0.43 to €0.63, in 2010 prices), in the decade since 2000 they dropped by 17%, from €0.64 to €0.53 per liter.[2] For example, in Italy taxes dropped by almost a third since 1995. Germany and the Netherlands held up better with less than 10% declines in real average tax rates.

Inflation is the most important reason for the decline. The 2000–2010 decade saw significant inflation rates, and fuel taxes were not corrected for this in many member states.

But another reason is the relentless shift to the cheaper fuel which is diesel. In 1980 almost two thirds of fuel used in Europe was petrol, nowadays two thirds

Fig. 7.1 Trend since 1980 in weighted, inflation-corrected tax on petrol and diesel in eight EU member states (representing two thirds of the EU market)

---

[2] Tax rates are inflation-corrected and weighted according to sales of petrol and diesel in each Member State; an average litre is defined as the average of a litre of petrol and a litre of diesel (roughly 34 MJ worth of fuel).

is diesel. This shift explains almost a quarter of the drop in fuel taxes in the last decade. We argued earlier that the lower diesel tax rate is not defensible from an environmental point of view.

The European Commission shares the view that fuel taxation should be fuel-neutral and be purely based on CO_2 impacts and energy content. A draft proposal lifts the minimum diesel tax rates by 20% in 2018, from €0.33 to €0.40 per liter, while leaving the petrol rate unchanged at €0.36 (in 2012 prices). The proposal is planned to be announced in spring 2011.

The car industry is strongly against such a move, claiming that more expensive diesel *fuel* will hurt sales of diesel *cars* and therefore put the achievement of the EU's 130 and 95 g/km CO_2 targets (see later in this chapter) in jeopardy.

However, as Fig. 7.2 shows, at present there is no correlation between the petrol-diesel tax differential on the one hand, and the share of diesel cars in new car sales on the other hand.

The main explanation for the share of diesel cars in the fleet is not fuel taxation, but rather vehicle purchase and circulation taxes, and company car taxation. Many member states (including the Netherlands) currently tax diesel cars more heavily to compensate for the lower tax on diesel fuel; levelling taxes on fuels could be used to also level taxes on cars, as the UK does. The UK has by far the most fuel- and technology-neutral system in place, with both vehicle and fuel taxes virtually independent on fuel and both strongly CO_2 based.

Chapters 6 and 9 in this book demonstrate that low diesel taxes cause strong rebound effects, cancelling out initial energy efficiency gains. And we should not forget trucks and vans, which together consume roughly half of all diesel used. Through the significant long-term price elasticity of –0.2 to –0.6 (Significance and CE Delft 2010) a 10% rise in diesel tax would reduce emissions from this segment of the road transport market with 2–6%.

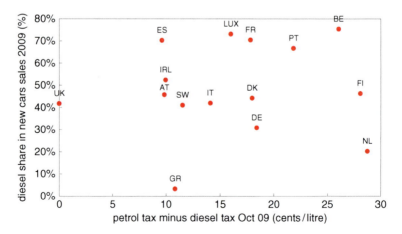

Fig. 7.2 Differences in taxes on petrol and diesel vs. the share of diesel in new car sales in 2009

7.2.2 Truck Diesel Tourism and What to Do About It

As mentioned, one of the greatest obstacles for increasing fuel taxes in Europe is diesel 'tourism' by trucks. Trucks can drive thousands of kilometers on a single tank and tend to criss-cross over Europe, giving great flexibility to fill up where diesel is cheapest. Hence smaller countries can raise total tax revenues by keeping diesel taxes low – an irresistible political proposition.

But such practices hold back the bigger member states from raising their diesel taxes further. It is not a coincidence that the highest diesel taxes can be found in the UK, France, Germany and Italy, countries with more limited evasion options. But even these countries lose part of the extra revenue from diesel tax increases to countries with lower taxes. For them the political proposition is difficult – full political pain of higher taxes, with part of the revenues leaking aboard.

Arguably the EU's strategy to introduce minimum fuel taxes has not been more than a stopgap measure to prevent the worst forms of abuse. It has not eradicated fuel tax havens; Luxembourg being the best example. It sells eight times as much diesel per head of population than its neighbors by keeping diesel tax 10–15 cents a liter lower. The fact that Luxembourg is closely following EU minimum tax rates strongly suggests that had the minimum rates not existed the situation would have been far worse. The Baltic countries and Spain are other examples of EU member states that closely follow minimum diesel tax rates.

The European Commission's intention is eventually to bind member states into fuel-neutral taxation, i.e. levy taxes in relation to carbon and energy content of the fuel. This would force member states, regardless of where their tax levels are in relation to the minimum, to impose roughly 10% higher taxes on a liter of diesel than on a liter of petrol. In theory, as we have seen, this is environmentally and economically the most efficient solution. But given the difficulty member states have in raising truck diesel taxes, it seems not unlikely that member states will take the easy route and lower petrol taxes instead. Obligatory fuel neutral taxation, as good as it is in theory, might hence backfire unless the EU forbids lowering petrol taxes too.

What else, besides minimum EU tax levels, can be done to avoid diesel tourism? Border checks are an option – checking whether hauliers are carrying excessive amounts of cheap fuel with them. In a borderless Europe, this only seems a realistic option at the EU's external border. And in fact EU legislation allows eastern member states to execute such checks. It is, however, rarely, if ever, done.

The definitive solution for diesel tourism is to move toward a European version of the International Fuel Tax Agreement (IFTA) in operation in North America (Mc Lure 2009).[3] This, interestingly, very old agreement completely eliminates the fuel tourism problem by taxing diesel on the basis of how much is *used* in a state, instead of how much is *sold*. It gives American and Canadian states fuel tax revenues based on distances travelled on their territory, average fuel efficiency of trucks,

---

[3] See also http://en.wikipedia.org/wiki/International_Fuel_Tax_Agreement [last accessed March 2011].

and their domestic fuel tax levels. Fuel purchases and miles driven in each state are recorded and tax payments either reimbursed or reimposed. This completely eliminates the incentive to fill up in the cheapest states, and hence also eliminates the incentive for states to keep taxes low to encourage truckers to fill up on their territory.

North America unfortunately administers the IFTA in a rather primitive and cumbersome way. But Europe can start from scratch and use the mandatory digital tachograph for trucks and its almost-ready GPS competitor Galileo to do the same at far lower cost. The only significant effort would be to record fuel sales, and of course to enforce the legislation by introducing a clearing house with powers to reimburse and bill hauliers.

An EU version of North America's IFTA is clearly the way forward for eliminating fuel tax tourism.

7.3 Europe's Cars and CO_2 Regulation

7.3.1 A Short History

The EU started thinking about car fuel efficiency in the early 1990s. A target to reduce average new car CO_2 emissions to 120 g/km was proposed by Germany at a meeting of European environment ministers in October 1994. It was presented as the ambition to lower fuel consumption of new petrol cars to 5 l/100 km and new diesel cars to 4.5 l/100 km. The target was formally announced in a 1995 European Commission communication (COM(95)689) and represented a 35% reduction over the 1995 level of 186 g/km.

Originally the target date was set for 2005. But before it became legally-binding, the target was postponed or weakened four times. Eventually, the original '120 g/km by 2005' evolved, via a failed voluntary agreement, into legislation to achieve a nominal '130 g/km by 2015' which in reality will be closer to 135–140 g/km (see below), a loss of about 15 g and 10 years.

On the upside, the law now finally adopted does offer a legally binding framework, including penalties, to deal with CO_2 emissions from cars. Significantly, it also adds a new 95 g/km target for 2020.

7.3.2 What the Law Says

The new law nominally strives to reduce the average CO_2 emissions from new cars to 130 g/km by 2015 (approx. 5.6 l/100 km for petrol cars and 5.0 l for diesel cars) and to 95 g/km by 2020. For comparison: the average for 2010 will be around 141 g/km.

The target is an average for all cars sold, not a fixed limit that no car may exceed. Manufacturers can average the CO_2 emissions from all cars they sell. They can also file for joint-compliance with other manufacturers, in order to average emissions over a larger pool of vehicles. This flexibility mechanism is called 'pooling'.

For the 2012–2014 period, the law features a so-called 'phase in', in which 65% (2012), 75% (2013) and 80% (2014) of cars from each manufacturer will have to comply. Carmakers are free to select 'compliance vehicles' and will therefore leave out the cars farthest from the target, i.e. the worst gas guzzlers. This has an enormous effect: 65% compliance effectively means 152 g/km on average; 80% means 146 g/km on average, i.e. a 16–22 g/km weakening. Hence it effectively postpones the 130 target from 2012 to 2015 (IEEP 2008).

Individual manufacturers' targets are differentiated on the basis of the weight of the cars they produce in the target year. For example, if a manufacturer's cars by 2015 are 100 kg heavier than the industry average (measured over the 2011–2013 period), they are allowed a 4.6 g/km higher CO_2 target (134.6 instead of 130 g/km CO_2 on average). Conversely, if their cars are lighter than average they get a tougher target.

Enforcement will take place through a system of fines. For every g/km a manufacturer exceeds its company target, it has to pay a €95 fine per vehicle sold, in principle.

Unfortunately the law features several loopholes:

- Up to 7 g/km credits for as yet undefined off-cycle credits (sometimes branded as 'eco-innovations') of unmeasured CO_2 that can be exchanged for measured reductions on the official test cycle. It is proving fiendishly difficult to implement these provisions in a way that guarantees reductions in real life;
- Counting low-CO_2 cars more than once, euphemistically dubbed 'supercredits'. This applies to cars with CO_2 emissions below 50 g/km (probably primarily electric and some plug-in hybrid vehicles). This allows manufacturers to count every such car as more than one car and would hence water down overall CO_2 reductions which are based on fleet averages. As an example: a 'supercredit' factor of 3 means that a manufacturer can sell three 260 g/km vehicles (essentially SUVs) for every electric car sold and still, in a legal sense, hit the 130 g/km target;
- Much lower penalties for missing the target by a few grams until 2018. The penalties for the first, second and third g/km over the target are only €5, €15 and €25 per g/km respectively instead of €95;
- Exemptions for carmakers with between 10,000 and 300,000 sales in the EU. They can apply for a default target of a 25% reduction compared with 2007 (Tata, the owner of Jaguar/Land Rover, and Porsche are likely applicants);
- Exemptions for carmakers with less than 10,000 sales in the EU, who can negotiate their own target with the Commission.

All these loopholes together in practice mean that the real target for fleet-average on-cycle CO_2 emissions 2015 is close to 140 g/km, rather than 130 g/km. On the positive side, the law says most loopholes will be phased out before 2020.

7.3.3 The 95 g/km Target and Its Review

Probably the most significant part of the new law is the addition of a 95 g/km target for 2020. This represents a 40% cut compared with 2007 levels, or 3.8% per year cut. Its 'modalities' and 'aspects of implementation' (i.e. not the target itself but possibly issues such as the weight-based utility parameter, see below) will have to be reviewed by the European Commission by January 2013.

7.3.4 Why Standards Should Be Footprint-Based, Not Weight-Based

In order for the cars CO_2 regulation to take the utility of different types of car into account, the European Commission decided to base CO_2 standards on weight. The idea being that bigger cars, that carry more people, are probably heavier.

The result, as we have seen, is that the 130 g/km standard for 2015 does not apply equally to all carmakers. In fact if carmakers manage to make their cars 100 kg lighter, their CO_2 target gets 4.6 g/km tougher.

This means carmakers are withheld 60% of the CO_2 credit from lightweighting technology, which is supremely inefficient in terms of incentives. Incomprehensibly, it was carmakers who lobbied furiously for weight-based standards.[4]

In fact, weight is not a utility parameter; consumers actually perceive car weight as a negative criterion, making handling and driving more difficult.

Footprint – wheelbase times track width – is a size-based metric, strongly related to interior space – and hence, unlike weight, strongly related to consumer utility. T&E commissioned a study from IEEP/TNO/CE Delft into footprint as an alternative parameter (IEEP, TNO and T&E 2008). Its conclusions were (quoted from the report):

- Technical analysis suggests that footprint performs at least as well as weight or pan area as a possible utility parameter, and in several important respects better.
- Using footprint avoids the problem that comes with using weight as the parameter, namely that the incentive for reducing vehicle weight – and thereby CO_2 emissions – is reduced or even eliminated.
- Footprint does not eliminate all perverse incentives, but as it is harder to increase the footprint (compared with increasing weight or pan area), it reduces the chances of cheap 'gaming' options.
- The overall cost of using footprint as a parameter in CO_2 reduction legislation is no greater than with weight or pan area, and could be less as the system would reward weight reduction. Also the impacts on individual companies would be about the same.

---

[4] See http://www.acea.be/index.php/news/news_detail/european_automobile_industry_united_in_approach_towards_further_reducing_co [last accessed March 2011].

- Insofar as US evidence is applicable to Europe, this seemed to suggest that footprint was associated with improved safety, whereas weight per se was associated with an increase in fatalities.

It is clear that the European Commission should propose to change the metric from weight to footprint as quickly as possible.

7.3.5 Progress to Date

In 2007 it became clear that the EU was finally getting serious about reducing CO_2 emissions from vehicles. Fleet average emissions for that year stood at 158 g/km. Fleet average emissions for 2010 were not yet known at the time of writing, but preliminary figures indicate an average of 141 g/km.[5] This is only 8% away from the nominal 130 g/km target for 2015, and even less when the above mentioned loopholes are taken into account.

This means that the average annual reduction in the 12 years before the regulation (1995–2007) was 1.4%, and the average reduction in the 3 years after (2008–2010) was 3.7% per year. This is a dramatic and very visible change of pace. Anyone who pays attention to car advertising will have noticed that CO_2 has become a central point in carmakers' marketing strategies.

7.3.6 Is Progress Due to Regulation Or to Other Factors?

Many observers have provided various explanations for the rapid progress on CO_2 emissions in the recent past. They point to the financial crisis of 2008–2009 and the ensuing so-called 'scrappage' schemes (subsidies for new cars provided an old one is scrapped) which have pushed people toward smaller and hence more efficient models (the scrappage allowances tended to be for fixed amounts that were a much larger proportion of purchase price for cheap cars than expensive ones). And they point to changes in national tax systems as a result of which low-CO_2 cars get more favorable treatment these days than was the case a few years back.

Starting with tax systems – it is true that they have played a big role. But it is also true that the fact that the EU was introducing car CO_2 legislation inspired and speeded up introduction of CO_2-based car taxes. The EU regulation changed the political dynamics. Before regulation, CO_2 based car taxes were a politically correct thing to do, but for industry a mixed blessing as they steered consumer demand toward generally less profitable low-CO_2 models. The EU CO_2 regulation turned CO_2-based car taxes into a tool to cut the cost of hitting the 130/95 g/km targets and hence became a great help for the car industry. Given this huge change in dynamics, it is no surprise that in the 2007–2010 period the number of countries which based

---

[5] Data from automotive data provider JATO, to be released at the time of writing.

part of their car taxes on CO_2 increased from 11 to 18, that the list of countries now includes Germany, and that many existing schemes were strengthened. Over 90% of car sales now take place in countries with CO_2-based tax systems.

The crisis and resulting 'scrappage schemes' had an impact too, but their effects on CO_2 emissions have been limited. Throughout 2009, 13 EU Member States adopted 'scrappage schemes' – such schemes were therefore in operation across 86% of the EU market in terms of sales. Almost €8bn was spent on direct payments plus overheads (Global Insight 2010). The result was that EU15 new car sales remained, at 13.3 million, virtually at the 1990–2009 average level.

Not only did these schemes boost car sales, they also changed the type of vehicle sold. As the schemes were typically designed as fixed cash paybacks, the effect on new prices was much larger for cheap cars than for expensive ones. This magnified the effect of people shifting to cheaper cars, a trend that could be expected during tough economic times. As a result, 2009 saw a strong increase in the market share of cheap cars, which are typically (but not always) lighter and less powerful than expensive cars, and typically (but not always) have lower CO_2 emissions per km. Therefore, even if technological progress would have halted completely in 2009, a decrease in average CO_2 emissions from newly sold cars in 2009 could have been expected.

T&E's 'How clean are Europe's cars' report (November 2010)[6] splits the 5.1% fleet average CO_2 reduction in 2009 into demand-side changes (smaller, less powerful cars) and supply-side changes (better technology). The analysis suggests that actually more than half of the reductions in 2009, or close to a 3% improvement, was achieved through better technology, and a bit over 2% could be explained by changes in demand. Figure 7.3 illustrates this.

Another way of assessing the impacts of the regulation is to look at responses by carmaker. Figure 7.4 below shows the gap carmakers had to close in 2008 to hit regulatory targets, and what they did to close the gap in 2009.

The graph shows that well-placed carmakers, those who needed to cut emissions by less than 12%, cut their emissions by less than 3% last year, whilst badly placed carmakers, those who needed to cut emissions by more than 12%, all made cuts of more than 3%. Those with more than 20% to go in 2008 cut by 5% or more in 2009. This again indicates that carmakers' actions were triggered by the regulation.

7.3.7 Distance to Regulatory Targets, by Carmaker

Figure 7.5 shows how in 2009 different carmakers were positioned to hit their specific target for 2015. The assumption is that the average weight per carmaker remains the same.

---

[6] http://www.transportenvironment.org/how_clean_are_europe-s_cars/ [last accessed March 2011].

7   The Right EU Policy Framework for Reducing Car CO$_2$ Emissions

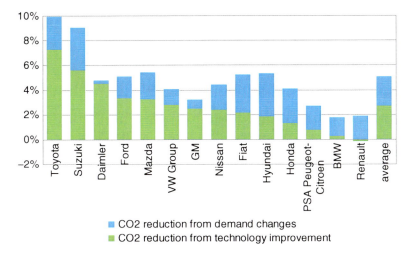

Fig. 7.3 Progress in sales-average CO$_2$ emission in g/km per carmaker in 2009 compared with 2008, and split between demand-side changes and technology changes. Carmakers are sorted on the basis of their technology-only performance

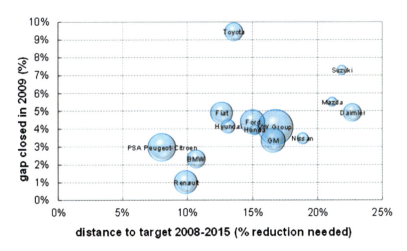

Fig. 7.4 What carmakers need to do between 2008 and 2015, and what they did in 2009

It should be borne in mind that (a) in 2010 carmakers are likely to report around 3% progress, and (b) that loopholes as previously described apply, so that in reality carmakers are even much closer to their targets than the graph suggests.

All available evidence points toward carmakers in Europe heading for very significant 'overcompliance' with the CO$_2$ regulation; hence they are likely to hit the target for 2015 years in advance.

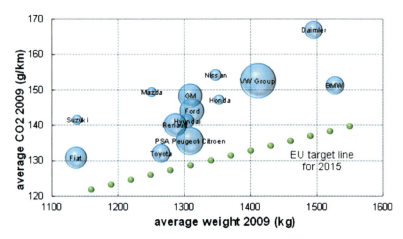

Fig. 7.5 Fleet-average weight and fleet-average CO_2 emissions by carmaker in 2009, compared with EU target curve for 2015

7.3.8 What Happened to the Promised 10 g/km?

Many questions have been asked about what actually happened to the 10 g/km that the European Commission promised to deliver when it weakened the 120 g/km target to 130 g/km. The regulation eventually adopted states that 'A further reduction of 10 g CO_2/km, or equivalent if technically necessary, will be delivered by other technological improvements and by an increased use of sustainable biofuels'. But a formal process to monitor and deliver these 10 g/km has never been established. What follows is our assessment of what happened.

7.3.8.1 Biofuels

As we have seen, unless indirect land use change impacts are accounted for, the 10% target for renewable energy in transport will increase, not decrease, GHG emissions, roughly by a factor of two compared with fossil fuels. If we assume cars will in 2020 run on 9% biofuels as estimated in the national action plans of EU member states, and their emissions are twice those of fossil fuels, that means car emissions by 2020 will rise by 9% which, multiplied by 130, yields roughly 12 g/km.

7.3.8.2 Tires

Standards and a label for low rolling resistance tires have been adopted (Official Journal of the European Union 2009a, b), and a regulation for tire pressure monitoring systems (TPMS) has almost been finalized. In total these measures could reduce emissions by some 3–4%, which is equivalent to some 4–5 g/km.

7.3.8.3 Light Commercial Vehicles (Vans)

A regulation on fuel efficiency on vans (European Commission 2009)[7] has been finalized, setting a 14% reduction over 2007 levels by 2017, which, excluding loopholes, is some 20 g/km net reduction. As vans constitute some 14% of emissions of cars, this is equivalent to roughly 3 g/km for cars.

7.3.8.4 Air Conditioning

A regulation on the energy efficiency of mobile air conditioners has been promised, but the process is severely delayed and is still far from being delivered.

The conclusion is that the approx. 7 g/km delivered through extra technological measures so far have been more than offset by the complete failure of biofuels policy to deliver emission cuts.

What is the lesson here? That the 'integrated approach' to reduce car emissions has, rather unsurprisingly, turned out to be a tool to exchange verifiable emissions reductions for unverifiable ones.

7.3.9 Have Carmakers Exaggerated the Difficulty of Making Cuts?

Over the past decades of environmental policymaking in transport, numerous reports have been produced that show that there is a systematic overestimation of costs to comply with environmental legislation. Partly this is due to poor analysis techniques (e.g. not accounting for learning effects, or mass-production effects, both of which reduce costs), but partly the bias is also caused by completely opposite incentives before and after introduction of legislation.

Before regulation has been adopted, carmakers (the most important source of cost data) have an interest in overestimating (stated) compliance costs – as high costs can make regulators decide to weaken or abandon intended regulations. Afterward, they engage in a furious race to reduce compliance cost to the minimum, usually quite successfully.

This paragraph makes a first assessment of whether such overestimation has again taken place in the case of CO_2 regulation. We do this on the basis of cost estimates used in the debate preceding Europe's regulation, and with some examples of recent 'best practice' cars.

Table 7.1 shows that 'best practice' small diesel cars have improved by 20 g/km, medium-sized ones have improved by roughly 30–35 g/km, and large ones by 40–70 g/km. The price premium of these cars over their less efficient predecessors is not easy to assess, particularly because often it concerns new models that are different in many ways from their predecessors, not just in fuel consumption. But it

---

[7] On 15 February 2011 the European Parliament voted through the compromise agreement, which sets standards of 175 g/km by 2017 and 147 g/km by 2020, weakening the Commission proposal.

Table 7.1 Improvement of 'best practice' diesel cars between 2007 and 2011. The basis for the data has been the 2007 and 2011 editions of the Dutch 'Brandstofverbruiksboekje' ('fuel consumption booklet')

Brand and model	CO_2 of best available diesel variant (g/km) 2007	2011	Improvement (g/km)	Fuel-saving program
Opel Corsa	115	94	21	Ecoflex
VW Golf	135	99	36	BlueMotion
Ford Focus	127	99	28	Econetic
BMW 118	150	119	31	Efficient Dynamics
Volvo S40	129	99	30	DrivE
VW Passat	151	109	42	BlueMotion
Volvo V70	172	119	53	DrivE
Citroën C5	142	120	22	Airdream
Mercedes C220	169	127	42	BlueEfficiency
Mercedes S	220	149	71	

is fair to say that none of the efficient 2011 models have been priced significantly above their predecessors, or less efficient models.

The graphs below demonstrate estimates of CO_2 reduction costs before the regulation was adopted. Firstly, we present in Fig. 7.6 industry estimates made in 2006.

According to this report a 20 g/km (approx. 12%) reduction of diesel car CO_2 emissions would cost €2,000, a 30 g/km reduction would cost €4,000. Apparently

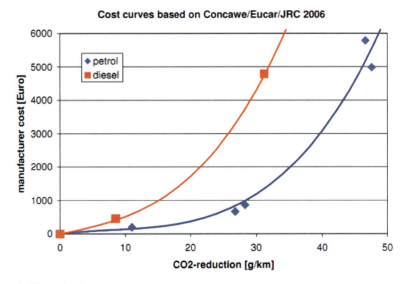

Fig. 7.6 CO_2 reduction cost curves
Source: Concawe Eucar JRC (2007)

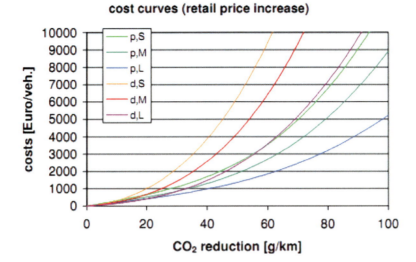

Fig. 7.7 CO_2 reduction cost curves
Source: TNO, IEEP and LAT (2006)

the industry found reductions beyond 40 g/km impossible in 2006, but that was proven false by the Passat, Volvo and Mercedes models highlighted above. This is a rather stunning illustration of what we said before – that before a regulation is adopted industry has an interest in exaggerating the costs (and even declare the requirements as 'impossible'), only to quickly and rather cheaply cut emissions once the law is in place.

Secondly, Fig. 7.7 presents the curves from the study the Commission used for its impact assessment (TNO, IEEP and LAT 2006).

The graph above shows that a 20 g/km improvement of a small diesel car would have cost roughly €1,000. A 30–35 g/km improvement on a medium-sized diesel car roughly €2,000. A 40–70 g/km improvement on a large diesel car should have cost roughly €2,000–5,000. While less obviously exaggerated than the industry figures, in particular the last two €2000+ per-car cost estimates still seem excessive given what has happened on the ground.

7.4 Real-World Vs. Official Fuel Consumption

7.4.1 A Big and Growing Problem

However strong CO_2 standards for cars are, what really matters is how much vehicles actually emit on the road, in everyday use.

And there is plenty of reason to worry on this front. Historically the gap between 'official' and 'real' CO_2 emissions and fuel consumption has always been assumed to be around 20%. While such a gap is not ideal, it is not a huge problem either if it is relatively predictable and stable, as consumers and policymakers alike know what to expect.

The problem is that in recent years the gap has grown quickly. Plenty of anecdotal reports in car magazines attest to this. But it was not until the first 'official' piece of research with a sufficiently large dataset came out, early 2010, that the spotlight was firmly put on the gap (TNO 2010). The research organization TNO analyzed fuel card (i.e. fuel sales) data for 140,000 vehicles, and plotted sales per km against the official fuel consumption – see Fig. 7.8. The work was carried out on behalf of the Dutch leasing company Travelcard.

This graph essentially shows that the lower the official CO_2 rating of a vehicle is, the larger the expected gap with real-world fuel consumption is likely to be. For sub-100 g/km CO_2 vehicles the gap can be as big as 45%.

This is a serious issue. Let us look at the EU regulation that is supposed to reduce fleet average emissions from 158 g/km in 2007 to 95 g/km in 2020, a 40% cut. According to the data, real world use for 158 g/km cars (average 2007) is about 21% higher or approx. 191 g/km, and real-world use of 95 g/km cars is according to the data roughly 140 g/km. So instead of a 40% cut, the 95 g/km would only achieve a 27% (1 − 140/191) cut compared with 2007 levels.

This means that if these finding still apply in 2020, a third of the 'on-paper' 40% reduction would not be delivered in real life because of the widening gap between official and real world fuel consumption. In order to achieve a 40% reduction in real life, an 'official' CO_2 target below 80 g/km would be needed in 2020.

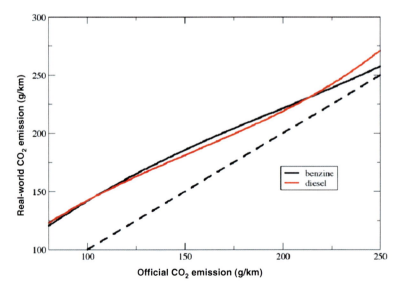

Fig. 7.8 Official CO_2 figures (*horizontal axis*) vs. real-world CO_2 emissions in 140,000 Travelcard vehicles
Source: TNO (2010)

The least we could wish for is that if we strengthen performance standards, that indeed performance is improved accordingly. But that's currently not the case.

7.4.2 How Can This Real-World Underperformance Be Resolved?

Before we can discuss possible solutions to this big problem, we need to find its root causes.

A first observation is that lately it has become extremely important for carmakers to lower their official CO_2 figures. CO_2 has become a matter of competition between carmakers, and now 18 EU Member states have car tax systems based on CO_2, it has become a major marketing issue. And let us not forget the EU fines for carmakers not complying with CO_2 limits. Both have made a gram of CO_2/km easily worth dozens of Euros if not a three-digit Euro figure per car. So the pressure to arrive at low official CO_2 rates is enormous, much bigger than before.

And it is this increased pressure that has laid bare, much more brutally than before, fatal flaws in the way CO_2 is measured.

A first measurement problem, often talked about, is the drive cycle, officially called the New European Drive Cycle or NEDC (although it is almost 40 years old now). The NEDC has woefully slow accelerations (26 s to arrive at 50 km/h...) and woefully short motorway driving (only 7 s at 120 km/h, out of a total of 1,180 s ...). All this reflects a period, long since passed, of less powerful vehicles and less ubiquitous motorways. The NEDC needs to be urgently updated, particularly because it overrewards 'urban' innovations such as start-stop systems, and underrewards innovations effective on motorways. But the cycle itself does not explain satisfactorily why suddenly the gap is widening so quickly. More is happening.

And that is how we arrive at the second problem, which is hardly talked about, but is probably at least as important. It is the lack of tight regulation of the way CO_2 is measured on the cycle. T&E has been closely involved in work to tighten up these procedures. We name a few shortfalls of the current system[8]:

- It allows switching off all auxiliary equipment (air conditioners, windows heaters, lights, seat heaters etc.), which can save more than 10% of emissions;
- It allows testing with a fully charged battery, so that the engine does not have to charge it during the test. That can save over 10% of emissions;
- It allows testing at a temperature window between 20 and 30°C (5° warmer means roughly 2% less emissions);
- It allows measurement of so-called 'road load' (the resistance programed in the roller bench) in a coast-down test under unrealistic conditions. For example, the test track can be up to 1% tilted, which, even though it has to be run in two

---

[8] Procedures are described in ISO standard 10521. Quantitative estimates have been partly derived from valuable work done by Schmidt and Johannsen (2010).

directions, gives a longer average coast down distance than a flat track. This is not theory; deliberately tilted test tracks exist.
- Air drag can be measured without 'optional' extras like right-side mirrors and roof racks;
- It allows the mass of the vehicle to exclude optional extras: air conditioning, sunroof and the like, to be rounded off when put in a mass category, and to be applied to a wide range of cars ('vehicle family' concept);
- It allows use of special low-friction lubricants that are unsuitable (and unaffordable) for daily use, and it allows to uncouple brakes, both measures lowering friction;
- And last but not least, it allows for a 4% downward 'administrative' correction of the CO_2 figure (6% for vans) to allow for calibration issues.

There is evidence that now that CO_2 is becoming so valuable for carmakers, ever more of these tolerances are actually being deployed, widening the gap between official and real-world CO_2. Type approval authorities are eager to attract business by offering the sharpest measurement conditions. And the lower the official CO_2 figure gets, the bigger the relative gap that these tolerances leave.

So the recipe for closing the gap is not all that complicated in principle: the drive cycle needs to be modernized, and measurement tolerances must be strongly reduced.

In the meantime, governments should make efforts to communicate better to their citizens the expected fuel consumption instead of the official one. If we wait too long for this, skepticism over CO_2 figures will only increase, which benefits no-one.

7.5 Conclusions

The key objective of this paper has been to demonstrate the key changes required to EU climate policy for cars and fuels in order to provide a framework to help rather than hinder the achievement of the EU's 60% CO_2 reduction target for transport by 2050.

In order to achieve this target, fuel economy standards for cars and low carbon standards for fuels should be tightened significantly, and minimum taxes on petrol and diesel should be raised. But the primary purpose of the paper is to make specific recommendations on how vehicle and fuel policy should be designed to facilitate such deep emissions cuts in the future.

The focus is EU policy, and therefore policy areas like urban transport get no attention. The paper focuses on areas where the EU has influence and where important changes are foreseen over the next years: climate regulations for vehicles and fuels, and on fuel taxation. The paper argues that, from a theoretical and a political economics standpoint, these instruments reinforce each other, are all needed, and that carbon prices should be much above those in the EU ETS and other sheltered sectors, in order to arrive at cost effective overall reductions of CO_2.

7 The Right EU Policy Framework for Reducing Car CO$_2$ Emissions 175

Fuel tax policy is the oldest and has had important effects. CO$_2$ regulation of cars has only recently been introduced but is clearly showing its benefits already, and at much lower costs than foreseen. Climate policy for fuels is still very messy and has so far been counterproductive, largely because of the promotion of biofuels regardless of true CO$_2$ emissions.

The paper draws the following conclusions on how the policy frame in all these three areas should be modified in order to be ready for the future:

- Much work should be done in order to close the ever-widening gap between 'official' CO$_2$ figures and real-world CO$_2$ figures. Measurement protocols need to be urgently cleaned up and tolerances tightened. Abuse of such tolerances is spreading rapidly now a gram of CO$_2$/km has become ever more important for a carmaker's bottom line;
- Targets per carmaker should be not be differentiated on the basis of the weight of the cars produced (this discourages lightweighting), but rather on their 'footprint' – the area between the four wheels;
- Targets for carmakers should not try to include the 'well to wheel' carbon footprint of fuels used by the cars produced, but mimic as well as possible an energy efficiency regulation. The former cannot be controlled by carmakers, while the latter can;
- Instead, targets for well to wheel carbon footprint of fuels should be imposed on fuel supplies ('low carbon fuel standard'). The EU fuel quality directive's 6% reduction target for the carbon footprint of transport fuels for 2020 is a good start. The 10% target for renewable fuels (mostly biofuels) should be scrapped because it judges fuels on their name (i.e. biofuels, second-generation biofuels), not on environmental performance, and hence does not provide incentives to improve fuels beyond minimum standards set;
- In order for this setup to work the carbon accounting of fuels should be strongly improved. Indirect land use change effects of biofuels should be accounted for, and there should be strong differentiation of values for fossil fuels (e.g. for tar sands and oil shale);
- Relatively high fuel taxes are a strong point of Europe but significant problems remain – in particular diesel taxes are still too low compared with petrol. This is partly for historical reasons, but partly also because diesel tourism with long-distance trucks rewards those member states that keep diesel taxes low. Luxembourg is an extreme case in point, with very low diesel taxes and excessive sales, but the principle is true for many small EU Member states, which in turn make it more difficult for the large ones to increase diesel tax;
- The European Commission has repeatedly proposed to raise minimum tax levels for diesel and is likely to continue doing so. While these efforts should be applauded and supported, they do not solve the problem that for many member states it remains attractive to keep diesel tax only just above this minimum;
- The definitive solution for truck diesel tourism is a system such as the International Fuel Tax Agreement, in operation in North America, that taxes diesel on the basis of fuel *used* in a state instead of diesel *sold*. With current information technology (i.e. satellite tracking) a cost-effective version of this system

could be introduced in Europe, which would completely eliminate the perverse incentive for many countries to keep diesel taxes low.
- Last but not least, Europe should drastically review the way it taxes private kilometers made with company cars. In almost all EU member states owning a company car is much cheaper than owning a private car, and private kilometers can often be made at zero cost. This arrangement, which increases overall CO_2 emissions from cars by 4 to 8%, should be ended urgently. Inspiration can be sought in other countries, such as the US.

References

Concawe Eucar JRC (2007) Well-to-wheels analysis of future automotive fuels and powertrains in the European context. Version 2c. http://ies.jrc.ec.europa.eu/uploads/media/WTW_Report_010307.pdf. Last accessed Mar 2011

Copenhagen Economics (2010) Company car taxation. Taxation Papers, Working Paper No. 22, European Commission, Luxembourg. http://ec.europa.eu/taxation_customs/resources/documents/taxation/gen_info/economic_analysis/tax_papers/taxation_paper_22_en.pdf. Last accessed Mar 2011

European Commission (2009) Proposal for a regulation of the European parliament and of the council: setting emission performance standards for new light commercial vehicles as part of the community's integrated approach to reduce CO_2 emissions from light-duty vehicles. COM (2009) 593 final, Brussels, 28.10.2009

Global Insight (2010) Assessment of the effectiveness of scrapping schemes for vehicles – economic, environmental, and safety impacts. Report prepared for the European Commission, DG Enterprise and Industry, March. http://ec.europa.eu/enterprise/sectors/automotive/files/projects/report_scrapping_schemes_en.pdf. Last accessed March 2011

IEEP (2008) The impact of phasing in passenger car CO_2 targets on levels of compliance. Institute for European Environmental Policy (IEEP), London, August. www.transportenvironment.org/Publications/prep_hand_out/lid:515. Last accessed Mar 2011

IEEP (2010) Anticipated indirect land use change associated with expanded use of biofuels in the EU, Institute for European Environmental Policy (IEEP), London, November. http://www.transportenvironment.org/Publications/prep_hand_out/lid/611. Last accessed Mar 2011

IEEP, TNO and T&E (2008) Footprint as utility parameter – a technical assessment of the possibility of using footprint as the utility parameter for regulating passenger car CO_2 emissions in the EU, TNO/IEEP/CE Delft, July. http://www.transportenvironment.org/Publications/prep_hand_out/lid/512. Last accessed Mar 2011

Mc Lure CE Jr (2009) Why tax commercial motor fuel in the EU member state where it's bought? Why not where it's consumed? CESifo Forum 2/2009, Munich, Germany. http://www.cesifo-group.de/portal/page/portal/DocBase_Content/ZS/ZS-CESifo_Forum/zs-for-2009/zs-for-2009-2/forum2-09-special1.pdf. Last accessed Mar 2011

Official Journal of the European Communities (2003) Council directive 2003/96/EC of 27 October 2003 restructuring the community framework for the taxation of energy products and electricity. O.J. L 283/51 of 31.10.2003

Official Journal of the European Union (2009a) Regulation (EC) No 661/2009 of the European parliament and of the council of 13 July 2009 concerning type-approval requirements for the general safety of motor vehicles, their trailers and systems, components and separate technical units intended therefore. O.J. L 200/1 of 31.7.2009

Official Journal of the European Union (2009b) Regulation (EC) No 1222/2009 of the European parliament and of the council of 25 Nov 2009 on the labelling of tyres with respect to fuel efficiency and other essential parameters. O.J. L 342/46 of 22.12.2009

Schmidt H, Johannsen R (2010) Future development of the EU directive for measuring the CO_2 emissions of passenger cars – investigation of the influence of different parameters and the improvement of measurement accuracy. Report for the German Federal Environmental Protection Agency (Umweltbundesamt), December. http://circa.europa.eu/Public/irc/enterprise/wltp-dtp/library?l=/wltp-dtp_procedures/meeting_brussels/wltp-dtp-labprocice-038/_EN_1.0_&a=d. Last accessed Mar 2011

Significance and CE Delft (2010) Price sensitivity of European road freight transport – towards a better understanding of existing results. Report 9012-1, Den Haag, June. http://www.cedelft.eu/?go=home.downloadPub&id=1130&file=4053_defreportASc_1297950058.pdf. Last accessed Mar 2011

TNO (2010) CO_2 uitstoot van personenwagens in norm en praktijk – analyse van gegevens van zakelijke rijders [CO_2 emissions from passenger cars in standard and practice – analysis of data from business drivers], TNO Report MON-RPT-2010-00114, January. http://www.tno.nl/downloads/co2_uitstoot_personenwagens_norm_praktijk_mon_rpt_2010_00114.pdf. Last accessed Mar 2011

TNO, IEEP and LAT (2006) Review and analysis of the reduction potential and costs of technological and other measures to reduce CO_2-emissions from passenger cars. Final Report to the European Commission, October. http://ec.europa.eu/enterprise/sectors/automotive/files/projects/report_co2_reduction_en.pdf. Last accessed Mar 2011

Part III
National Policies

Debating whether local and national efforts to reduce greenhouse gas emissions undermine global efforts or whether global efforts generate net costs rather than net benefits produces a lot of hot air but not necessarily better solutions... Acknowledging the complexity of the problem, as well as the relatively recent agreement among scientists about the human causes of climate change, leads to recognition that waiting for effective policies to be established at the global level is unreasonable. Rather than only a global effort, it would be better to self-consciously adopt a polycentric approach to the problem of climate change in order to gain the benefits at multiple scales as well as to encourage experimentation and learning from diverse policies adopted at multiple scales.

Elinor Ostrom, A Polycentric Approach for Coping with Climate Change. Background Paper to the 2010 World Development Report, Policy Research Working Paper 5095, The World Bank, Washington, DC, 2009, pp. 31–32

Chapter 8
CO_2-Based Taxation of Motor Vehicles

Nils Axel Braathen

Abstract This chapter describes CO_2-related tax rate differentiation currently applied in one-off or recurrent motor vehicle taxes in OECD countries. It also calculates the tax rates applied, measured in Euros per tonne of CO_2 emitted over the lifetime of a vehicle. For subsidies to low-emission vehicles in one-off vehicle taxes, the cost per tonne CO_2 'saved' is also calculated. The chapter ends with a discussion of the current practices, *inter alia* in the context of the policy measures applied to combat climate change in other parts of the economy.

8.1 Introduction

Many countries have for many years levied special taxes on motor vehicles. There are two main categories of such taxes: *One-off* taxes that are levied when the vehicles are first registered; and *recurrent* taxes that one has to pay (e.g., annually) in order to be allowed to use the car.

From a purely economic efficiency perspective, the arguments for special vehicle taxes are weak. One should in principle rather seek to tax the various negative externalities that car purchases and use entail (emissions of greenhouse gases and air pollutants, congestion, etc.) more directly, and avoid taxing vehicles differently than other goods and services.[1,2]

Rather than economic efficiency, it is probably revenue-raising considerations and/or income distribution that have motivated the introduction of current vehicle taxes. Especially back at the time when these taxes were first introduced, the poorer

---

Disclaimer: The views expressed are those of the author and do not necessarily reflect the views of OECD or its member countries.

[1] See OECD (2009a) for an in-depth discussion.

[2] Sallee (2010) discusses taxation of the fuel economy of vehicles in the United States and Canada and emphasizes that while fuel economy taxation does have an impact on fleet fuel economy; such taxation is a less efficient policy for reducing fuel consumption than would be direct taxation of motor vehicle fuels.

N.A. Braathen (✉)
OECD, Environment Directorate, F-75775 Paris, France
e-mail: Nils-Axel.Braathen@oecd.org

parts of the populations would generally not own cars at all, so taxing motor vehicles was a way to tax the rich, while avoiding taxing the poor.

With an increasing focus in many countries on the threat of major climate change, a number of countries have in recent years modified their motor vehicle taxes, or introduced new ones, to take into account the amount of CO_2 each vehicle category on average emits per kilometer (km) driven. Section 8.2 gives a description of the current use of such CO_2-related tax rate differentiation in both *one-off* and *recurrent* vehicle taxes in OECD countries. Section 8.3 provides some discussion of current practice, including comparisons with incentives given to abate CO_2 emissions in other sectors of the economy.

8.2 Use of CO_2-Related Tax Rate Differentiation in Motor Vehicle Taxes

Drawing on OECD (2009b) and OECD (2010), this section will provide a detailed description of the use of CO_2-related tax rate differentiation in motor vehicle taxes in OECD member countries as of 1 January 2010.[3] While in some countries the tax rate differentiation is based directly on the certified CO_2 emissions of a given vehicle type, in other countries the tax rate depends on the certified fuel efficiency of the vehicles. Given that there currently are no possibilities for cleaning the CO_2 emissions that a certain amount of fuel use of a motor vehicle entails, the two approaches are in this context equivalent – if a correction is made for the differences in the amounts of CO_2 emitted from the combustion of 1 l of petrol and 1 l of diesel.[4]

In some countries, the tax rate depends on the price of a given vehicle. For illustration purposes, this chapter uses pre-tax prices of 10,000 EUR and 25,000 EUR as examples.

8.2.1 CO_2-Related Tax Rate Differentiation in One-Off Taxes on Motor Vehicles

Figures 8.1 and 8.2 illustrate the tax rates levied per vehicle, as a function of the amount of CO_2 they emit per km driven, in the OECD countries that apply

---

[3] CO_2-related tax rate differentiation of motor vehicle taxes is also used in some non-OECD member countries, including some of the EU member states that are not OECD members. As of 1 September 2010, South Africa also introduced a CO_2-related purchase tax on motor vehicles, with a tax rate of 75 ZAR (South African currency) per gram of CO_2 emitted per km, above 120 g of CO_2 per km, for passenger vehicles. For small goods-transporting vehicles, the tax rate is 100 ZAR per gram of CO_2 emitted, above 175 g/km. In October 2010, one ZAR equalled about 0.1 EUR. Hence, e.g., for a vehicle emitting 180 g of CO_2 per km driven, the tax rate is about 450 EUR.
[4] Whereas the combustion of 1 l of petrol causes 2.343 kg of CO_2 to be emitted, the combustion of 1 l of diesel causes emissions of 2.682 kg of CO_2.

8 CO_2-Based Taxation of Motor Vehicles

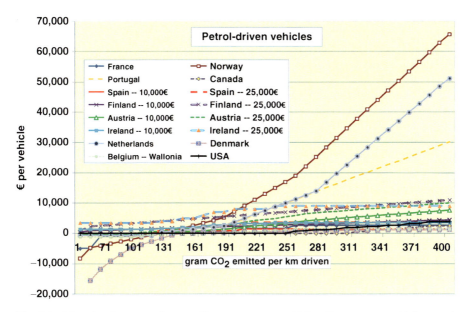

Fig. 8.1 CO_2-related tax rates in one-off taxes on motor vehicles. Tax rates per vehicle, petrol-driven vehicles

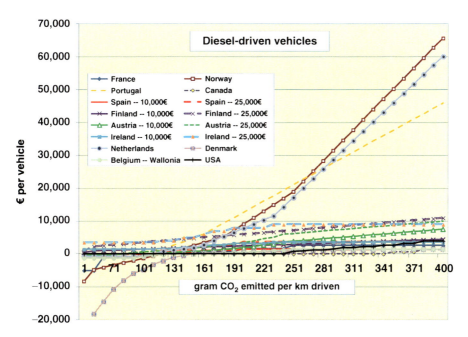

Fig. 8.2 CO_2-related tax rates in one-off taxes on motor vehicles. Tax rates per vehicle, diesel-driven vehicles

CO_2-related tax rate differentiation in one-off motor vehicle taxes.[5] As some countries vary their tax rates between petrol- and diesel-driven vehicles, Fig. 8.1 shows the tax rates for petrol-driven vehicles, while Fig. 8.2 illustrates the same for diesel-driven vehicles.[6] As the 'outliers' in the two graphs can make it difficult to see the differences across the other countries, Fig. 8.3 'zooms in' on the middle range of the tax rates, in the case of petrol-driven vehicles.

The three graphs clearly show that there are large differences in the tax rates applied per vehicles, with Norway, the Netherlands and Portugal applying the highest tax rates for (very) high-emission vehicles. One can also see that Denmark, France and Norway are applying negative tax rates – i.e., provide subsidies – for vehicles with low CO_2 emissions per km driven.

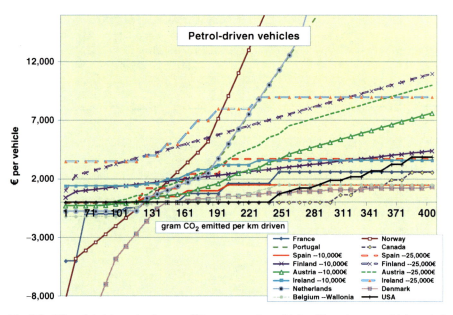

Fig. 8.3 CO_2-related tax rates in one-off taxes on motor vehicles. Tax rates per vehicle, petrol-driven vehicles, selected tax rate range

---

[5] The tax rate differentiation in the vehicle purchase tax in Belgium only applies to the province of Wallonia. Sallee (2010) indicates that the vehicle models affected by the 'Gas Guzzler Tax' in the United States had a market share of 0.7% in 2006 – and that the taxed models are overwhelmingly made by foreign manufacturers. Light trucks, including Sports Utility Vehicles (SUVs) are not covered by the tax.

[6] It is emphasized that the comparisons in this chapter only take *CO_2-related* tax rates into account. In addition to the CO_2-related element, the *Motor Vehicle Tax* in Norway also contains a cylinder volume part and a kW-based part – that each apply to the same vehicles. The total tax that a car purchaser in Norway has to pay is thus *significantly* higher than what is described in this chapter.

In the countries where the CO_2-related tax rate varies with the price of the vehicle (Austria, Finland, Ireland and Spain), the tax rate per vehicle is always higher for the more expensive vehicles.

8.2.2 CO_2-Related Tax Rate Differentiation in Recurrent Taxes on Motor Vehicles

Figures 8.4 and 8.5 illustrate CO_2-related differentiation in *recurrent* (annual) taxes on motor vehicles in OECD countries as of 01.01.2010, for petrol-driven and diesel-driven vehicles respectively, as a function of the amount of CO_2 the vehicles emit per km driven. One can notice that France applies different tax rates for company-owned cars and for 'big polluters' among cars owned by others. The highest tax rates per year in the OECD countries are applied for company-owned cars in France, and in Ireland, followed by Denmark and Luxembourg. One can, however, notice that in Ireland, the tax rate culminates at 225 g of CO_2 emitted per km – vehicles emitting (even) more are not subject to a higher tax rate per year. In France, the tax rate for company-owned cars increases quite rapidly with increasing CO_2 emissions per km a vehicle is driven, reaching very high levels for the vehicles with the largest emissions.

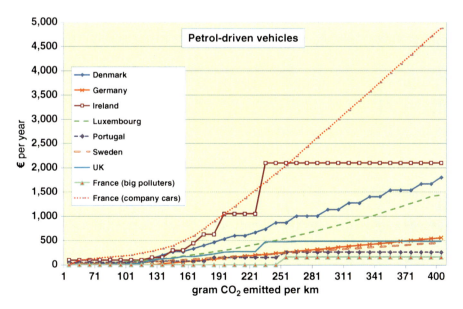

Fig. 8.4 CO_2-related tax rates in recurrent taxes on motor vehicles. Tax rates per year, petrol-driven vehicles

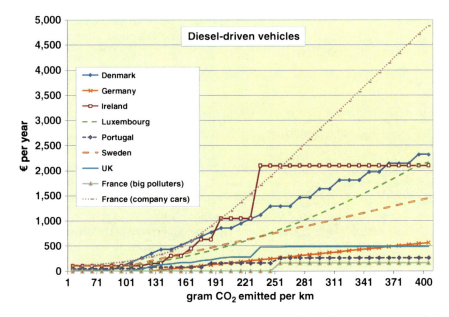

Fig. 8.5 CO_2-related tax rates in recurrent taxes on motor vehicles. Tax rates per year, diesel-driven vehicles

8.2.3 Calculation of Average Tax Rates per Tonne Emitted Over a Vehicle's Lifetime

Knowing the tax rates per vehicle, or per year for the recurrent motor vehicle taxes, it is easy to calculate the tax rate per tonne of CO_2 the vehicle will emit over its lifetime once two assumptions are made: (i) the total number of km a vehicle is driven over its lifetime; and, (ii) in the case of recurrent taxes, the total duration of the lifetime of a vehicle. In this section, the tax rates shown are *averaged* across the number of grams each vehicle emits per km it is driven. *Marginal* tax rates are illustrated in Section 8.2.4.

It is assumed that each vehicle is driven *200,000 km* over its lifetime,[7] and that the lifetime is *15 years*.[8] The subsequent calculations would not *qualitatively* be changed much if these parameters in fact were somewhat different for all vehicles, but there *would* be some bias in the calculations if – e.g. – vehicles with high emissions per km driven were driven more km over their lifetime than vehicles with lower emissions per km.

---

[7] According to Sallee (2010), in the calculations made of fuel savings in relation to the Hybrid Vehicle Tax Credit in the United States it is assumed that each vehicle is driven 120,000 miles over their lifetime – which equals 193,000 km.

[8] Sallee (2010) indicates that the average lifespan of cars in the United States is 14 years.

8 CO$_2$-Based Taxation of Motor Vehicles

As there is one million grams per tonne; when one knows the number of grams a car emits per km, and if the vehicle will be driven 200,000 km in total, one just has to multiply the number of grams emitted per km by 0.2 to find the number of tonnes emitted over the lifetime of the vehicle.

With the assumptions mentioned, Figs. 8.6 and 8.7 illustrate the *average* tax rate per tonne of CO$_2$ emitted over the lifetime of a vehicle in *one-off* motor vehicle taxes in OECD countries, for petrol and diesel-driven vehicles respectively.

Several countries provide subsidies in their one-off vehicle taxes for vehicles with low CO$_2$ emissions. In addition to looking at the magnitude of these subsidies per tonne CO$_2$ these vehicles (nevertheless) will emit over their lifetime, it can be of relevance to calculate the size of the subsidies per tonne of CO$_2$ emissions avoided – for example compared to the emission levels of the lowest-emitting vehicles that do not receive any subsidies. Comparisons of such estimates are a bit complicated, as the level of emissions at which subsidies start varies significantly across countries, and between different vehicle categories within a given country. A comparison is nevertheless made in Table 8.1– and it can be seen that the subsidies per tonne CO$_2$ 'saved' vary a lot, and are very large in some cases. At the extreme, the subsidies can approach 1,000€ per tonne of CO$_2$ emissions avoided.

Figures 8.8 and 8.9 present the average tax rate per tonne of CO$_2$ emitted over the lifetime of the vehicles in *recurrent* motor vehicle taxes. The annual tax payments are for simplicity just multiplied by 15, to take into account that the taxes in question have to be paid every year.

A possible objection to the estimates regarding recurrent taxes is that tax payments made some 10–15 years from now matter less to consumers than tax payments they have to make right away. There is hence a valid argument for discounting future tax payments – the question is 'only' which discount rate to use. Figure 8.10 illustrates the same cases as Fig. 8.8, but – as an example – a 7% discounting per year of future tax payments is used. The overall picture – the 'ranking' of different countries – is (of course) not affected, but all the tax rates per tonne of lifetime CO$_2$ emissions are lowered.

Having expressed the CO$_2$-related tax rates in both one-off and recurrent motor vehicle taxes in terms of taxes per tonne of CO$_2$, one can add the two together in order to compare the total use of such tax rates in different countries. This is done in Fig. 8.11, for selected levels of CO$_2$ emissions per km driven – without any discounting of future payments of recurrent taxes.[9] One can notice that while some countries provide significant subsidies for vehicles 'only' emitting 100 g of CO$_2$ per km driven, in France (for company-owned vehicles) and in Ireland, a tax rate of more than 100 EUR per tonne of CO$_2$ is applied. For vehicles emitting 150 g of CO$_2$ per km, in these two cases, the tax rate exceeds 200 EUR per tonne of CO$_2$.

---

[9] As in any case, the choice of a discount rate would be somewhat arbitrary, for simplicity, no discounting is used in this graph.

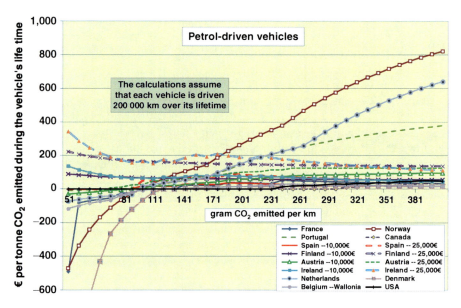

Fig. 8.6 CO_2-related tax rates in one-off taxes on motor vehicles, per tonne CO_2 emitted over the lifetime of a vehicle, petrol-driven vehicles

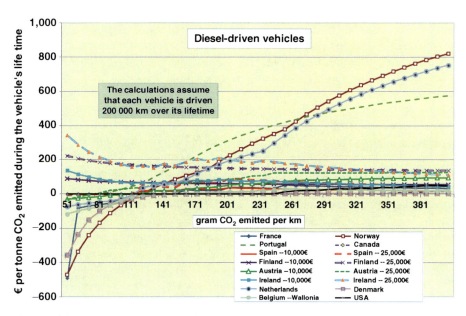

Fig. 8.7 CO_2-related tax rates in one-off taxes on motor vehicles, per tonne CO_2 emitted over the lifetime of a vehicle, diesel-driven vehicles

8 CO$_2$-Based Taxation of Motor Vehicles

Table 8.1 Subsidies in one-off vehicle taxes per tonne CO$_2$ 'saved', calculated based on emission reductions compared to the lowest-emitting vehicles not being subsidized

Country	Vehicle type	# g CO$_2$ per km at which subsidies start		125	100	75	50
Austria	10,000€ vehicle, diesel	81	Subsidy per vehicle	0	0	60	300
			g CO$_2$ saved per km driven	0	0	6	31
			Tonnes CO$_2$ saved over lifetime	0	0	1.2	6.2
			Subsidy per tonne CO$_2$ saved	**0**	**0**	**50**	**48.4**
	10,000€ vehicle, petrol	109	Subsidy per vehicle	0	80	260	300
			g CO$_2$ saved per km driven	0	9	34	59
			Tonnes CO$_2$ saved over lifetime	0	1.8	6.8	11.8
			Subsidy per tonne CO$_2$ saved	**0**	**44.4**	**38.2**	**25.4**
	25,000€ vehicle, diesel	74	Subsidy per vehicle	0	0	0	300
			g CO$_2$ saved per km driven	0	0	0	24
			Tonnes CO$_2$ saved over lifetime	0	0	0	4.8
			Subsidy per tonne CO$_2$ saved	**0**	**0**	**0**	**62.5**
	25,000€ vehicle, petrol	88	Subsidy per vehicle	0	0	200	300
			g CO$_2$ saved per km driven	0	0	13	38
			Tonnes CO$_2$ saved over lifetime	0	0	2.6	7.6
			Subsidy per tonne CO$_2$ saved	**0**	**0**	**76.9**	**39.5**
Belgium-Wallonia	All vehicles	126	Subsidy per vehicle	200	800	1,200	1,200
			g CO$_2$ saved per km driven	1	26	51	76
			Tonnes CO$_2$ saved over lifetime	0.2	5.2	10.2	15.2
			Subsidy per tonne CO$_2$ saved	**1000.0**	**153.8**	**117.6**	**78.9**
Denmark	Petrol	138	Subsidy per vehicle	1,074	3,761	8,058	16,654
			g CO$_2$ saved per km driven	13	38	63	88
			Tonnes CO$_2$ saved over lifetime	2.6	7.6	12.6	17.6

g CO$_2$ emitted per km driven

Table 8.1 (continued)

Country	Vehicle type	# g CO$_2$ per km at which subsidies start		g CO$_2$ emitted per km driven				
				125	100	75	50	
France	Diesel	142	Subsidy per tonne CO$_2$ saved	**413.3**	**494.8**	**639.6**	**946.3**
			Subsidy per vehicle	1,074	4,298	8,596	17,191
			g CO$_2$ saved per km driven	17	42	67	92
			Tonnes CO$_2$ saved over lifetime	3.4	8.4	13.4	18.4
			Subsidy per tonne CO$_2$ saved	**316.0**	**511.6**	**641.5**	**934.3**
	All vehicles	121	Subsidy per vehicle	0	500	1,000	5,000
			g CO$_2$ saved per km driven	0	21	46	71
			Tonnes CO$_2$ saved over lifetime	0	4.2	9.2	14.2
			Subsidy per tonne CO$_2$ saved	**0.0**	**119.0**	**108.7**	**352.1**
Netherlands	Petrol	121	Subsidy per vehicle	0	750	750	750
			g CO$_2$ saved per km driven	0	21	46	71
			Tonnes CO$_2$ saved over lifetime	0	4.2	9.2	14.2
			Subsidy per tonne CO$_2$ saved	**0.0**	**178.6**	**81.5**	**52.8**
	Diesel	118	Subsidy per vehicle	0	580	750	750
			g CO$_2$ saved per km driven	0	18	43	68
			Tonnes CO$_2$ saved over lifetime	0	3.6	8.6	13.6
			Subsidy per tonne CO$_2$ saved	**0.0**	**161.1**	**87.2**	**55.1**
Norway	All vehicles	120	Subsidy per vehicle	0	1,394	3,137	4,879
			g CO$_2$ saved per km driven	0	20	45	70
			Tonnes CO$_2$ saved over lifetime	0	4	9	14
			Subsidy per tonne CO$_2$ saved	**0.0**	**348.5**	**348.5**	**348.5**
Portugal	Petrol	94	Subsidy per vehicle	0	0	68	100
			g CO$_2$ saved per km driven	0	0	19	44

8 CO$_2$-Based Taxation of Motor Vehicles

Table 8.1 (continued)

Country	Vehicle type	# g CO$_2$ per km at which subsidies start		125	100	75	50
				\multicolumn{4}{c}{g CO$_2$ emitted per km driven}
	Diesel	80	Tonnes CO$_2$ saved over lifetime	0	0	3.8	8.8
			Subsidy per tonne CO$_2$ saved	**0.0**	**0.0**	**17.9**	**11.4**
			Subsidy per vehicle	0	0	76	100
			g CO$_2$ saved per km driven	0	0	5	30
			Tonnes CO$_2$ saved over lifetime	0	0	1	6
			Subsidy per tonne CO$_2$ saved	**0.0**	**0.0**	**76.1**	**16.7**

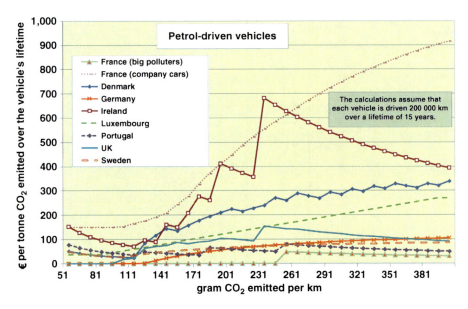

Fig. 8.8 CO_2-related tax rates in recurrent taxes on motor vehicles, per tonne CO_2 emitted over the lifetime of a vehicle, petrol-driven vehicles

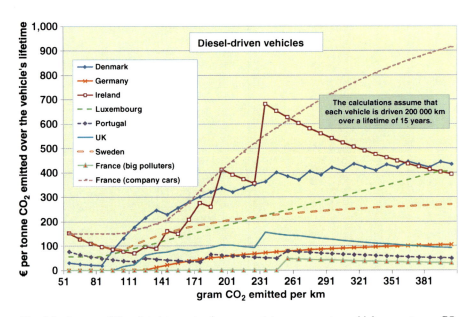

Fig. 8.9 Average CO_2-related tax rates in recurrent taxes on motor vehicles, per tonne CO_2 emitted over the lifetime of a vehicle, diesel-driven vehicles

8 CO_2-Based Taxation of Motor Vehicles

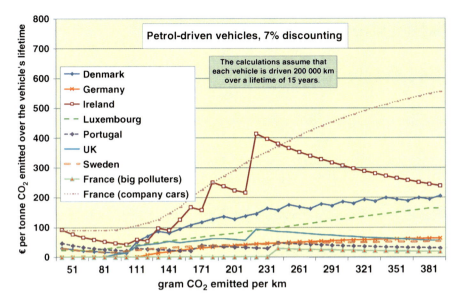

Fig. 8.10 CO_2-related tax rates in recurrent taxes on motor vehicles, per tonne CO_2 emitted over the lifetime of a vehicle, petrol-driven vehicles; 7% discounting

For vehicles with very high CO_2 emissions per km driven (here examples of 230 or 330 g/km are used), the tax rates per tonne of CO_2 emitted over the lifetime exceeds 300 EUR in a number of cases, and even 700 EUR in some cases. One can notice that in Ireland, the tax rate per tonne of CO_2 is in fact much higher for vehicles emitting 230 g/km than for vehicles emitting 330 g of CO_2 per km. This is because tax payments per year do not increase any further as a function of emissions per km driven, beyond 225 g of CO_2 per km.

8.2.4 Calculation of Marginal Tax Rates per Tonne of CO_2 Emitted Over a Vehicles' Lifetime

The estimates of tax rates per tonne of CO_2 presented thus far have been *averaged* across all the grams of CO_2 a given vehicle emits per km it is driven. It can also be of interest to look at the *marginal* tax (per tonne of lifetime CO_2 emissions) impact of choosing a vehicle that emits 1 g of CO_2 more each km it is driven. This is done in Fig. 8.12, for one-off vehicle taxes in selected countries and vehicle categories. In several of the countries *not* included in the graph, the marginal tax rate is for most emission levels zero – because the amount of tax due does not increase if one chooses a vehicle with marginally larger emissions. This corresponds to cases where the curves shown in Fig. 8.1 above are horizontal.

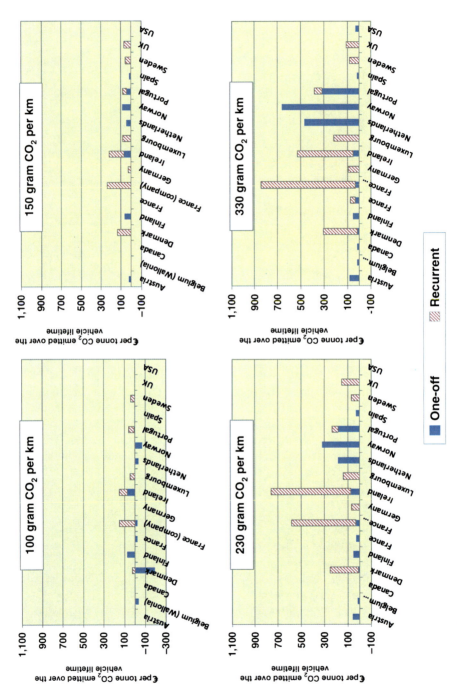

Fig. 8.11 Total CO_2-related tax rates in taxes on motor vehicles, per tonne CO_2 emitted over the lifetime of a vehicle, petrol-driven vehicles

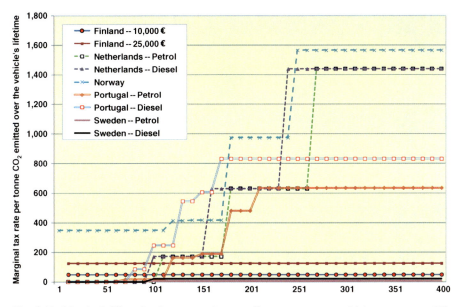

Fig. 8.12 Marginal CO_2-related tax rates in one-off taxes on motor vehicles, per tonne CO_2 emitted over the lifetime of a vehicle, selected countries

Figure 8.12 illustrates that the marginal tax 'punishment' of choosing a vehicle model with slightly higher emissions of CO_2 per km driven can be very high in some countries – well over 1,000€ for each tonne of CO_2 that the vehicle will emit over its lifetime in the Netherlands and Norway for high-emitting vehicles. One can also notice that even if Norway provides subsidies in its one-off tax to low-emitting vehicles, the *marginal* tax rate for such vehicles is strictly positive.[10]

On the other hand, Finland and Sweden provide relatively modest and constant marginal incentives to choose a vehicle with slightly lower CO_2 emissions, at all levels of emission per km driven.

8.3 Discussion of Current CO_2-Related Tax Rate Differentiation Practice

The comparisons made above make it clear that the tax rates applied per tonne CO_2 emitted over a vehicle's lifetime vary significantly between countries – with France, Norway, the Netherlands, Portugal, Ireland and Denmark having particularly high tax rates for high-emission vehicles. While there are good reasons to increase the tax

---

[10] The same is, for example, the case in Denmark.

rates *per vehicle* (in the case of one-off taxes) or *per year* (in the case of recurrent taxes) with increasing CO_2 emissions per km driven, it is more difficult to see good reasons why the tax rate *per tonne lifetime CO_2 emissions* should increase with increasing emissions per km driven.[11] A *given tonne* of CO_2 emitted into the atmosphere causes the same environmental damage, regardless of whether it is emitted from a vehicle emitting much or little per km it is driven.

Given the much lower marginal abatement costs for other CO_2 emission mitigation options in the respective countries (e.g., as regards industry – cf. the CO_2 emission allowance prices in the European Union's Emission Trading System, of about 15–20 EUR per tonne of CO_2), one can ask why *such* strong incentives are provided in these countries to abate CO_2 emissions from some motor vehicles – not least when taking the marginal tax rates into account? Part of the answer can probably be found in political economy issues that often are seen as obstacles to stricter regulation of emissions from industrial sources, for example due to a fear of loss of sectoral competitiveness. There is no risk of 'carbon leakage' in relation to the use of the vehicle taxes discussed in this chapter.

One can argue that the allowance prices in the EU ETS do not currently provide sufficiently strong abatement incentives to avoid serious human-induced climate change in the future. However, the tax rates per tonne of lifetime CO_2 emissions in some of these countries are also rather high compared, for example, to the development over time of the price that emitters would need to be charged for each tonne of GHG emissions to induce them to reduce emissions enough to keep global emission trends on track with a scenario that OECD recently elaborated, that stabilizes the CO_2 concentration in the atmosphere at 450 ppm, cf. e.g. OECD (2008) and (2009c).

One should also keep in mind that CO_2-related tax rate differentiation in motor vehicle taxes comes *on top of* the often already quite strong CO_2 abatement incentives stemming from taxes on motor vehicle fuels, especially in European countries. Taxes on petrol and diesel are, of course, levied for a number of other reasons,[12] and one should not count all of them as 'carbon taxes'. Nevertheless, it is the full rate of tax that will influence the extent to which CO_2 will be emitted.

It is noticeable that several countries apply different tax rates per tonne of CO_2 emitted over the lifetime of petrol- and diesel-driven vehicles, with the higher rates being applied for the latter category. From a *local air pollution* perspective, it makes good sense to tax diesel-driven vehicles more heavily than petrol-driven ones. However, a given tonne of CO_2 does the same harm regardless of the fuel-type it is stemming from. Hence, one can again question the rationale for some of the current practices.

---

[11] One possibility could in principle be that vehicles with high CO_2 emissions per km driven tend to be driven longer distances over their lifetimes than vehicles causing lower emissions per km they are driven. The differences in km driven would, however, have to be very large in order to 'compensate' for some of the differences described in this chapter.

[12] For example, West and Williams (2007) found that motor fuels are complements in consumption to leisure. Taxing motor fuels hence make it possible to indirectly tax leisure, thus correcting a distortion otherwise difficult to address regarding the choice between work and leisure.

8 CO$_2$-Based Taxation of Motor Vehicles

The price of a vehicle is not a good indicator of the environmental harm it causes. Hence, from an environmental point of view, the arguments for applying a higher tax rate per tonne of CO$_2$ emitted from an expensive vehicle than from a cheaper one seem weak. It is probably equity concerns that explain the use of such tax rate differentiation in several countries.

Some people might say that the cost to society of reducing CO$_2$ emissions by taxing emissions from high-emitting vehicles is very low, because people simply can buy low-emitting vehicles instead. However, there *could* be a significant loss of 'consumer surpluses' involved in such policies. In the absence of any climate policies, the people that would buy high-emitting vehicles would often be willing to pay (substantially) more for those vehicles than they would be willing to pay for vehicles that emit less.[13] One should consider carefully whether the benefits to society of doing so actually exceed this welfare loss.[14]

As mentioned in the introduction, in a 'perfect world', with no market- or policy-failures, one can question the need for CO$_2$-related differentiation of motor vehicle taxes. As there is a direct link between the carbon content in the motor fuels used and the CO$_2$ emissions of a given vehicle, it could be more environmentally effective and economically efficient to just apply a tax reflecting the carbon content of different fuels.

As emphasized in Sallee (2010), taxes that promote purchases of more fuel efficient vehicles will tend to increase the car-use externalities that are linked to the number of km driven (rather than to the amount of fuels used), such as accidents and congestion. Fuel taxes would, however, also be rather imprecise instruments to address congestion externalities, as the magnitude of these externalities depends strongly on where and when the driving takes place.

Regarding environmental effectiveness, differentiation of a tax on vehicle *purchases* only affects (directly) the decisions of those that *buy* a new vehicle, and it has no (or little) impact on how much the vehicle is used. Differentiation of *recurrent* vehicle taxes can affect the decision to continue to *own* both new and old vehicles, but will also have no (or little) impact on how much the vehicle is *used*.

There is also a problem related to any policy that is based on the certified emissions or fuel efficiency of a given vehicle type[15]: Car producers have a strong

---

[13] A counterargument could be the quite high willingness to pay for emission reductions that some people show in buying hybrid and other low-emission versions of some vehicle models.

[14] There can be co-benefits related to reducing motor vehicle CO$_2$ emissions, in the form of reductions in emissions of local air pollutants, possibly noise, etc. It is, however, not evident that this is a good argument for taxing a given tonne of CO$_2$ emitted from a high-emission vehicle (much) more than a tonne emitted from a low-emission vehicle. For further discussion of co-benefits from climate policies, see Bollen et al. (2009).

[15] This point is just as valid for policies obliging car producers to meet certain average fuel efficiency standards as it is for tax rate differentiation of motor vehicle taxes.

incentive to 'fit' the characteristics of the vehicles they send for testing to the specificities of the test-cycles use, without necessarily causing lower emissions in actual use of the vehicles.[16]

Sallee (2010) points out that there can be important interactions between a tax related to the fuel efficiency of motor vehicles and regulations addressing the average fuel economy of vehicles, such as the CAFE standards in the United States and the coming rules on vehicle fuel efficiency in the EU, described in Chapters 4 and 6 of this book. Sallee (2010) finds that 'in the future, the interaction between fuel economy taxes and CAFE regulation is likely to create a situation in which fuel economy taxation will often influence the sales of particular vehicles without having any net impact on fleet fuel economy'.

However, the world is not quite perfect. For example, it is possible that consumers to some extent are 'myopic', and don't take future fuel consumption much into account when buying a new car.[17] This could argue in favor of *some* tax incentives to promote the sales of low-emission vehicles, especially if it would prove possible to develop test-cycles that well reflect actual vehicle use.

It can also be 'politically easier' to introduce CO_2-related tax rate differentiation in vehicle taxes than to introduce (only) 'sufficiently high' motor fuel taxes, in part, because this can be done in a revenue-neutral way.[18] New, or higher, taxes are rarely popular, but it can perhaps help in the implementation process if the tax can be 'sold' as not raising additional revenue.[19]

---

[16] Sallee (2010) documents that vehicles in the US have been 'tweaked' to fit in to tax-preferred categories of the 'Gas Guzzler Tax' applied there. These adjustments were said to entail negative welfare impact equal to three times the positive welfare impact that can be expected from an ideal Pigouvian fuel tax.

[17] A study by Vance and Mehlin (2009) of the German car market does, contrary to some earlier evidence, indicate that consumers there in fact to a considerable extent do take future tax payments into account in their purchasing decisions. Their results suggest that recurrent motor vehicle taxes and fuel costs significantly determine market shares of different car categories, and hence may serve as effective instruments in influencing the composition of the car fleet and associated CO_2 emissions.

[18] *All* use of motor vehicles contributes to some negative externalities – for example, congestion and accidents. One can hence question the practice of providing subsidies for low-emission vehicles, as these contribute to increasing the total number of vehicles on the roads. If the policy context is one where it is deemed 'necessary' to provide subsidies to stimulate car sales, in order to 'save' the national motor vehicle sector, it can nevertheless be useful to include a stimulus for low-emission vehicles in these subsidies, cf. OECD (2009a) for a further discussion.

[19] In the aftermath of the financial and economic crisis in recent years, most OECD countries will need to go through a period of fiscal consolidation, with increases in tax revenues and/or reductions in public spending. It *might* prove easier to 'sell' the introduction of higher taxes on fossil fuels in such a context – where such taxes might be seen by the electorate as the lesser of several 'evils'.

8.4 Conclusions

While applying carbon-differentiated motor fuel taxes and road charging to address other externalities related to motor vehicle use would be the first-best approach, *some* CO_2-related tax rate differentiation of motor vehicle taxes can possibly play a useful role if, for example, political economy constraints make it difficult to put in place an 'ideal' system.

A priori, this should be valid for both developed and developing countries. In the latter category, taxes on motor vehicles (and on motor vehicle fuels)[20] would tend to progressive, as the poorest part of the population would generally not own motor vehicles at all. Regarding one-off taxes on (new) motor vehicles, only rich people would be able to buy vehicles with high CO_2 emissions. Regarding recurrent vehicle taxes, some middle-income people might own old cars with relatively high CO_2-emissions, but the poorest parts of the population should not be much affected.

While *some* CO_2-related tax rate differentiation in motor vehicle taxes can be useful, the degree of tax rate differentiation applied in *certain* OECD member countries at present can seem to be out of proportion to the CO_2 abatement incentives they provide elsewhere in their economies. This is of relevance both as regards the 'punishment' sometimes facing high-emission vehicles and as regards the 'encouragement' sometimes provided for low-emission vehicles.

Acknowledgements The author thanks Nick Johnstone, Sverre Mæhlum and Theodoros Zachariadis for helpful comments.

References

Bollen J, Guay B, Jamet S, Corfee-Morlot J (2009) Co-benefits of climate change mitigation policies: literature review and new results. Economics Department Working Papers No. 693, OECD, Paris. http://www.oecd.org/officialdocuments/displaydocumentpdf/?cote=eco/wkp(2009)34&doclanguage=en

OECD (2008) Climate change mitigation: What do we do? OECD, Paris. www.oecd.org/dataoecd/31/55/41751042.pdf

OECD (2009a) The scope for CO_2-based differentiation in motor vehicle taxes – in equilibrium and in the context of the current global recession. OECD, Paris. www.oecd.org/officialdocuments/displaydocument/?doclanguage=en&cote=env/epoc/wpnep/t(2009)1/final

OECD (2009b) Incentives for CO_2 emission reductions in current motor vehicle taxes. OECD, Paris. www.oecd.org/officialdocuments/displaydocument/?doclanguage=en&cote=env/epoc/wpnep/t(2009)2/final

---

[20] Low-income households can to some extent be affected by increases in prices of public transport if motor fuel taxes are increased. However, with the additional revenues that such a tax increase would raise, it ought to be possible to provide targeted subsidies to relevant public transport schemes to offset the most negative distributive impacts. A full analysis of the distributional impacts of a fuel tax increase should also take into account the distribution of relevant changes in e.g. local air quality stemming from the tax increase.

OECD (2009c) The economics of climate change mitigation: policies and options for global action beyond 2012. OECD, Paris. www.oecd.org/document/56/0,3343,en_2649_34361_43705336_1_1_1_1,00.html
OECD (2010) Taxation, innovation and the environment. OECD, Paris. www.oecd.org/env/taxes/innovation
Sallee J (2010) The taxation of fuel economy. NBER Working Paper Series. Working Paper 16466. National Bureau of Economic Research. Cambridge, MA. www.nber.org/papers/w16466
Vance C, Mehlin M (2009) Tax policy and CO_2 emissions – an econometric analysis of the German automobile market. Ruhr Economic Papers #89, Rheinisch-Westfälisches Institut für Wirtschaftsforschung, Essen. www.econstor.eu/dspace/bitstream/10419/26854/1/593498305.PDF
West SE, Williams RC III (2007) Optimal taxation and cross-price effects on labor supply: estimates of the optimal gas tax. J Public Econ 91:593–617

Chapter 9
Fuel Taxation in Europe

Jessica Coria

Abstract Based on the European experience, this chapter highlights the very important role of motor fuel taxes for carbon emission mitigation. Fuel demand and CO_2 emissions would have been much higher in the absence of fuel taxes of the level currently implemented in Europe. Moreover, given the close link between fuel consumption and CO_2 emissions, fuel taxes are likely to be more effective reducing CO_2 emissions than alternative policies such as fuel-efficiency standards. Nevertheless, in order to reduce the broad range of external costs of road transportation, additional instruments are necessary. The fuel demand elasticities and the role of fuel taxes reducing CO_2 emissions are discussed, and policy recommendations are provided.

9.1 Road Transportation Externalities and Fuel Prices in Europe

In Europe, as well as in many other countries around the world, there is an increasing concern about the social costs of transport. Some of the major externalities that transport causes are traffic congestion, traffic accidents, local pollution and global warming.[1]

Congestion arises from the mutual disturbance of users competing for limited transport system capacity, and its main consequence is the increase in the travel time. The economic costs of congestion are determined by a number of parameters regarding the value of travel time, the speed-flow relationship describing the effect of an additional vehicle on the transport system, and demand elasticities describing the likely reaction of users to the internalization of the external costs of congestion. Several studies have estimated the marginal social congestion cost based on different models for urban areas and rural roads. The results vary significantly due to different model settings, aggregation levels, local characteristics and traffic conditions; however, bandwidths of marginal social costs in Europe range from €0.25

---

[1] Other externalities caused by road transportation include oil dependency, noise, highway maintenance costs, urban sprawl and improper disposal of vehicles and vehicle parts. See Parry et al. (2007) for a related discussion.

J. Coria (✉)
Department of Economics, University of Gothenburg, SE 405 30 Gothenburg, Sweden
e-mail: Jessica.Coria@economics.gu.se

to €2.00 per vehicle kilometer in large urban areas and from €0.25 to €0.3 per vehicle kilometer in small and medium urban areas (CE Delft 2008).

Road users impose also accident risks on other road users; some of the most important accident cost categories are material damages, medical costs, production losses and the willingness to pay of the victim and of the relatives and friends of the victim to avoid an accident (the so-called warm-blooded costs). In the determination of the marginal external accident costs, there are two main problems. Firstly, one needs to determine the relationship between the number of road users and the number of accidents. Secondly, there is the determination of that part of the accident costs which is internalized in each road-user's decision process. Empirical estimates of the accident costs can be obtained in different ways; the direct economic costs are relatively easy to estimate as they are directly observable while the only way to determine the warm-blooded costs is on the basis of revealed or stated preferences for risk reductions.

Road transportation is responsible for the emissions of several air pollutants, as particulate matter, nitrogen oxides, sulfur oxide, ozone and volatile organic compounds (Michaelis 1995). These pollutants cause damages to humans, biosphere, soil, water, buildings and materials at the local and regional level. For instance, particulate matter contributes to asthma attacks and other respiratory problems (McCubbin and Deluchi 1999) while carbon monoxide elevates the hospitalization rate of young children (Neidell 2004).

Finally, road transportation has also a large share in the total emissions of carbon dioxide. Though carbon dioxide does not have direct adverse health effects, it is one of the main sources of global warming. In 2007, the transport sector produced nearly one-fifth of global CO_2 emissions, of which roughly three fifths can be attributed to private automobiles. Moreover, the emissions of this sector have increased at a faster rate than global emissions and they continue to grow strongly: 45% versus 38% average between 1990 and 2007 (International Energy Agency 2009; see also Chapter 3 of this book).

The emissions of CO_2 caused by driving a kilometer in a vehicle are mainly dependent on the fossil carbon content of the total fuel consumed. However, other variables such as the modal choice, the efficiency of the equipment, the distance travelled, and the efficiency of the overall transportation system and infrastructure have also an impact. Formulating appropriate policies to tackle the externalities caused by automobiles is not that straightforward due to the varied range of externalities associated with road use, and the complex interactions between modes of transport. Ideally, policies should take account of the combined effect of different externalities and induce appropriate choices regarding the use of competing modes of transport. Generally speaking, one can make a distinction between pricing, regulation and infrastructure policies (Sterner 2002, Mayeres 2003). Each type of instrument encompasses a wide range of options. For example, pricing includes economic instruments such as fuel taxes, taxes on vehicle ownership and road pricing. Regulatory measures include emission standards, traffic rules and rationing of car use, while infrastructure policy refers to, for example, spatial planning. One instrument does not preclude the use of others. They are often complementary, since the optimal policy mix requires the use of a number of instruments in combination.

9 Fuel Taxation in Europe

As part of its efforts to reach the targets of the Kyoto Protocol, the European Commission enacted in 2009 new regulations to reduce emissions of CO_2 per km of newly registered automobiles. As explained in Chapter 4 of this book, this legislation was motivated by two main considerations: the fact that the transport sector has not been so far included in the EU Emissions Trading Scheme (EU ETS), and the fact that voluntary commitments of the European Automobile Manufacturer's Association (ACEA) to reduce emissions have not been very successful (see also Frondel et al. 2010). Several studies have shown that there is a significant positive association between increased fuel economy and increased driving, and a significantly negative fuel price elasticity. Taken together, these results suggest that fuel taxes are likely to be a more effective policy measure in reducing emissions than fuel efficiency standards. In this chapter, we focus on the role played by fuel taxes discouraging automobile use and reducing CO_2 emissions from road transportation in Europe.

The vast majority of vehicles in Europe are still powered by internal combustion engines fuelled with distillates of petroleum, although in recent years alternative fuels have attracted increased attention. Most countries levy specific excise taxes upon motor petroleum fuels. In addition, the value added tax (VAT) is generally applied over and above the excise tax. The combined effect of the motor fuel taxes plus VAT is that motor fuels are taxed more heavily than other goods and other sources of energy (Crawford and Smith 1995).

European domestic fuel prices vary considerably among countries, primarily due to differences in tax rates. In spite of the fact that there is a minimum fuel tax mandated level[2] in the EU, fuel-related fiscal measures are determined at a national level and hence there are several fuel regimes across EU members (Ryan et al. 2009). Rietveld and Van Woudenberg (2005) use cross-section data to explain fuel price differences between European countries. They find that fuel taxes increase with per capita income and per capita government expenditure in Europe. For instance, in the 15 countries that were EU members before 2004, the level of fuel prices and taxes is about 20% above the level in the Eastern European countries that became EU members from 2004 onward (EEA 2010). In addition, small European countries tend to be more aggressive than large countries, charging lower fuel taxes in order to attract consumers from neighboring countries. For example, Luxembourg has lower fuel taxes than its neighbors so that substantial cross-border fuelling and shopping trips take place.

Table 9.1 summarizes fuel taxes in EU Members. In all Member States (except for the UK) diesel is taxed less than petrol, leading to lower prices (even though the external costs of diesel vehicles are on average higher than those of petrol vehicles). Table 9.1 shows that excise taxes on diesel in the EU can be as much as 40% less than on gasoline. This has contributed to a shift from petrol to diesel vehicles in recent decades, as illustrated in Table 9.2. In France, Belgium and Austria, for

---

[2] This minimum level is equal to €0.359/l in the case of unleaded gasoline and €0.33/l for diesel. The implementation of the minimum rates implied an increase of the energy tax in Member States that joined the EU more recently, as in most of them only transport fuels were taxed and at a lower rate.

Table 9.1 Fuel taxes and fuel prices in selected countries

	Fuel taxes[a]		Fuel prices[b]		kg CO_2/capita road transport[c]	Ratio fuel tax/price[d]	
	Gasoline	Diesel	Gasoline	Diesel		Gasoline	Diesel
US	0.126	0.139	0.725	0.79	5057	17%	18%
Canada	0.309	0.252	1.051	0.967	3859	29%	26%
Japan	0.637	0.366	1.42	1.3	1680	45%	28%
EU	0.648	0.497	1.64	1.48	1814	39%	34%
Austria	0.569	0.447	1.56	1.45	2725	37%	31%
Belgium	0.790	0.525	1.86	1.54	2284	43%	34%
Bulgaria	0.451	0.395	1.33	1.29	924	34%	31%
Cyprus	0.462	0.425	1.35	1.31	2469	34%	32%
Czech Republic	0.650	0.554	1.63	1.57	1714	40%	35%
Denmark	0.729	0.497	1.89	1.60	2382	39%	31%
Estonia	0.544	0.506	1.44	1.42	1667	38%	36%
Finland	0.807	0.502	1.86	1.49	2330	43%	34%
France	0.781	0.551	1.75	1.50	1958	45%	37%
Germany	0.862	0.625	1.84	1.59	1712	47%	39%
Greece	0.862	0.530	1.92	1.66	1732	45%	32%
Hungary	0.572	0.464	1.56	1.49	1258	37%	31%
Ireland	0.699	0.578	1.72	1.61	3171	41%	36%
Italy	0.726	0.544	1.78	1.60	1963	41%	34%
Latvia	0.489	0.425	1.41	1.38	1523	35%	31%
Lithuania	0.559	0.353	1.55	1.34	1384	36%	26%
Luxembourg	0.598	0.403	1.51	1.30	13421	40%	31%
Malta	0.591	0.454	1.57	1.34	1312	38%	34%
Netherlands	0.919	0.542	1.95	1.52	2061	47%	36%
Poland	0.581	0.389	1.47	1.37	1067	40%	28%
Portugal	0.750	0.469	1.78	1.51	1703	42%	31%
Romania	0.448	0.377	1.36	1.31	556	33%	29%
Slovakia	0.662	0.474	1.63	1.45	1013	41%	33%
Slovenia	0.624	0.550	1.55	1.49	2491	40%	37%
Spain	0.547	0.426	1.50	1.40	2242	36%	30%
Sweden	0.481	0.548	1.76	1.65	2299	27%	33%
United Kingdom	0.808	0.808	1.83	1.87	1961	44%	43%

[a]*Source*: European Commission (2010), The American Petroleum Institute (2010), The Ministry of Natural Resources Canada (2010) and The National Tax Agency Japan (2010). Fuel taxes are expressed in US dollars/liter in 2010
[b]*Source*: European Union: European Environmental Agency (2010). Fuel prices are expressed in US dollars/liter in 2010. United States, Canada and Japan: Transport Policy Advisory Services (2009). Fuel prices are expressed in US dollars/liter in 2008
[c]*Source*: CO_2 emissions from fuel combustion 2009, International Energy Agency
[d]This ratio corresponds to the fraction of fuel tax in final consumer price

instance, diesel cars had a share close to or larger than 50% in 2006. Moreover, in 1980, petrol accounted for about 70% and diesel for 30% of the fuel sales, while the share of diesel rose to 58% and unleaded petrol dropped to 42% more recently (EEA 2010). In recent years, however, the price difference between diesel and gasoline has

Table 9.2 Share of diesel cars in the car stock 1995 and 2006

Country	1995	2006
France	22%	47%
Germany	14%	22%
Spain	14%	38%
Italy	12%	30%
Netherlands	11%	16%
Belgium	33%	52%
UK	8%	19%
Austria	24%	53%

Source: Schipper and Fulton (2008)

been narrowing in most countries, partly because the increased demand for diesel has lead to higher retail prices (Schipper and Fulton 2008).

According to Sterner (2002), several reasons explained the lower tax on petrol. Historically, diesel was preferred over gasoline because it is generally more energy efficient and was believed to create less toxic exhaust emissions. At the time, energy efficiency had a greater weight than health issues in policy making. Today, however, concerns over the health effects of particulate matter have increased. On the other hand, recent findings indicate that although diesel cars in Europe may provide significant fuel savings to individual drivers, they do not provide significant national energy or CO_2 emissions savings, since they are in average larger in size and are driven 40–100% more per year than gasoline cars (Schipper and Fulton 2008).

In addition to excise taxes and VAT, motor fuels are subject to a number of other 'special' taxes in some Member States. For instance, a number of countries are now applying some form of CO_2 related tax rate differentiation in their taxes on either the purchase or the use of motor vehicles. There are also a number of countries where the motor vehicle tax rates depend on the fuel efficiency of the vehicles in question – which is closely linked to the CO_2 emissions caused. For instance, France and Norway subsidize the purchase of vehicles with relatively modest CO_2 emissions per kilometer driven. In France, the subsidies gradually decrease, and increasing taxes are levied for vehicles with emissions larger than 160 grams CO_2 per kilometer (g/km). In the Norwegian case, rapidly increasing taxes are applied to petrol-driven vehicles with CO_2 emissions exceeding 120 g/km. More details on such taxation are provided in Chapters 6 and 8 of this book.

As shown in Table 9.1, fuel prices to consumers vary among European countries not only because of variations in the level of the excise tax and VAT, but also because of differences in distribution and retailing costs. On average, fuel taxes in the EU represent around 39% of the retail price of gasoline and 34% of the retail price of diesel.

Most fuel taxes regimes in the world have been designed with revenue generation motivations. In comparison with many other revenue instruments, fuel taxes are unusually efficient when it comes to collection costs; they are collected from a reasonably small number of fuel wholesalers or at the refineries, with the charges being passed along to retailers and ultimately to consumers. This keeps the costs low and also reduces prospects for fraud or evasion (Wachs 2003).

In practice, however, fuel taxes do not only have the potential to raise revenues, but also to improve environmental quality (Sterner 2007). Lower fuel prices in the United States have contributed to a trend toward the use of larger vehicles, while in Europe higher fuel prices have helped encourage improved fuel economy. In 2005, for example, on-road fuel economy in United States was slightly above 8.9 km/l, while in Japan, Germany, the U.K. and France it was 9.4, 12.3, 13.2 and 13.6 km/l respectively. In addition, although vehicle ownership and car use in Europe has continued to grow, there is some evidence of saturation, as more and more families that acquire a second car do not use it as much as they use the first one, and kilometers driven lag behind GDP growth (Schipper 2011).

As a result of the differences in fuel economy and automobile use, there is a large variation in the average fuel consumption as well as per capita emissions of CO_2 by road transportation across OECD member countries (Rietveld 2007). For instance, Table 9.1 shows that per capita CO_2 emissions from road transportation in the USA are more than twice as high as in the EU (5057 versus 1814 kg CO_2/capita). Conversely, per capita CO_2 levels in the EU and Japan are comparable.

9.2 Demand Elasticities

For forecasting and evaluation of the environmental effects of fuel taxation, a critical consideration is the responsiveness of total fuel consumption to changes in fuel prices. This is characterized by the price elasticity of demand for fuel.

Gasoline demand can be expressed as the product of (i) fuel efficiency (gasoline use per kilometer), (ii) mileage per car and (iii) car ownership. Changes in gasoline price affect total demand through all these channels, although the responses depend on the timing. In this regard, the short-run and long-run responses to gasoline taxes are expected to differ for a number of reasons. In the short run, the stock of cars is relatively constant. Responses might take the form of a curtailment of travel, substitution to public transportation, carpooling of individual car use, improved automobile maintenance to enhance fuel efficiency, and a shift in the use of different vintages within the existing fleet, which increases the relative use of newer automobiles (that tend to be more fuel efficient and less polluting). Instead, since in the long run the stock of automobiles will vary in size and composition, there will be a shift in demand (due to both replacement and the expansion generated by increased population and income) from less fuel-efficient to more fuel-efficient cars. There might be also a change in land use and lifestyle, resulting in new living patterns that economize on travel and fuel consumption, such as living closer to work.

Several econometric studies determining the role of fuel prices and income in fuel consumption have been conducted over the years, particularly during the 1970s and the early 1980s when fuel prices were high and concerns about energy conservation and energy security were strong. Current concerns about global warming have re-ignited interest in understanding the demand for gasoline, particularly in explaining cross-country differences in gasoline consumption and automobile driving and in predicting the impact of fuel tax changes on driving and fuel consumption. Many gasoline demand studies have also been motivated by an interest in the role of income in gasoline demand and the distributional impacts of gasoline taxes.

A wide variety of models have been estimated, using different functional forms and estimation techniques, covering different time periods and different parts of the globe. In all cases, the main dependent variable is gasoline demand. For instance, static models are usually based on cross-sectional data and are used to estimate a single price and income elasticity parameters. Instead, lagged endogenous models are estimated using time series or panel data, and they contain a lagged dependent variable to distinguish between short and long run elasticities.

Reviews by Dahl and Sterner (1991), Espey (1998), Graham and Glaister (2002) and Brons et al. (2008) provide qualitative and quantitative summaries of gasoline demand research and estimates of price elasticities. As expected, the evidence shows important differences between the long and short run price elasticities of fuel consumption. Long run price elasticities typically range between −0.6 and −0.8, while short run elasticities range between −0.2 and −0.3 (Graham and Glaister 2002).

Evidence also shows that simple static models that only include income and gasoline price as explanatory variables measure an intermediate price elasticity that is close to the long run elasticity (Dahl and Sterner 1991). Instead, models that include vehicle characteristics, vehicle ownership and fuel efficiency capture the 'shortest' short run elasticities by effectively measuring the influence of price and income changes on driving only (Espey 1998).

The price elasticity of gasoline (ε_G) can be decomposed as follows:

$$\varepsilon_G = -\varepsilon_{FE} + \varepsilon_{\frac{KM}{C}} + \varepsilon_C \tag{9.1}$$

where ε_{FE}, $\varepsilon_{\frac{KM}{C}}$ and ε_C stand for the point elasticity of fuel efficiency, mileage and car ownership with respect to gasoline price, respectively. That is, they indicate the response in fuel efficiency, mileage per car and car ownership to a change in the price of gasoline. Johansson and Schipper (1997) study the effects of fuel efficiency on gasoline price elasticity. Their fuel use data for 12 OECD countries over the period 1973–1992 allows them to conduct separate estimations for vehicle stock, mean fuel intensity and mean annual driving distance. They estimate a long run fuel price elasticity of approximately −0.7, in which the largest fraction (approximately 60%) is due to changes in fuel intensity.

Increased fuel efficiency is not only explained by changes in the demand for new vehicles toward fuel-efficient ones: increased gasoline prices also induce a fuel-efficient vehicle to stay in service longer while a fuel-inefficient vehicle is more likely to be scrapped. In this regard, Li et al. (2008) analyze the effects of gasoline prices on the survival probability of old cars (with more than 10–15 years of service) in the United States; they find that an increase in gasoline price would prolong the life of vehicles with fuel efficiency higher than 29 miles per gallon (12.3 km/l), while it would shorten the lifetime less fuel-efficient vehicles.

Brons et al. (2008) analyze whether the set of elasticities in the literature can be combined by making use of the linear relationship between the point elasticities in equation (9.1). For this purpose, they develop a meta-analytical estimation approach based on a seemingly unrelated regression (SUR) model that allows them to combine observations of elasticities from different primary studies and thus increase their sample size. Their results show that the estimated mean short run price elasticity of gasoline demand is –0.34, which is somewhat higher in absolute value that the estimates found in previous studies. This value can be deconstructed into estimates for the price elasticity of fuel efficiency (–0.14), mileage per capita (–0.12) and car ownership (–0.08). Thus, like Johansson and Schipper (1997), they find that the response in demand resulting from a change in gasoline price is mainly driven by responses in fuel efficiency; mileage per car and car ownership affect the response to a lesser extent.

Their estimate of the long-run price elasticity of gasoline demand is –0.84 (again, higher than the estimates found in previous studies). This value can be decomposed into estimates for the price elasticity of fuel efficiency (–0.31), mileage per capita (–0.29) and car ownership (–0.24). Thus they conclude that in the long run, the response in demand resulting from a change in gasoline price is driven equally strongly by responses in fuel efficiency, mileage per car and car ownership.

They also find that consumers' demand for gasoline became more price-sensitive between 1949 and 2003, which can be explained by the increased consumption of gasoline in this period – since its share in total expenditures has increased, so has the price sensitivity. However, other studies have suggested that fuel consumption has become more price-inelastic over time, and that this is increasingly explained by changes in fuel efficiency rather than in the amount of driving, but also by rising incomes in many countries in the world and falling real fuel prices (see, for instance, Small and Van Dender 2007 and Hughes et al. 2008).[3]

Estimates indicate also that there are regional variations in the elasticity of demand. For instance, Espey (1998) shows that gasoline consumers in European countries seem to be more income-sensitive than consumers in the US both in the short and in the long run. Certainly, this might be explained by a series of reasons related to model specification and the nature of the data. However, an important reason for the existence of higher elasticity values in European countries is the availability of alternative means of transport. In several countries in Europe, the share of non-motorized transport modes in passenger transport reaches 40–50%, whereas in the USA and Canada it is between 10 and 20%. In the United States, the absence of alternative means of transportation coupled with an uneven income distribution makes fuel taxation particularly sensitive for low-income groups. Conversely, in many European countries, public transportation offers reasonable alternatives and shelters low-income households somewhat from the cost of higher fuel prices. This is also explained by urban architecture; most U.S. cities have population densities

---

[3] These two studies refer to the US and use date of recent decades only, so this may explain the different results.

much lower than those in Europe. Therefore, it is not surprising that fuel consumption in the dispersed U.S. cities is several times as high as in typical European cities. Clearly, the lack of alternatives reduces the capability of U.S. consumers to react to increased prices and increases the political resistance to fuel taxes (Sterner 2002).

Price elasticities also differ among various socio-economic groups. Wadud et al. (2010) employ semi-parametric econometric techniques to accommodate the possibility of differences in responses among households. They model the heterogeneity in price and income elasticities through interacting price and income with demographic variables. In the context of the United States, they find that price elasticity decreases with higher income. Quite the opposite, multiple vehicles and multiple wage earner households are more sensitive to price changes, as well as households located in urban areas. This could be related to the ease with which they can switch to a more efficient second vehicle or to alternative transport modes when price increases.

Gately (1992) and Dargay and Gately (1997) examine the price-reversibility of fuel demand for road transport. They use econometric models based on price-decomposition techniques to measure separately the effects of different types of price increases and decreases. Their results suggest that consumers do not necessarily respond in the same fashion to rising and falling prices, or equivalently to sudden and substantial price rises as to minor price fluctuations. In fact, consumers have reacted more strongly to the price rises of the seventies than to other price rises, and that the resulting reductions in fuel use have not been totally reversed as prices return to lower levels. This means that when prices rise above some previous maximum level, the long run demand changes, so that subsequent price declines will not totally undo the demand reductions caused by the initial price rise. Clearly, the existence of price asymmetries has important implications for the estimation of price elasticities and for transport policy. In terms of price elasticities, reversible models underestimate the impact of certain price rises and overestimate the effect of other smaller price rises and of price cuts. In terms of transport policy, real fuel prices, and thus fuel taxes, would have to be increased substantially to be successful at reducing fuel demand. However, once a reduction in demand is attained, it will not be fully reversed if real prices fall again; although, given the effect of income growth eroding the effects of price increases, price will need to rise more rapidly than income if fuel demand is to remain at certain level.

Recent increases in the number of diesel cars in most European countries might imply that some estimates are overstated if studies on gasoline demand do not distinguish between gasoline and diesel-powered cars. Pock (2010) uses a panel data set from 14 European countries over the period 1990–2004 to produce separate estimates of price elasticities with respect to the diesel-powered cars in a dynamic gasoline demand equation. His price elasticity estimates are lower than those in previous studies, which might be interpreted as a signal that car owners react to increasing fuel prices by gradually replacing their gasoline-powered cars with diesel-powered ones.

When it comes to income elasticities, Dargay and Gately (1999) projected the growth in the car and total vehicle stock up to the year 2015, for OECD countries and a number of developing economies with widely varying per capita income. Their projections are based on econometric models that explain the growth of the car/population ratio and vehicle ownership (based on annual data for 26 countries over the period 1960–1992) as a function of per capita income. They estimated both short and long run income elasticities. They find that income is by far the most important variable explaining vehicle ownership and that the relationship between vehicle ownership and income in each country tended to be nonlinear and S-shaped; income elasticities may reach a value of about two, for low- and middle-income levels (that is, ownership grows twice as fast as income). However, when income levels increase, vehicle ownership increases only as fast as income, and decreases down to zero as ownership saturation is approached for the highest income levels. Moreover, Dargay (2001) finds that car ownership responds more strongly to rising than to falling income – there is 'stickiness' in the downward direction. These results have clearly negative implications when it comes to the scale and scope of the future problems associated with road transportation, as vehicle stock in developing countries increases and so fuel consumption.

9.2.1 Demand Elasticities and CO_2 Emissions

The use of specific data or methodological approaches can certainly create crucial differences in the magnitude of elasticity estimates. However, evidence from most surveys suggests that long run price elasticities are high enough to play a significant role in moderating CO_2 emissions. For example, Sterner (2007) provides some estimates of the environmental effects of gasoline taxes by calculating the hypothetical effect on OECD carbon emissions from transport that would have occurred if all OECD countries had applied (for a long period) the tax policies pursued by the European countries with the highest tax levels (for example Netherlands, Italy and the UK). For a price elasticity of –0.8 he concludes that the whole OECD emissions of carbon from transport would have been 44% lower[4]; in contrast, they would have been 30% higher if all OECD countries had as low taxes as the US. Thus, in terms of global carbon emissions, the effect of gasoline taxes is sizeable. In the former case, the difference in gasoline consumption would have been around 270 million tons of fuel per year. If these savings were added over a decade, they would lead to roughly 8.5 billion tons of CO_2, which implies that the atmospheric carbon content would have been 1 ppm higher than it is today if gasoline taxes had not been used the way they have in Europe.[5]

---

[4] The numbers are slightly smaller when assuming smaller price elasticities. For an elasticity of demand equal to –0.7 the reduction is equal to 40% while it is equal to 36% if the price elasticity is equal to –0.6.

[5] In an earlier and somehow oppositional study, Sipes and Mendelsohn (2001) examine whether charging higher taxes would result in significant emission reductions in Southern California and

Davis and Kilian (2010) have recently argued that the sensitivity of gasoline consumption to changes in price is not appropriate for evaluating the effectiveness of gasoline taxes in reducing CO_2 emissions because of two reasons. First, most of these studies do not address the endogeneity of the price of gasoline. Since increases in the demand for gasoline cause the price of gasoline to increase, there is a spurious correlation between the price and the regression error that biases the estimates of the price elasticity toward zero. Second, since price changes induced by tax changes are more persistent than other price changes, gasoline taxes may induce larger behavioral changes. In addition, gasoline tax increases are often accompanied by media coverage that may have an effect on its own. The authors explore a variety of alternative econometric methods designed to account for the endogeneity of gasoline prices and to exploit the historical variation in US federal and state gasoline taxes. Their results indicate that short run gasoline consumption is more sensitive to gasoline taxes than suggested by previous studies; by using instrumental variables and restricting attention to dates of nominal state tax increases, they find a statistically significant short run price elasticity of −0.46. Nevertheless, their estimates also imply that a gasoline tax increase of the magnitude currently contemplated by policy makers would have only a modest short run impact on carbon emissions. For example, a 10 cent per gallon increase in gasoline taxes would decrease US carbon emissions from the transportation sector by about 1.5% and decrease total US carbon emissions by about 0.5%, which is roughly equal to one-half of the typical annual increase in U.S. carbon emissions.[6] This estimate captures only the short run response. The long run response is likely to be considerably larger as drivers substitute toward more fuel-efficient vehicles. They point out that one could clearly induce larger emission decreases with larger gasoline tax increases. Indeed, some of the policies currently being proposed would increase the gasoline tax by as much as 1 US dollar per gallon or more. However, such an increase would be far larger than any gasoline tax increase in US history; therefore, there is reason to doubt the accuracy of predictions generated from linear econometric models, and it is not possible to estimate such nonlinear effects from historical data.

However, even if the impact of fuel taxes on emissions is not that sizeable, it could be still argued that they are more cost-effective than alternative command-and-control policies in place since taxes both reduce kilometers driven and influence vehicle choice, while standards would operate only through the latter channel. For instance, in the US, regulation targeted toward the reduction of gasoline consumption was introduced following the 1973 oil crisis, in the form of the corporate

---

Connecticut. They use both experimental survey data and actual behaviour to explore whether people would change their driving behaviour in response to higher gasoline prices. They find that imposing environmental surcharges on gasoline will result in only a small reduction in driving and thus only a small improvement on the environment.

[6] The percentage change in total carbon dioxide emissions in the U.S. is calculated by multiplying the gasoline consumption effect by 0.338, the fraction of carbon dioxide emissions in the United States derived from the transportation sector.

average fuel economy (CAFE) standards.[7] These standards impose a limit on the average fuel economy of the vehicles sold by a particular company in each year, with separate limits for passenger cars and light duty trucks. A number of studies have considered fuel economy standards in the context of comparing alternative policy instruments. According to Kleit (2004) and Austin and Dinan (2005), the welfare cost of CAFE standards were much higher than if a corresponding fuel tax had been used, while some studies even doubt that the standards had any aggregate fuel-saving effect at all.

Another factor affecting the effectiveness of fuel efficiency standards is the 'rebound effect' (Espey 1997). As the fuel efficiency of automobiles improves, it become less costly to drive a mile which provides an incentive to increase the miles driven. That is, total fuel consumption changes less than proportionally to changes in the fuel efficiency standard. This effect is usually quantified as the extent of the deviation from proportionally. In the case of CAFE, estimates indicate that it is approximately 22.2% for the long run (Small and Van Dender, 2007).

Moreover, (Parry et al. 2007) divide automobile related externalities into those arising from gasoline use and those from miles driven, showing that gasoline taxes reduce a greater number of important externalities than do CAFE standards (for example, CAFE does not directly address automobile accidents or congestion). A further discussion on the appropriateness of using command-and-control policies as a supplement to fuel taxation has been provided in Chapter 6 of this book.

9.3 The Political Economy of Fuel Taxes and Distributional Concerns

If fuel taxes are such a good instrument to improve the environment and to reduce automobile-related gasoline consumption and oil dependency, why are they not used more universally? Sterner (2007) points out several explanations for the prevalence of low gasoline taxes in many countries. Firstly, it is the political lobbying. Although in theory policies are designed to maximize welfare, in practice they are shaped by economic interests and the higher the dependence on motoring among the (electorate) population the more difficult it is politically to raise fuel taxes. Indeed, several studies have reported that tax levels are sensitive to a variety of political and economic conditions. Hammar et al. (2004), for instance, investigate the determinants of gasoline tax rates across a panel of Western European countries, the US and New Zealand, finding that while low taxes encourage greater gasoline consumption, high levels of consumption lead to substantial pressure against tax rate increases. This seems to be particularly the case in the US, where the oil industry, along with construction and automobile manufacture, form the core of the 'highway lobby' supporting policies that favor motor vehicle transportation (Parry and Small

---

[7] Substantial increases in the stringency of U.S. fuel economy standards are planned for the period 2010 to 2020, corresponding to the growing salience of concerns associated with gasoline use.

2005). Hammar et al. (2004) also find that other governmental variables influence tax rates, such as the level of government debt. In the same line, Decker and Wohar (2006) analyze the determinants of US state excise taxes levied on diesel fuel. They find that the freight trucking industry's contribution to total state employment is a highly significant determinant of a state's diesel tax rate, consistently suggesting that the greater this contribution the lower the tax rate.

Secondly, it is often argued that since poor families spend more on transportation than higher income families (as a proportion of their income) and since they drive vehicles that pollute more, fuel taxes impose a greater economic burden on the poorer than on higher-income families. This burden, however, may be mitigated to some extent by lower vehicle ownership rates and higher price responsiveness among poor households[8] (Poterba 1991, Santos and Catchesides 2005, Walls and Hanson 1999, West 2004a, b, 2005, West and Williams 2004). There is also the argument that if fuel prices rise, then all other prices in the economy will rise as a result of the increased cost of transportation, and this might be particularly detrimental to poor households.

The evidence for the regressivity of gasoline taxes comes primarily from cross-sectional surveys, which show that low-income families spend a larger fraction of their annual income on gasoline than high income families (Chernick and Reschovsky 1997). However, more recent research has shown that regressivity cannot be taken for granted, and that the choice of methodology has proven to be of great importance for the distributional outcome. A general conclusion arising from this literature (e.g. Caspersen and Metcalf 1994, Davies et al. 1991, Lyon and Schwab 1995, Rogers 1995) is that consumption-based taxes are less regressive when incidence calculations are based on lifetime income as opposed to annual income; the argument is that if most people with low income are only temporarily poor, and if gasoline consumption decisions tend to be made on the basis of lifetime income, calculating tax burdens based on data from a single year will yield substantially higher burdens than those calculated on the basis of lifetime or permanent income.[9]

As suggested by Sterner (2010), the distributional impacts of fuel taxes might be also affected by different social conditions that differ among countries. For example, in the US, a car is often a necessity even for low-income earners due to the lack of public transportation. On the contrary, in developing countries, cars and gasoline are luxury goods. Thus, there might be a tendency to progressivity in low income countries, but regressitivy in high income countries. Therefore, it is important to study this issue in countries with different characteristics.

Sterner (2010) analyzes the distributional effects of taxing transport fuel in seven European countries – France, Germany, Italy, Serbia, Spain, Sweden and the UK.

---

[8] Higher price responsiveness increases, however, the consumer surplus loss among poor households.

[9] Although the use of lifetime incidence is highly appealing, there are a number of practical issues with this approach – see Chernick and Reschovsky (1997) for a discussion.

He considers both the direct effect of gasoline taxes throughout fuel purchases and the indirect effect through the indirect use of fuel in public transport. His results are quite mixed when the tax burden is measured as the share of total annual disposable income. In many countries the tax appears close to proportionality, though there is some overall regressivity in Sweden and the UK, while in Germany middle-income earners seem to bear the largest burden. The author also finds that there are clear differences in the level of burden across countries, which might be partly explained by the level of taxation. For example, in France, households seem to bear a relatively lower burden than in the UK. Taking the national averages across all deciles, France has an average burden of 0.78% while in the UK this is approximately 2.20%. Instead, when the tax burden is measured as the share of total expenditure, results turn out to be more progressive.

Distributional impacts are also very dependent on how the additional revenues from the tax increase are recycled. Bento et al. (2009) analyze the distributional effects of a permanent increase in gasoline taxes in the US under alternative recycling methods: – 'flat' recycling, where revenues are returned in equal amounts to every household, 'income-based' recycling, where revenues are allocated to households based on each household's share of aggregate income, and 'vehicle miles travelled – based' recycling, where revenues are allocated based on each household's share of vehicle miles travelled (VMT) in the baseline. They find that under flat recycling, lower-income groups receive a share of the tax revenues that is considerably larger than their share of gasoline tax; thus, flat recycling might more than fully offset the potential regressivity of fuel taxes. Instead, the pattern of impacts is U-shaped under income-based recycling: since rich households have the lowest ratio of miles travelled to income, they are the only group that experiences welfare gains under this scheme. Finally, although the pattern of impacts across income distribution is fairly flat under VMT-based recycling, higher-income households benefit the least from this scheme since they drive cars that are more fuel-efficient; therefore, the ratio of gasoline taxes to VMT travelled is larger for richer countries.

9.4 Double Dividend and Fuel Taxes

It has been suggested that the substitution of externality-correcting taxes in place of distortionary taxation may lead to a 'double dividend' in the sense that it will lead to both environmental benefits and lower welfare costs of raising public revenues. This 'double dividend' argument promotes the use of environmental taxes since, unlike the other principal instruments of environmental policy (direct regulation and freely allocated tradable permits), they provide revenues that can be used to reduce other taxes and the distortions those taxes cause. Nevertheless, several studies have stressed the inefficiencies caused by adding (or raising) environmental taxes to an already distorting tax system, concluding that the distortions associated with environmental taxes are quite significant (Bovenberg and de Mooji 1994, Bovenberg and Goulder 1996, Goulder et al. 1997, Goulder et al. 1999). The double dividend argument shown to be wrong in the following sense: If the utility function of a representative consumer depends on the consumption of a polluting good and leisure – and there is a distortionary tax on labour – by driving up the price of the polluting good relative to leisure, environmental taxes would lead consumers to reduce labour supply, producing a negative welfare impact called the tax-interaction effect. Thus, on the one hand, environmental taxes correct the externality; if the revenues are recycled through cuts in the pre-existing labour tax, they also reduce the distortion created by the pre-existing tax and increase welfare. However, on the other hand, the tax interaction effect exacerbates the distortion imposed by the labour tax.[10] The final effect will depend on the magnitude of the cross-price elasticity between the polluting good and leisure, and will be positive only if the polluting good is a weak substitute for leisure.

In a recent study, West and Williams (2007) estimate the cross-price elasticity between gasoline and leisure, along with the optimal second-best gasoline tax. They find that gasoline is a leisure complement, and that the second-best gasoline tax exceeds the marginal external damage associated with gasoline consumption by about 35%. Thus, these results suggest that the efficiency gains from increasing the gasoline tax would be even larger than what a first-best analysis would indicate. Of course, the practical relevance of these results may be limited by political constraints. On the other hand, as discussed in the following section, gasoline taxes are only an imperfect instrument for dealing with some of the externalities caused by road transportation.

9.5 Limits to Fuel Taxes

As discussed previously, setting appropriate taxes toward road transportation is not an easy task due to the large number of externalities associated with transport. The first-best policy requires that drivers are charged the full marginal cost of marginal vehicle use; but it is difficult to restructure existing fuel taxes so as they reproduce exactly the first-best solution for a series of reasons. Firstly, the success of fuel taxation depends on the link between fuel consumption and the external effect. In the case of global warming, fuel consumption and the potential of vehicle usage are closely related variables. However, other externalities – such as local air pollution and congestion – vary dramatically with respect to the exact time, the location and the density of exposed population. Emissions of local air pollutants also depend on factors such as fuel choice, driving style and weather conditions. For instance, a significant share of the emissions of volatile organic compounds and carbon monoxide

---

[10] The previous argument disregards the effect of improved environmental quality on individuals' labour supply decisions. If pollution affects labour productivity and health, regulation produces an additional benefit-side tax-interaction effect, whose magnitude would greatly depend on the effects of the pollutant in question. Williams III (2002) shows that the benefit-side interaction effect can be of the same magnitude as the cost-side interaction effect.

is typically emitted during the first kilometer driven, since catalytic converters do not work properly when cold.

If a Pigovian tax were available, it would induce households to internalize all the negative effects, by driving less (and less aggressively), buying fuel-efficient cars, using cleaner fuels, installing abatement technologies and avoiding cold start-ups (Fullerton and West 2002). Nevertheless, in spite of technological advances in this area (as on-board diagnostic equipment and remote sensing), technologies to measure the emissions of each vehicle in a cost-effective and reliable way are still not available (Fullerton and West 2002). Fuel taxes can, therefore, only provide an approximate reflection of the environmental marginal social costs of transport decisions, unless they can be differentiated according to fuel characteristics, mileage, and vehicle and driver characteristics (which should be observable) at the pump station. Such a tax would be essentially equivalent to an emission tax; like an emission tax it is not feasible since it requires much information and it would be very difficult to monitor since drivers could easily roll back their odometers to affect the payment (Fullerton and West 2002).

There has been a good deal of attention defining first- and second-best tax policies. In the European context, Jansen and Denis (1999) examine tax and other policies for reducing both CO_2 emissions and conventional pollutants. They find that the best policy mix to reduce CO_2 emissions consist of fuel taxes – based on the carbon content of fuel – that are combined with differentiated purchase taxes to encourage the switching to increased fuel efficiency vehicles. In spite of its relatively high fuel efficiency, the share of diesel cars decreases in their simulations because the taxes imply a high proportional burden on diesel. This is in contrast to the current situation, where diesel is subject to relatively low taxes. When it comes to reducing local air pollutants, the authors find that the best mix includes an emissions-based kilometer tax combined with a new vehicle purchase tax based on the emissions equipment of the vehicle. Finally, joint optimization of several external effects gives a greater weight to instruments that target driving behavior, such as improved public transportation, road pricing and traffic management. Even if these instruments are not the best options when considering emissions in isolation, in a joint optimization they are likely to play an important role.

Parry and Small (2005) develop an analytical framework to assess the second-best optimal level of gasoline taxes, considering local and global pollution, congestion externalities and the interactions with the fiscal system. They illustrate their framework by calculating the second-best optimal taxes for the US and the UK. They show that the optimal gasoline tax in the US is more than twice the current rate. Quite the opposite, the optimal gasoline tax in the UK is slightly less than half the current rate. The congestion externality seems to be the largest component of the optimal fuel tax, followed by the Ramsey component – which reflects the balance between excise taxes and labour taxes financing the government's budget. The next most important components are accidents and local air pollution. Global warming only plays a minor role and seems to be the only component for which the fuel tax is approximately the right instrument.

9.6 Conclusions

It seems unlikely that the global demand for transport would decrease in the future. Estimates indicate that transport will grow by 45% by 2030 (IEA 2009). Clearly, fuel taxes can play an important role in reducing fuel consumption and CO_2 emissions, especially in developing countries where the fleet of automobiles in growing rapidly; in practice they have already had an important effect constraining the emissions of CO_2 from road transportation in many countries. However, there are some limits. The evidence available in the literature indicates that as fuel prices rise, fuel consumption falls by a less than proportional amount and that there is a 'stickiness' in the downward direction, which implies that further reductions will require much higher increases in fuel prices. On the other hand, if we consider that other externalities caused by road transportation – like congestion and local air pollution – are associated with higher marginal external costs than climate change, it is clear that policies to deal directly with these issues also deserve attention.

The political economy and distributional concerns also impose limits to use of fuel taxes. The argument that fuel prices rises might be particularly detrimental for poor households creates strong resistance against fuel taxes.

Acknowledgment The author thanks Thomas Sterner and Theodoros Zachariadis for helpful comments.

References

Austin D, Dinan T (2005) Clearing the air: the costs and consequences of higher CAFE standards and increased gasoline taxes. J Environ Manage 50(3):562–582

Bento A, Goulder LH, Jacobsen MR, von Haefen RH (2009) Distributional and efficiency impacts of increased U.S. gasoline taxes. Am Econ Rev 99(3):667–699

Bovenberg AL, de Mooji RA (1994) Environmental levies and distortionary taxation. Am Econ Rev 84(4):1085–1089

Bovenberg AL, Goulder LH (1996) Optimal environmental taxation in the presence of other taxes: general-equilibrium analyses. Am Econ Rev 86(4):985–1000

Brons M, Nijkamp P, Pels E, Rietveld P (2008) A meta-analysis of the price elasticity of gasoline demand. a SUR approach. Energy Econ 30(5):2105–2122

Caspersen E, Metcalf G (1994) Is a value added tax regressive? Annual versus lifetime incidence measures. National Tax J 47(4):731–746

CE Delft (2008) Handbook on estimation of external costs in the transport sector, Commissioned by European Commission DG TREN. http://ec.europa.eu/transport/sustainable/doc/2008_costs_handbook.pdf

Chernick H, Reschovsky A (1997) Who pays the gasoline tax? National Tax J 50(2):233–259

Crawford I, Smith S (1995) Fiscal instruments for air pollution abatement in road transport. J Transport Econ Policy 29(1):33–51

Dahl C, Sterner T (1991) A survey of econometric gasoline demand elasticities. Int J Energy Syst 11(2):53–76

Dargay J (2001) The effect of income on car ownership: evidence of asymmetry. Transportation Res Part A: Policy Practice 35(9):807–821

Dargay J, Gately D (1997) Vehicle ownership to 2015: implications for energy use and emissions. Energy Policy 25(14–15):1121–1127

Dargay J, Gately D (1999) Income's effect on car and vehicle ownership, worldwide: 1960–2015. Transportation Res Part A: Policy Practice 33(2):101–138

Davis L, Kilian L (2010) Estimating the effect of a gasoline tax on carbon emissions. Forthcoming in J Appl Econometrics. http://onlinelibrary.wiley.com/doi/10.1002/jae.1156/abstract

Davies J, St.-Hilare F, Whalley J (1991) Some calculations of lifetime tax incidence. In: Atkinson AB (ed) Modern public finance, vol 1. International library of critical writings in economics, no. 15. Aldershott, UK and Brookfield, pp. 238–254

Decker C, Wohar M (2006) Determinants of state diesel fuel excise tax rates: the political economy of fuel taxation in the United States. Annu Regional Sci 41:171–188

Espey M (1997) Pollution control and energy conservation: complements or antagonists? A study of gasoline taxes and automobile fuel economy standards. Energy J 17:49–60

Espey M (1998) Gasoline demand revisited: an international meta-analysis of elasticities. Energy Econ 20:273–295

European Commission (2010) Taxation and customs union: excise duties on alcohol, tobacco and energy, Part II, July 2010. http://ec.europa.eu/taxation_customs/taxation/excise_duties/energy_products/rates/index_en.htm

European Environmental Agency (EEA) (2010) Fuel prices (TERM 021). http://www.eea.europa.eu/data-and-maps/indicators/fuel-prices-and-taxes/fuel-prices-and-taxes-assessment-1. Acceded Sept 2010

Frondel M, Schmidt CM, Vance C (2010) A regression on climate change policy: the European Commission's legislation to reduce CO_2 emissions from automobiles. Forthcoming in Transportation Research Part A. http://www.sciencedirect.com/science/article/pii/S096585640900127X

Fullerton D, West SE (2002) Can taxes on cars and on gasoline mimic an unavailable tax on emissions? J Environ Econ Manage 43:135–157

Gately D (1992) Imperfect price – reversibility of oil demand: asymmetric responses of gasoline consumption to price increases and declines. Energy J 13(4):163–182

Goulder LH, Parry IWH, Burtraw D (1997) Revenue-raising vs. other approaches to environmental protection: the critical significance of pre-existing tax distortions. Rand J Econ 28:708–731

Goulder LH, Parry IWH, Williams RC III, Burtraw D (1999) The cost-effectiveness of alternative instruments for environmental protection in a second-best setting. J Public Econ 72:329–360

Graham D, Glaister S (2002) The demand for automobile fuel: a survey of elasticities. J Transport Econ Policy 36:1–26

Graham D, Glaister S (2004) Road traffic demand: a review. Transport Rev 24:261–274

Hammar H, Löfgren Å, Sterner T (2004) Political economy obstacles to fuel taxation. Energy J 25:1–17

Hughes JE, Knittel CR, Sperling D (2008) Evidence of a shift in the short-run price elasticity of gasoline demand. Energy J 29(1):93–114

International Energy Agency (IEA) (2009) CO_2 emissions from fuel combustion: highlights. http://www.iea.org/co2highlights/CO2highlights.pdf. Acceded Sept 2010

Jansen H, Denis C (1999) A welfare cost assessment of various policy measures to reduce pollutant emissions from passenger road vehicles. Transportation Res D 4:379–396

Johansson O, Schipper L (1997) Measuring the long-run fuel demand for cars. J Transport Econ Policy 31(3):277–292

Kleit AN (2004) Impacts of long-range increases in the corporate average fuel economy (CAFE) standard. Economic Inquiry 42:279–294

Li S, von Haefen R, Timmins C (2008) How do gasoline prices affect fleet fuel economy. NBER Working Paper No. 14450

Lyon A, Schwab RM (1995) Consumption taxes in a life-cycle framework: are sin taxes regressive? Rev Econ Statistics 77(3):389–406

Mayeres I (2003) Taxes and transport externalities. Public Finance Manage 3(1):94–116

McCubbin DR, Deluchi M (1999) The health costs of motor-vehicle related air pollutant. J Transport Econ Policy 33(3):253–286

Michaelis L (1995) The abatement of air pollution from motor vehicles. J Transport Econ Policy 29(1):71–84

Neidell MJ (2004) Air pollution, health and socio-economic statues: the effects of outdoor air quality on childhood asthma. J Health Econ 23(6):1209–1236

Parry IWH, Small K (2005) Does Britain or the United States have the right gas tax? Am Econ Rev 4:1276–1289

Parry IWH, Walls M, Harrington W (2007) Automobile externalities and policies. J Economic Literature 45(2):373–399

Pock M (2010) Gasoline demand in Europe: new insights. Energy Econ 32(1):54–62

Poterba JM (1991) Is the gasoline tax regressive? In: Bradford D (ed) Tax policy and the economy, vol. 5. MIT Press, Cambridge, pp. 145–165

Rietveld P (2007) Transport and the environment. In: Tietenberg T, Folmer H (eds) The international yearbook of environmental and resource economics 2006/2007. Edward Elgar Publishing

Rietveld P, van Woudenberg S (2005) Why fuel prices differ. Energy Econ 27(1):79–92

Rogers D (1995) Distributional effects of corrective taxation: assessing lifetime incidence from cross sectional data. In: Proceedings of the eighty – sixth annual conference on taxation. National Tax Association, Columbus, pp. 192–202

Ryan L, Ferreira S, Convery F (2009) The impact of fiscal and other measures on new passenger car sales and CO_2 emissions intensity: evidence from Europe. Energy Econ 31(3):365–374

Santos G, Catchesides T (2005) Distributional consequences of gasoline taxation in the United Kingdom. Transportation Res Record 1924:103–111

Schipper L (2011) Automobile use, fuel economy and CO_2 emissions in industrialized countries: encouraging trends through 2008? Transport Policy 18(2):358–372

Schipper L, Fulton L (2008) Disappointed by diesel? The impact to the shift to diesels in Europe through 2006. http://www.metrostudies.berkeley.edu/pubs/reports/004_trb_diesel.pdf

Sipes KN, Mendelsohn R (2001) The effectiveness of gasoline taxation to manage air pollution. Ecol Econ 36(2):299–309

Small K, van Dender K (2007) Fuel efficiency and motor vehicle travel. The declining rebound effect. Energy J 28(1):25–51

Sterner T (2002) Policy instruments for environmental and natural resource management. RFF Press, Washington, DC

Sterner T (2007) Fuel taxes: an important instrument for climate policy. Energy Policy 35:3194–3202

Sterner T (2010) Distributional effects of taxing transport fuel. Forthcoming in Energy Policy. http://www.sciencedirect.com/science/article/pii/S0301421510001758

The American Petroleum Institute (2010) State gasoline tax report. http://www.api.org/statistics/fueltaxes/index.cfm

The Ministry of Natural Resources Canada (2010) Report on retail fuel prices. http://www2.nrcan.gc.ca/eneene/sources/pripri/price_map_e.cfm#allprices

The National Tax Agency Japan (2010) Tax national statistics. http://www.nta.go.jp/foreign_language/statistics/tokei-e/menu/kansetu2/h11/02.htm

Transport Policy Advisory (2009) International fuel prices 2009. http://www.gtz.de/de/dokumente/gtz2009-en-ifp-full-version.pdf

Wachs M (2003) A dozen reasons for raising gasoline taxes. Public Works Manage Policy 7(4):235–242

Wadud Z, Noland RB, Graham DJ (2010) A Semiparametric model of household gasoline demand. Energy Economics 32(1):93–101. http://www.sciencedirect.com/science/article/pii/S0140988309001054

Walls M, Hanson J (1999) Distributional aspects of an environmental tax shift: the case of motor vehicle emissions taxes. National Tax J 52(1):53–65

West SE (2004a) Distributional effects of alternative vehicle pollution control policies. J Public Econ 88:735–757

West SE (2004b) Equity implications of vehicle emissions taxes. J Transport Econ Policy 39(1):1–24
West SE (2005) The cost of reducing gasoline consumption. Am Econ Rev 95(2):294–299, Papers and proceedings of the one hundred seventeenth annual meeting of the American Economic Association
West SE, Williams RC (2004) Estimates from a consumer demand system: implications for the incidence of environmental taxes. J Environ Econ Manage 47(3):535–558
West SE, Williams RC (2007) Optimal taxation and cross-price effects on labor supply: estimates of the optimal gas tax. J Public Econ 91(3–4):593–617
Williams R III (2002) Environmental tax interactions when pollution affects health or productivity. J Environ Econ Manage 44(2):261–270

Chapter 10
Passenger Road Transport During Transition and Post-transition Period: Residential Fuel Consumption and Fuel Taxation in the Czech Republic

Milan Ščasný

Abstract The main trends in the determinants of residential motor fuel consumption during the transition and post-transition period in one former centrally-planned country, the Czech Republic, are examined. We show that passenger car ownership in Czech households has been increasing and that these cars are increasingly equipped with stronger engines. However, the age structure of the fleet has not changed as much. The environmental burden has therefore increased, mainly due to increasing domestic car ownership and the vehicles' technical attributes. Richer and larger households, households with children, and those living in smaller rather than larger cities are the household segments with larger passenger car penetration, and toward which a policy maker might specifically target a policy aimed at changing transportation patterns. Interestingly enough, expenditures on motor fuel showed less change during the analyzed period 1993–2009, and on average remained relatively the same across all income deciles. Fuel taxation, as measured by the Suits and the Jinonice indexes, was quite even too, and this might lead one to conjecture about an even rather than an uneven effect of further fuel taxation across income deciles. Indeed, utilizing a micro-simulation model embodied with price and income responses of fuel demand as estimated for several household segments, we support this conjecture. However, our conclusion about relatively even distribution of the burden from higher fuel taxation across Czech households does not hold, if we analyze the incidence for household segments as defined according to several transportation-relevant household and housing characteristics. Should certain differences due to higher fuel pricing across household segments appear, the overall effect of quite large fuel price increases would be relatively small and the effect on distribution would not be large either. Identifying a household segment that would be affected relatively more even after recycling revenues, we recommend targeting mitigating measures to households of pensioners and families with children and those living in smaller towns or villages.

M. Ščasný (✉)
Environment Center, Charles University Prague, 162 00 Prague 6, Czech Republic
e-mail: milan.scasny@czp.cuni.cz

10.1 Institutional Background

Transportation in all former communist countries in Central Europe was fully centrally planned as were all other segments of their economies. The attention of the social planner was concentrated on the enhancement of the economic production of centrally-planned firms. The consumer choice set was fully determined by supply, which was set by the planner. For example, if a household wished to buy a new car, it was not unusual for the household to have to be placed on a 'waiting list' to get a car, and that car would have to be, of course, produced in one of the countries that belonged to the communistic block. The vehicle choice set was in this way regulated both through the total quantity of vehicles supplied to the 'market' and through the technical and qualitative characteristics the car was equipped with. Queuing for petrol was not unusual, and the price of fuel was not at all cheap. Once in possession of a car and fuel one was allowed to drive freely, but only within the area of the communistic block. Special permission was needed to cross the border, a permit that was rarely provided. On the other hand, public means of transport generally worked well and was cheap, especially in urban areas, with relatively dense public transport networks developed both within and outside urban areas. This system of transportation formed specific behavior patterns among consumers. All of this changed after 1989, when the so-called Velvet Revolution kick started the economic and political transformation of all communist countries in Europe toward democracy and a market system, and this change also affected the transportation system.

The main aim of this chapter is to look at the main incentives for the use of passenger vehicles by households during the transformation and post-transformation period in one of the former communist countries. Specifically, we examine fuel consumption and expenditures of Czech households during the period 1993–2009. Because direct fuel consumption is conditional – with some exceptions – on having a car, we also analyze household ownership of passenger cars. The main interest here is to discuss policy-relevant issues that are relating to the individual road transportation of households, some of which have also been raised in Chapter 9 of this book that discusses the usefulness and public acceptability of fuel taxation. We therefore start with a brief review of the institutional background specific to the Czech Republic.

Since the beginning, three specific instruments have been introduced to regulate individual passenger road transport in the Czech Republic. Excise tax on propellants is the first and most effective one. The rates of excise tax on petrol and diesel have been set at quite high levels since 1993 when the current Czech tax system was established (8.20 Czech crowns (CZK) per liter of petrol, 6.95 CZK per liter of diesel, at an exchange rate in early 2011 of approximately 25 CZK per Euro). During the following years, nominal rates for both these fuels were increased several times and reached levels of 12.84 CZK and 10.95 CZK per liter of petrol and diesel respectively in 2010. A circulation road tax is the second instrument introduced in the area of individual road transportation. Road tax is, however, levied only on business vehicles and thus has not had any direct effect on household behavior.

The third, highway tolls, was introduced in mid-1990s with a fee that has increased several times and is paid according to highway use.[1]

Each of these three instruments was introduced mainly for fiscal reasons, i.e., to collect public revenues either for the government, or for the State Fund of Transport Infrastructure in order to finance infrastructure projects. We argue that none of these had a strong effect on fuel savings. Let's consider our first instrument, fuel tax. As will be shown later in Fig. 10.2, all increases in nominal rates during the entire period 1993–2009 were not sufficiently large to even balance the effect of inflation. As a result of this policy, the real final price of motor fuels declined. European Union Directive 2003/96/EC on the taxation of energy products and electricity did not bring any impetus for increasing the tax rates, and thus, fuel savings, either. In fact, Czech tax rates were actually above the minimal rates as set in the Directive before the period when Czech law had to comply with the requirements of the *acquis communitaire*.[2] To conclude, the excise tax on motor fuels was not a strong motive for fuel savings and/or for increasing the fuel efficiency of the passenger car fleet. Indeed, as we show later, the age structure of the fleet remained almost unchanged for about 20 analyzed years, during which time the share of cars with stronger engines was continuously increasing.

The regulation of fuel consumption is based on an argument for correcting negative externalities (see Verhoef 1994, 2002). In fact, driving a car generates external costs including a health impact associated with airborne pollution, a wide range of impacts due to climate change, or from the depletion of non-renewable resources. Considering pollution, for example, mobile sources are responsible, in the Czech Republic, for 47% of total nitrogen oxide (NOx) emission, 28% of volatile organic compounds (VOCs), 23% of particulate matters and 16% of carbon dioxide (CO_2). Emissions attributable to passenger car use by households contribute about one fourth (NOx, PM) to one third (VOC) of the part attributable to all mobile sources.[3]

Despite the fact that Czech authorities introduced quite effective instruments[4] to control emission levels from stationary emission sources during the 1990s, emissions from mobile sources were continuously increasing. Figure 10.1 describes the trend in emission levels and compares these levels for stationary, area and mobile sources from 1995 to 2008. For instance, while emissions of particulate matter decreased by more than 90% for stationary sources, PM emissions from

---

[1] The toll was later replaced by a mileage charge for large vehicles used for business.

[2] The minimal rates set in EU Directive 2003/96/EC are 359 and 330 Euros per 1,000 l of petrol and diesel respectively. Moreover, recent (2010–2011) Czech rates of tax on petrol and diesel that are 688 Euros, or 513 Euros per toe (assuming 25 CZK per Euro) are already above the minimum rates as proposed in a proposal for Revision of this Directive for the year 2018 (that are set at 460 Euros, or 466 Euros per toe respectively).

[3] Emissions are taken from Czech Register of Emission Sources, REZZO and they describe the status in the year 2005.

[4] Emissions from stationary sources were regulated by the Air Pollution Act introduced in 1991, which required fulfilling certain emission limits on each large emission source until 1999.

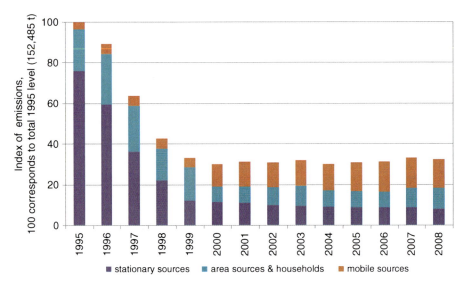

Fig. 10.1 Emission of particulate matter, CZE 1995–2008
Source: REZZO database (CHMI 2010)

mobile sources remained constant (e.g. during 1993–1999), or in some periods even increased slightly (especially during 2003–2006).

Due to negative environmental externalities generated from fuel use, the Czech government has discussed several instruments for the regulation of transport emissions, especially within so-called environmental tax reform.[5] Apart from nominal increases in rates of petrol and diesel tax and the provision of tax rebates on environmentally-friendly fuels, no stricter regulation of fuel use by households was introduced. The negative effect of such regulation on the competitiveness of Czech firms (on the business side) and on distribution and equity (on the household side) were pronounced strongly to be the main policy obstacles to higher fuel and/or vehicle taxation. We therefore later examine whether this is true in the case of households. Specifically, we analyze what might be the distributional effect of a fuel price increase on Czech households.

The structure of the rest of this chapter is as follows. First we examine the use of motor fuel by Czech households (Section 10.2). Then we move on and look in detail at household vehicle ownership (Section 10.3). Next, Section 10.4 focuses on distributional aspects; first we examine household fuel expenditures

---

[5] During 2000–2001, a first proposal on Environmental Tax Reform (ETR) was based on higher energy taxation with rates as proposed in the so-called 'Monti Proposal' (EC 1997). The next proposals being discussed during 2004–2008 relied on higher energy or carbon taxation, using information about unit external costs as quantified by ExternE method. One of the ETR proposals also included a new circulation vehicle tax with a rate based on average carbon intensity (new cars) or engine size (older cars).

across several household segments, then we predict the effect of fuel pricing policy on household expenditures and welfare. The last section summarizes policy-relevant recommendations.

10.2 Residential Fuel Consumption

We start by looking at average household budget shares on propellants. Because fuel use is conditional on having a car, we look first at fuel expenditures of those households that owned a car. On average, fuel expenditures remained quite constant across all years analyzed, with the mean ranging between 13,000 and 15,000 CZK per annum (all at 2005 real prices). Fuel expenditures of households that did not have a car, and most likely used a car belonging to a relative or friend (but still excluding households with zero fuel consumption), are significantly lower with a mean between 2,000 to 4,000 CZK per year (see the right side of Table 10.1 that provides details and also reports statistics for both these groups merged together).

As a result of continuously growing income in years 1993–2009, the fuel budget share of households with a car were, on average, decreasing and ranged between 5.2% and 5.8% of their total expenditures, or between 4.8% and 5.5% of their total net incomes respectively. The fuel budget share of households without a car was much smaller and because of that the budget shares for those two household groups merged together are slightly smaller.

Why did expenditures on fuel (the bolded line in Fig. 10.2 below) remain almost at a constant level even while the net incomes of Czech households were increasing over the period? To answer this question, one would need an accurate estimate of household demand (see later). Figure 10.2 can at least provide first insights into the reason. At first, although the Czech government increased the rate of excise tax on motor fuels several times after 1993, this change was not enough to increase the real price of propellants and as a result, except for a few years (2001 and 2005–2007), the real price of fuel was decreasing (bottom line in Fig. 10.2).[6] Second, the absolute magnitude of price and income elasticity of fuel demand – as estimated by Brůha and Ščasný (2006) for Czech households – is similar (see Section 10.4), which means that any price effect would be counterbalanced by the wealth effect. As a consequence of a continuously increasing income and a slightly decreasing real fuel price, average fuel consumption has remained at a constant level since 1999, and about 1.4 times above the 1993 level. The escalation in the price of oil in 2001 was the only time fuel consumption decreased; once the real price of fuel reached its previous level, the consumption of fuel quickly returned to its previous levels (see double-lined curve of Fig. 10.2).

---

[6] We compute fuel prices as a weighted average of the price of petrol and of diesel with a ratio of 80:20.

Table 10.1 Fuel consumption and expenditures (households with zero fuel expenditures excluded)

	Only households with own car ($N = 29{,}398$)				Households with a car or with fuel expenditures ($N = 34{,}675$)			
	Consumption [l per year]	Fuel expenses [CZK(2005) a year]	Fuel expenses [% of net incomes]	Fuel expenses [% of total expenditures]	Consumption [l per year]	Fuel expenses [CZK(2005) a year]	Fuel expenses [% of net incomes]	Fuel expenses [% of total expenditures]
1993	386	13,583	5.88	6.10	336	11,837	5.20	5.40
1994	428	13,840	5.93	6.10	367	11,882	5.14	5.31
1995	476	13,666	5.51	5.79	403	11,568	4.72	4.97
1996	517	13,647	5.26	5.51	444	11,739	4.57	4.80
1997	540	13,935	5.28	5.45	476	12,280	4.70	4.85
1998	508	12,866	4.95	5.20	447	11,333	4.41	4.64
1999	537	13,209	4.95	5.16	472	11,621	4.44	4.63
2000	588	14,641	5.59	5.89	511	12,719	4.91	5.17
2001	471	14,075	5.25	5.65	417	12,468	4.70	5.05
2002	470	13,110	4.85	5.24	411	11,475	4.30	4.64
2003	538	13,510	4.85	5.21	474	11,921	4.32	4.65
2004	566	13,956	4.87	5.34	505	12,468	4.38	4.82
2005	559	14,750	5.23	5.81	497	13,116	4.70	5.23
2006	544	15,029	5.12	5.68	480	13,267	4.58	5.07
2007	527	14,795	4.82	5.47	456	12,783	4.21	4.77
2008	567	14,756	4.73	5.42	490	12,759	4.16	4.76
2009	525	14,044	4.55	5.15	450	12,042	3.95	4.46

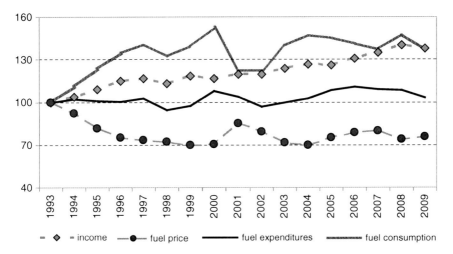

Fig. 10.2 Fuel consumption and its drivers (1993 levels=100)

10.3 Household Ownership of Passenger Vehicles

Despite a quite stable average level of fuel consumption and expenditures of Czech households, especially since 2003, the environmental burden associated with personal car use has been increasing over time. This was mainly driven by two factors: the technical characteristics of the fleet and changes in the stock of passenger cars.

We examine at first the former. In the early 1990s the average car had a small engine and ran on petrol. However, both these characteristics of vehicles have changed significantly in the past 15 years. Until 1998, diesel cars presented only a small fraction of the stock – up to 8% of the fleet. Since then, diesel cars have been replacing petrol fuelled cars and, according to vehicle register (CDV 2009), their share reached about 25% in 2009. This change was caused by two factors: the relatively lower price of diesel compared to the price of petrol,[7] and the increased wealth that allowed the purchase of relatively expensive diesel cars.

The composition of the passenger car fleet based on engine size changed significantly too, and moved in particular toward cars with stronger and thus more fuel-intensive engines (see the right side of Fig. 10.3). Specifically, while the share of cars with the smallest engines (up to 1,200 cubic centimeters or cm^3) and of cars with engines above 1,600 cm^3 was 37% and 20% of the fleet in 2000, these shares reversed and became 21% and 34% respectively by 2009. In other words, about 15% of the fleet switched from the smallest cars, up to 1,200 cm^3, into the category of

---

[7] Petrol that was used by households cost about 20% more than diesel until 1999, and then about 13% more until 2003. Since then diesel was cheaper only by less than 1%, in some periods it was even slightly more expensive than petrol due to obligations to fulfil a minimum quota on bio-diesel share.

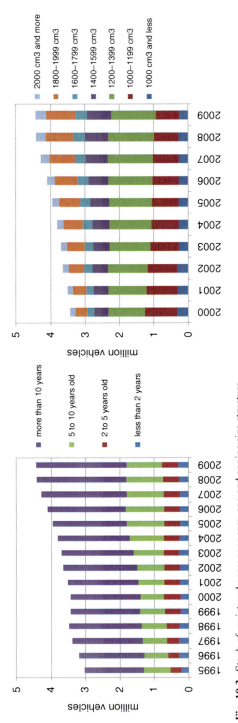

Fig. 10.3 Stock of registered passenger cars, age and engine size structure
Source: Transport Yearbook, various years; CDV (2009)

stronger cars, with more than 1,600 cm³, during 2000–2009 (see the right side of Fig. 10.3). A continuation of this trend for the replacement of energy-saving cars by cars with stronger engines might present a strong driver for boosting further fuel use and increasing the burden on the environment from passenger car transportation.

Let's now look at the stock. Opening the market in the Czech Republic in 1990 also brought new opportunities for households to purchase new cars and more cars. While in 1993 about 2.8 million passenger cars were registered (270 per 1000 inhabitants), the stock of passenger vehicles became 20% larger in 1998, increased by another 10% in 2003 and now is almost 60% higher than the 1993 stock level (see Fig. 10.3).[8] We highlight that the Register records all passenger cars owned by households as business cars.

Because households are of our special concern in this Chapter, we utilize microdata from two surveys regularly conducted by the Czech Statistical Office, the HBS[9] and the EU-SILC,[10] to examine this trend in the household sector. Indeed, both of these datasets confirm an increasing trend in car ownership. The increasing percent of car-owning households in the EU-SILC, from 60 to 66% during 2005–2009, clearly shows an increasing penetration of passenger vehicles within Czech households. This trend is also supported by HBS data: the share of households having at least one car increased from about 53% in 1993–1996 to 62% since 2003; the number of households with a second car was increasing as well, specifically from about 2% in 1993 to 7% in 2009, while there have been very few who owned three or more

---

[8] The increasing number of households which own or use firm-owned cars over time i8 s also reported in two available individual-level datasets – Household Budget Surveys and EU-SILC (see Table 10.2). In the HBS, and on average, while less than 53% of households owned a car in the early 1990s, there are more than 60% of households with a car after 2001. The share of households with a car is also increasing in EU-SILC data – 59% of households owned a car in 2005 while there are almost 66% such households in 2009. Households that could use a firm-owned car for their family represent only a small portion of the HBS population, about 3% to 4%. Both of these datasets, however, suffer from a lack of consistent information about passenger car ownership over the entire analyzed period. The first, the HBS, reports a number for passenger cars owned by household. However, it uses a quota sampling and some changes in its sampling strategy in 2006 make consistent comparison over time difficult. The second, the EU-SILC, applies random sampling, but its surveys only started in 2005.

[9] The Household Budget Survey (HBS) is regularly collected by the Czech Statistical Office and its database includes information about household annual expenses on several hundred consumption items, income from various sources, possession of durable goods, home characteristics and other socio-economic data of household members. Households included in the survey are selected using the non-probability quota sampling technique and the annual samples have on average 2,700 to 3,000 observations. The variable PKOEF reflects how each household is represented in the entire Czech population allowing us to compute aggregate statistics and make country-wide predictions. Our dataset covers the period 1993–2009 and includes more than 46,596 observations.

[10] The EU-SILC is an EU-wide survey of family statistics on incomes and living conditions. This survey has been conducted annually by the Czech Statistical Office since 2005 (before this, Microcensus surveys were conducted in 1996 and 2002). Households are selected using random sampling and the size of its samples range between 4,000 and 11,000 households each year. Except housing expenditures, the EU-SILC does not include any information about expenditures of households, or more detailed information about durables such as passenger vehicles.

Table 10.2 Passenger car ownership and the age of the fleet

	Own car (EU-SILC)	Own car (HBS)			Business-cars (HBS)		
	Ownership	Ownership	Average age	Median age	Possibility to use	Average age	Median age
1993		53.0%	10.36	9			
1994		51.0%	10.81	10			
1995		51.5%	11.24	10			
1996		53.4%	11.26	10			
1997		55.6%	11.34	11			
1998		56.0%	11.44	11			
1999		58.1%	11.01	10	2.9%	4.87	3
2000		58.6%	11.04	10	3.0%	4.94	3
2001		60.9%	11.36	11	3.4%	5.32	4
2002		60.6%	11.52	11	3.3%	5.83	5
2003		61.6%	11.62	11	3.1%	6.58	5
2004		63.1%	10.95	9	4.4%	7.06	6
2005	59.4%	62.3%	11.37	10	3.9%	7.63	7
2006	61.0%	61.7%	11.62	10	3.9%	7.19	7
2007	62.9%	61.2%	11.37	10	4.2%	7.47	7
2008	64.3%	60.8%	11.07	10	4.5%	6.64	6
2009	65.6%	62.1%	10.62	10	4.2%	6.80	6

Source: Household Budget Surveys and EU-SILC (weighted by PKOEF)

cars. About 3–4% of households could use a business car, and this share has been slowly increasing since 1999 when HBS started to report this data. However, one should be aware of the fact that the HBS is not based on random sampling and also that some changes in sampling strategy do not allow consistent comparisons across later years.[11]

Increases in the vehicle stock did not, however, have a larger effect on the age structure of vehicles; indeed, almost 60% of passenger cars are older than 10 years and less than 20% are younger than 5 years during the entire period of economic transformation and post-transformation period (see Fig. 10.3). An old passenger car fleet is also supported by HBS data (see the left side of Table 10.2). If we approximate the age of the fleet by the year of purchase that the HBS has been reporting since 1993,[12] the average age of cars owned by households was around 11 years during the entire period. The average age of firm-owned cars used by households – also reported in the HBS data – is in fact younger, being about 6–7 years. The younger age of firm-owned cars is a result of the legally set depreciation period for cars, which motivates a faster renewal of business-car stock.

The relatively stable age of the fleet, with a simultaneous increase in its stock over the period, was partly a result of the large number of purchases of passenger vehicles in the second-hand market, a common behavioral pattern in almost all transition countries in Central and Eastern Europe. In fact, while in the HBS data in a given year about 5–6% of Czech households reported the purchase of a car within the past 5 years, only 11–20% of them bought a new car and the rest, i.e., 80–89% of buyers, chose to buy a second-hand car. Purchases of second-hand cars slowed down increases in the stock and resulted naturally in fixing the average age of the fleet.

10.4 Distributional Aspects of Fuel Use

Consumers are as different in their behavior as in their tastes. In this section we examine how car ownership and fuel expenditures vary across some household segments. Specifically, in Section 10.4.1, we describe firstly how many households in a given household segment own a car. Then, we econometrically estimate the probability of owning a car while controlling for the effect of certain household characteristics and the size of the household's place of residence. Then we move to examine expenditures on motor fuels across household segments, paying attention to the progressivity of fuel tax payments. In the next section, we review estimates of household responsiveness to price and income changes, i.e. the key parameters of household demand that are needed for proper welfare measurement of (tax)

---

[11] We use HBS data rather to analyze the main determinants of car ownership among Czech households later.

[12] The approximation of vehicle age by the year of purchase might underestimate the average age of the fleet, particularly because second-hand cars are usually older than the age given by the year of their purchase.

incidence (Section 10.4.2). And using these estimates, we predict the distributional effect of fuel pricing policies (Section 10.4.3). All analysis presented here is based on household-level data from the Household Budget Surveys, as in the previous sections.

10.4.1 Fuel Use and Car Ownership: Ex Post Assessment Across Household Segments

Firstly, we examine car ownership across income deciles, which are defined by net total income per household member. As expected, lower income deciles own relatively fewer cars than upper deciles do (see the first column of Table 10.3). The share of car owners is increasing monotonically, but with fewer car owners in middle income deciles; this can be explained through detailed inspection of the HBS dataset, which reveals that middle income deciles include a higher share of families with retired persons who exhibit lower passenger car penetration In fact, our conjecture is supported by following econometric analysis.

Next, Table 10.4 displays the same descriptive statistics for household segments as defined by the size of residence (small, medium, or large respectively), the social status of the family head (i.e., farmer, retired, or with an economically active person) and whether or not there are children in the family ('+' sign denotes families with children). A simple comparison shows that the number of car owners increases across the period in all household segments. Families of retired persons have fewer cars, as do households with only one economically active person. Households living in larger places of residence are similarly equipped with fewer cars, indicating

Table 10.3 Car ownership and fuel expenditures across household income deciles

	Car ownership, % of households				Zero fuel expenditures, % of households				Fuel expenditures as % of income[a]			
	1993	1998	2003	2008	1993	1998	2003	2008	1993	1998	2003	2008
whole sample	0.53	0.56	0.62	0.61	0.37	0.33	0.29	0.26	5.35	4.45	4.41	4.22
decile 1	0.50	0.56	0.59	0.50	0.41	0.31	0.29	0.36	5.28	4.60	4.60	5.33
decile 2	0.38	0.45	0.57	0.51	0.49	0.42	0.36	0.35	5.08	4.34	5.16	4.38
decile 3	0.42	0.55	0.55	0.60	0.48	0.33	0.37	0.30	5.36	4.21	4.68	4.00
decile 4	0.44	0.48	0.57	0.57	0.46	0.42	0.34	0.31	5.49	4.62	4.37	4.36
decile 5	0.52	0.50	0.56	0.54	0.38	0.43	0.31	0.32	5.37	4.50	4.01	4.16
decile 6	0.54	0.60	0.60	0.65	0.35	0.32	0.30	0.20	5.58	4.63	4.46	3.93
decile 7	0.58	0.53	0.67	0.65	0.32	0.36	0.25	0.21	5.19	4.15	4.61	3.89
decile 8	0.61	0.64	0.69	0.67	0.33	0.27	0.22	0.18	5.31	4.92	4.29	4.55
decile 9	0.67	0.66	0.68	0.68	0.21	0.23	0.22	0.18	5.51	4.56	4.28	4.06
decile 10	0.65	0.62	0.68	0.70	0.28	0.24	0.20	0.17	5.25	3.96	3.79	3.81

[a]Households with zero expenditure on fuel are excluded

Table 10.4 Car ownership and fuel expenditures across other household segments

	1995			2005		
	Car-owner	Zero fuel expenses	Fuel expenses as % of incomes[a]	Car-owner	Zero fuel expenses	Fuel expenses as % of incomes[a]
farmer_small	0.79	0.12	0.06	0.82	0.10	0.06
farmer_large	0.74	0.11	0.05	0.82	0.12	0.06
retired_small	0.34	0.53	0.05	0.48	0.35	0.05
retired_medium	0.28	0.65	0.05	0.44	0.42	0.05
retired_large	0.28	0.63	0.04	0.35	0.56	0.04
EA1_small	0.25	0.60	0.07	0.43	0.40	0.07
EA1+_small	0.62	0.24	0.07	0.82	0.14	0.07
EA2_small	0.69	0.11	0.07	0.84	0.07	0.07
EA2+_small	0.74	0.06	0.05	0.95	0.03	0.07
EA1_large	0.22	0.68	0.07	0.35	0.51	0.05
EA1+_large	0.46	0.36	0.05	0.65	0.28	0.05
EA2_large	0.68	0.14	0.05	0.83	0.11	0.05
EA2+_large	0.72	0.16	0.05	0.84	0.11	0.05

[a]Households with zero expenditure on fuel are excluded
Small (large) describes the size of community in which a household is living, in this case a municipality with less (more) than 20,000 inhabitants. For the pensioners, we delineate three community sizes: fewer than 20,000, more than 50,000 and in between these two numbers (medium-sized municipality)

better (public) transport alternatives in cities. Households with children and families of farmers have more cars than similar childless households.

Using the HBS household-level data for the years 1994–2009 with 43,674 observations in total, we specifically estimate a logit model[13] to analyze the probability of there being at least one passenger car in a household. Our fairly simple model shows (Table 10.5) that household structure matters when a car-purchase decision is taken and the effect we found is similar to results found elsewhere (see, e.g., Dargay 2006 or Johnsotone et al. 2009 for the literature review on the determinants of car ownership and car use). We found that the more adults there are living in a family, the more likely they are to have a car, while being a single male counterbalances the effect of the former control variable and being a single female almost doubles the adult effect. Families of retired persons are indeed less likely to own a car. Having children increases the probability of car ownership, but only for younger children, and especially where the children are younger than 5 years old. Having children older than 10 years has no significant effect on car ownership. This indicates that families may buy a car – we conjecture most likely their first car purchase – either when they

---

[13] Because only a small share of households owned two cars, and very few three and more, we rely on bivariate logit rather than using a count model or multinomial logit model.

Table 10.5 Probability of owning a car, logit model

Variable name	Variable description	Coeff. estimate	Marginal effect	t stat
Intercept		−3.6956		−21.59
adult	continuous: number of adults in the family	1.5132	0.2398	45.36
singleM	dummy: =1 if single male	−1.4318	−0.2127	−13.74
singleF	dummy: =1 if single female	1.2651	0.1880	24.79
retired	dummy: =1 if household of retired	−0.2926	−0.0435	−8.18
child5	count; number of children younger than 5	0.1198	0.0178	3.99
child69	count; number of children of age 6–9	0.0958	0.0142	2.94
child1014	count; number of children of age 10–14	0.0259	0.0038	0.94
child15plus	count; number of children older than 15	0.0083	0.0012	0.33
income	continuous; net annual household income in 1,000 CZK	0.0067	0.0010	36.98
fuelprice(−1)	continuous; lagged real (2005) price of liter propelant	−0.0188	−0.0028	−3.16
city500	dummy; residence with less than 500 inhabitants	1.3701	0.2036	23.4
city2000	dummy; residence with 500–2,000 inhabitants	0.8652	0.1285	21.17
city5000	dummy; residence with 2,000–5,000 inhabitants	0.6974	0.1036	14.46
city10k	dummy; residence with 5,000–10,000 inhabitants	0.5010	0.0744	8.93
city50k	dummy; residence with 10,000–50,000 inhabitants	0.5259	0.0781	15.18
city100k	dummy; residence with 50,000–100,000 inhabitants	0.2760	0.0410	7.05
N. obs.		43,674		
LogLik		−21,183		
McFadden's LRI		0.2625		
Adjusted Estrella		0.3294		
AIC		42,400		

have a child, or while their child is quite young, a change in circumstances that usually leads to changes in consumption patterns and time allocation within a family. The probability of car ownership also monotonically increases the smaller the size of community (with the lowest probability of owning a car in Prague, a capital city, which is a reference value in the specification of our model). Income increases the probability of car ownership just because richer families can afford to buy one. The effect of fuel prices was not significant, however, unless the fuel price is lagged by 1 year, in which case the price effect, though still small, is negative, as one would intuitively expect.

Now we move to fuel expenditures. If we analyzed the whole HBS sample, we would find a similar pattern as in other studies on distributional analysis, i.e., the budget share for fuels increases slightly with income (see e.g. Sterner 2011). Similarly, the fuel budget shares of Czech households are on average the smallest for the first five income deciles, except for households in the lowest decile, which have relatively larger expenses, while the sixth to the ninth deciles have slightly larger expenses. However, if we concentrate on fuel users only, we hardly find any difference in the fuel budget share across income deciles and across years (see the last part in Table 10.3). This observation indicates that any differences in fuel expenditures across deciles are driven by the differences in the share for non-users across deciles. Indeed, the ratio of those with no fuel expenditures declines toward richer households, except for the first income decile, indicating that there are relatively fewer non-users of fuel in the least rich households compared to other lower deciles (see the middle column of Table 10.3). We revealed in our previous analysis (Ščasný 2006) that single females with children are relatively more represented in the lowest deciles; single female is the household segment and having children is one of the characteristics that we found from our econometric analysis is likely to have a higher probability of having a car, and thus having some fuel expenditures.

We support our conclusion about the relatively even distribution of fuel expenditures by measurement of the distribution of fuel tax payments. Specifically, we use the Suits index (Suits 1977) and the Jinonice index (Brůha and Ščasný 2008b, Ščasný 2011) to examine the progressivity of fuel tax payments.

The Suits index compares cumulated percents of total income and cumulated percentages of the total tax burden.[14] The Suits Index basically aims at evaluating whether the payments of certain taxes are distributed equally among households. In its graphical presentation, it examines the percentage of tax payments paid by households holding x percent of incomes. Negative values of the Suits Index indicate regressivity in tax payments, and any reductions in the Suits index indicate increased regressivity of a tax. The diagonal in the graphical representation describes a flat-tax rate, i.e., the case where each dollar is taxed evenly, where the Suits index has a value of zero.

---

[14] We approximate the integral by a trapezoid rule and weight each observation by a specific variable pkoef, which corresponds to how a particular household recorded in the HBS is represented in the entire Czech population; see details in Brůha and Ščasný (2008b), or Ščasný (2011).

Similarly to Brůha and Ščasný (2008a) or Ščasný (2011), statistical inference and the confidence interval of the Suits index, as well as the Jinonice index we describe later, are computed by using a wild bootstrap.

Our *ex post* assessment of progressivity of a fuel tax supports our previous findings: the fuel tax is relatively even. The value of the Suits index, which is around −0.1 for almost all years, indicates only a small regressivity of fuel taxes (Fig. 10.4). Or, to illustrate further, the least rich households, holding one half of the total income, paid about 60% of the cumulative tax payments in 2005.

Brůha and Ščasný (2008a) proposed an alternative index to measure tax progressivity, which they call the Jinonice Index. The Jinonice index combines the main features of the Suits Index and the Gini approach based on the Lorenz curve, with units ordered according to income on the *x*-axis (rather than cumulative incomes on the *x*-axis as in the case of Suits). In contrast to the Suits index, which compares the cumulated percentages of total income and cumulated percentages of the total tax burden, the Jinonice Index compares cumulated percentages of units (households), ranked by their income, and cumulated percent of the total tax burden. As in the Suits index, the value of the Jinonice index is bounded by −1 and +1, but the diagonal and zero value of the Jinonice index indicates a lump-sum tax instead.

The Jinonice index for motor fuel taxes has a value of about +0.1 and indicates a small progressivity of fuel tax. The slight differences in the progressivity lie in the composition and interpretation of the indexes. While the progressivity measurement based on the Suits index relates tax payments to the cumulative of incomes held, the Jinonice index relates tax payments to the cumulative of units (households). Using our illustrative example, the first half of less rich households paid about 42% of

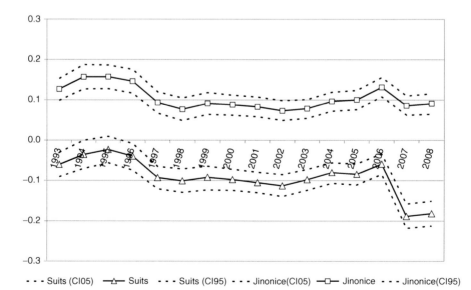

Fig. 10.4 Ex post measurement of progressivity of fuel taxes using the Suits and Jinonice index

fuel taxes. Both of our results for the fuel tax progressivity measurements for the period 1993–2008 suggest that any further increase in fuel taxation might be evenly distributed, having possibly only a small effect on equity.

The distribution of the no fuel-users and expenditures on fuels is even less for household segments that are defined according to those household and demographic characteristics which are more likely to determine car use, such as having a child, living in a small versus larger municipality, or with a retired head of family. Looking at the average numbers in Table 10.4 we find that, for instance, while there are only about 10% of households of farmers with no fuel use in 2005, there are more than half such households in the segment of families with economically active persons living in a large town or city. The number of non users of fuel decreases for all these segments, and the number of households with a car increases.

There is also a larger variance in the fuel budget share across these segments compared to the variance in the budget shares for income deciles. For instance, the budget shares of fuel users are lowest for households with pensioners (between 4% and 5%), quite small for households with economically active persons living in a large town (about 5%), and the largest for similar households living in a small town or village (about 7%). Interestingly enough, the budget share remained the same across these segments and years.

Distributional analysis based on income deciles may not uncover important differences in consumption patterns across different household segments within the decile. We therefore suggest performing an incidence analysis of fuel or car pricing policy at the level of properly defined household segments, defined by transport-relevant household and/or demographic characteristics, and not only on the level of income deciles.

10.4.2 *Responsiveness of Czech Households to Price and Income Changes*

Computing the effect of any fuel pricing policy by multiplying the before-policy consumption levels with the new price vector would be a quite naive approach because one can hardly imagine that under new conditions such as a new price, consumers will consume the same amount of regulated goods as in the before-policy situation. If this is true, any evaluation of the distributional effect of policy should consider the behavioral responses of consumers to price changes and, if relevant, to changes in income.

Such estimates of behavioral responses for Czech households are provided, for instance, by Brůha and Ščasný (2006), who estimate price and income elasticity for motor fuel demand, but also elasticities for several transportation services. Their 2006 study addresses several shortcomings of their previous estimate based on an annual macro-data and time series based estimation technique. Specifically, they rely on a coherent demand system, namely the Almost Ideal Demand System (Deaton and Muelbauer 1980), which consists of motor fuel, use of buses, rail, and urban public means of transport. Using Household Budget Surveys for the years 2000–2004, with a sample size of almost 15,000 observations, they estimate the demand system from micro-data rather than from macro aggregates in order to capture heterogeneity in behavior (see e.g. Halvorsen (2006) for a critique of estimation approaches based on the macro data). Further, their AID system is estimated separately for 13 distinct household groups as defined already above, in order to derive segment-specific parameters of demand. To avoid biased estimates, the problem arising from zero expenditures is treated with the two-stage Heckman style correction (1979), and the authors did not consider the budget share equation for segments for which certain services are not available (e.g., expenditures on urban public means of transport for the four household segments living in small communities where this kind of transportation in reality does not exist).

Income elasticity of motor fuel demand was found to be about +0.7 with the strongest income response for households with economically active person(s) with children. The estimates of own price elasticities have all the expected signs, and the weighted average is about –0.5. The strongest response with respect to fuel price is found for households with one economically active person and families of pensioners living in medium-sized towns. Households of farmers living in larger towns and cities and, surprisingly, the remaining households of pensioners, have the weakest response on own price of fuel. A few estimates of cross price effects have unexpectedly negative signs, but they are quite small in absolute value. Overall, Brůha and Ščasný indeed document large heterogeneity in behavioral responses across analyzed household segments in the Czech Republic (see Table 10.6 for detailed results).

Table 10.6 Estimates of price and income elasticity for motor fuel demand

Household group	Income elasticity	Own price elasticity	Cross price elasticity Bus	Rail	Urban public means
farmer_small	0.70	−0.51	0.00	0.22	n.a.
farmer_large	0.63	−0.06	−0.03	0.06	0.20
retired_small	0.60	−0.44	0.32	0.27	n.a.
retired_medium	0.60	−0.67	−0.04	0.11	0.01
retired_large	0.57	−0.44	0.04	0.11	0.04
EA1_small	0.66	−0.59	0.18	0.38	n.a.
EA1+_small	0.82	−0.55	0.28	−0.07	0.20
EA2_small	0.64	−0.55	0.29	0.01	n.a.
EA2+_small	0.78	−0.52	0.00	−0.01	n.a.
EA1_large	0.66	−0.60	0.28	0.00	0.10
EA1+_large	0.82	−0.62	0.10	0.24	0.11
EA2_large	0.69	−0.51	0.20	−0.25	0.25
EA2+_large	0.74	−0.49	0.38	0.02	0.12
Weighted average	0.71	−0.52	0.21	0.07	0.12

Source: Brůha and Ščasný (2006); also in Ščasný (2011)

10.4.3 Incidence of Fuel Taxation on Households

Price and income elasticities as being estimated for several transport goods were then entered into a micro-simulation model, DASMOD, in order to predict the impact of increases in the price of fuel. The DASMOD model specifically simulates the effect of various price-based policies on several variables, such as expenditures for specific households. In this way, the model predicts the effect on expenditures and consumption of several non-durable consumption items including motor fuel or energy, taking into consideration respective demand elasticities.

To assess the incidence of a policy, DASMOD predicts the effects both on household expenditures and on welfare. While any change in the former indicator may inform a policy maker about the expected environmental effect and fiscal impact, the latter measure provides information about welfare loss or benefit induced by that policy.[15] Incidence analysis undertaken by this model also allows assuming revenue recycling options, such as using additional revenues from fuel taxation for labor taxation cuts, or provision of compensation to mitigate any adverse social effect.

Using predictions about fuel consumption and thus paid fuel tax, about labor taxes in the case of revenue recycling, and about the provision of social benefits, we can simulate the effect on total public revenues and/or dead-weight loss.

Simulation by DASMOD is performed at the lowest possible level, i.e. the model simulates the impact for each household included in the model database. The effect as being predicted for each individual household might be then aggregated for household segments such as income deciles, or for the entire Czech population.

The DASMOD model was built on the micro-level household data taken from the Household Budget Survey and a policy evaluation presented just below particularly utilizes HBS 2005 data. We also plug in the values of segment-specific price and income elasticity as they were estimated by Brůha and Ščasný (2006) for 13 household segments (with the values reported in Table 10.6).[16]

Our reference scenario describes the year 2008; i.e., a final price of 28.4 CZK per liter of fuel and parameters for labor taxation as valid in that year. We predict the effect for the following five policy scenarios (see Table 10.7):

- *Scenario 1* assumes the rate of excise tax on motor fuels and the VAT rate as valid in the year 2010–2011. Both tax changes would increase the final price of fuel by about 5%. We do not further assume any change in labor taxation.
- *Scenario 2* introduces a tax that would increase the final price of fuel to about 35 CZK per liter, which is the actual price Czech households faced at the beginning of the year 2011;.

---

[15] In a recent version of DASMOD, the welfare changes are approximated by the geometric mean between the Paasche and Laspeyres cost-of-living indices, although these indices can only provide an approximation of theoretically proper measures of welfare changes (see more e.g. Hausman 1981, or Vartia 1983).

[16] Detailed description of the model provided, for instance, Ščasný (2011).

Table 10.7 Definition of policy scenarios

	Before policy	Scen 1	Scen 2	Scen 2a	Scen 2b	Scen 2c
Fuel pricing						
pre-tax price [CZK per l]	12.4	12.4	12.4	12.4	12.4	12.4
excise tax [CZK per l]	11.5	12.5	17.0	17.0	17.0	17.0
VAT [%]	19%	20%	19%	19%	19%	19%
final price [CZK per l]	28.4	29.8	35.0	35.0	35.0	35.0
fuel price change		5.1%	23.2%	23.2%	23.2%	23.2%
Labor taxation						
PIT	15%	15%	15%	14.22%	15%	15%
SSC	12.5%	12.5%	12.5%	12.5%	11.57%	12.5%
tax credit [CZK a year]	24,840	24,840	24,840	24,840	24,840	26,187

- Scenarios 2a, 2b and 2c are just the same as Scenario 2 with respect to fuel taxation, however, we recycle all additional revenues from the fuel tax via lowering personal income tax rate (*Scenario 2a*), lowering the rate of obligatory paid social security contributions, SSC (*Scenario 2b*), or via increasing tax credit of personal income tax (PIT) (*Scenario 2c*). Each of these three scenarios keeps a revenue-neutrality principle, i.e. the net effect on public revenues at the macro level is zero.

Table 10.8 reports the effect of one of our policy scenarios – *Scenario 2a* – on certain variables for average households in a given household segment (as described in its first column) in Czech crowns per year. For instance, on average, if the price of fuel was 35 CZK per liter of fuel (i.e. it would be increased by 23% compared to its 2005 level), a household in the first income decile would spend 890 CZK more annually on motor fuels [*Column 3*]. *Columns 4* and *5* report increases in fuel tax payments, and changes in the consumption of fuel respectively. Assuming the cross price effect on use of public transport, we also predict the effect on expenses on public transportation, as aggregated across all three kinds, in *Column 2*. If additional revenues from fuel taxation (*Column 8*) are recycled via lowering labor taxation, the effect on labor tax payments (including payments for obligatory social security contributions) are reported in *Column 6*. The negative numbers there indicate that that a household is paying less tax on labor. The effect on household welfare is reported in *Column 7*, when we consider the effect due to price changes and income from revenue recycling. A recent version of the model does not include benefits from improvements in environmental quality or from avoiding other externalities. The final two columns report the net effect on public revenues and dead-weight loss attributable to a given household. Dead-weight loss is computed as a difference between the welfare effect and the net change in public revenues. The effect on each of those variables as expressed in Czech crowns per year are at first reported for each income decile, then for 13 household segments as defined earlier. At the bottom of Table 10.8 we report the aggregated effect as summed over all households

Table 10.8 Effect of *Scenario 2a* on several household segments and public finances

In CZK per year and household	Expenses on public transport	Fuel expenses	Fuel tax	Fuel use (l)	Paid labor taxes	Welfare	Additional revenues from fuel taxation	Net change in public revenues	DWL
decile 1	37	890	1,344	−36.7	−1,271	−604	1,198	168	435
decile 2	30	857	1,272	−33.6	−1,248	−506	1,132	121	385
decile 3	21	706	1,057	−28.4	−989	−475	941	140	335
decile 4	14	734	1,078	−27.8	−848	−625	958	271	353
decile 5	24	736	1,075	−27.3	−968	−494	954	170	324
decile 6	19	816	1,183	−29.7	−1,109	−495	1,050	152	343
decile 7	50	991	1,453	−37.4	−1,631	−355	1,289	−32	388
decile 8	64	1,053	1,551	−40.3	−1,958	−167	1,375	−211	378
decile 9	78	1,223	1,798	−46.5	−2,283	−181	1,594	−255	436
decile 10	102	1,359	2,005	−52.3	−2,838	87	1,776	−523	435
farmer_small	95	1,229	1,781	−44.7	−1,844	−570	1,575	81	489
farmer_large	34	2,873	3,035	−13.1	−1,882	−1,127	2,572	1,047	80
retired_small	4	607	837	−18.6	−30	−1,063	740	716	347
retired_medium	22	313	600	−23.2	−29	−928	548	524	404
retired_large	−20	354	488	−10.8	−29	−607	434	410	197
EA1_small	53	624	1,000	−30.4	−1,046	−405	894	47	358
EA1+_small	8	1,178	1,793	−49.7	−1,323	−1,192	1,604	532	660
EA2_small	−81	1,480	2,249	−62.1	−2,501	−651	2,021	−4	655
EA2+_small	24	1,908	2,785	−70.9	−2,589	−1,205	2,478	381	825
EA1_large	49	936	1,394	−37.1	−1,610	−317	1,239	−65	382
EA1+_large	22	726	1,218	−39.7	−1,512	−299	1,099	−125	424
EA2_large	98	1,367	1,974	−49.0	−2,698	32	1,743	−442	411
EA2+_large	92	1,586	2,239	−52.7	−2,930	−43	1,974	−399	442
Entire population bln. CZK	0.18	3.93	5.80	−151.2	−6.36	−1.60	5.15	0.00	1.60

Note: Aggregated effect on fuel consumption is reported in million liters and properly weighted to provide the impact on the entire Czech population; this aggregated annual effect is expressed in billion CZK.

Let's look first at the consumption of fuel by households at its aggregate; *Scenario 1*, if the fuel price increased by 5%, this would reduce aggregated consumption by 2.6%. Scenario 2, which introduces a 23% increase in the fuel price, would then reduce its consumption by about 11.2%. Revenue-recycling slightly decreases the reductions in fuel use as delivered without revenue recycling with a net reduction of 10.8%. The effect on fuel consumption is similar across all three considered revenue recycling options. On average, *Scenario 2* would reduce the use of fuels by 30 to 50 liters per annum across income deciles. If we analyze the effect on fuel consumption across our 13 household segments, the reductions in household fuel use vary widely and are about 10 to 70 l and year.

The burden of higher fuel pricing – as measured by changes in fuel expenses – increases along income deciles toward the highest one (from 700 to 1,400 CZK a year), however, because the ratio of employed persons across income deciles varies, a negative effect on welfare – including recycling the revenues – increases toward the lowest decile. The welfare effect of a proposed price increase is negative in all deciles, except one – the highest income decile – which enjoys net positive welfare due to fuel pricing thanks to benefits from a reduction in their labor tax. The net effect on both tax payments is positive in the four highest income deciles, while in the rest of the income deciles they pay more taxes even after labor taxation cuts. If we looked at our 13 segments, policy would not bring a negative net effect on welfare only to households with more economically active persons without children and living in larger cities (the effect on households with the same characteristics but with children would be quite small). Households of pensioners, households of farmers living in large communities and families with one or more economically active persons with children and living in smaller municipalities ('EA1+_small' and 'EA2+_small') would lose the most among all analyzed segments.

Predicted effects on expenditures, welfare or fuel use vary across households widely if we analyze incidence for household segments as defined by social status and the size of place of residence. For instance, fuel use is now, on average, reduced by between 10 and 70 l, expenditures are increased between 300 and 3000 CZK, and the welfare impact ranges between +30 and –1,200 CZK.

Overall, Table 10.9 shows that a quite large increase in fuel price, as considered in all scenarios 2, would have a relatively small effect on both fuel expenditures and welfare. Recycling revenue from additional fuel taxation via labor taxation cuts would mitigate negative direct effects on households, especially in those segments where more persons are employed in the labor market. However, after revenue recycling, the negative effect of higher fuel taxation on welfare would remain the same for households of (economically non-active) pensioners who can't simply benefit from labor taxation cuts. Still, on average, the welfare impact as a financial effect on pensioners would be quite small, in absolute terms up to 0.4–0.7% of their total net incomes. Families with children and living in smaller municipalities present the

Table 10.9 Effect of fuel taxation on household expenditures and welfare

	Change in fuel expenditures, % of total net incomes					Change in welfare, % of total net incomes				
	Scen 1	Scen 2	Scen 2a	Scen 2b	Scen 2c	Scen 1	Scen 2	Scen 2a	Scen 2b	Scen 2c
decile 1	0.10%	0.39%	0.41%	0.41%	0.42%	−0.21%	−0.87%	−0.28%	−0.31%	−0.09%
decile 2	0.09%	0.38%	0.40%	0.39%	0.40%	−0.20%	−0.81%	−0.23%	−0.25%	−0.15%
decile 3	0.08%	0.33%	0.35%	0.35%	0.35%	−0.18%	−0.72%	−0.24%	−0.22%	−0.20%
decile 4	0.09%	0.35%	0.36%	0.36%	0.36%	−0.18%	−0.72%	−0.31%	−0.30%	−0.30%
decile 5	0.08%	0.33%	0.34%	0.34%	0.34%	−0.16%	−0.67%	−0.23%	−0.22%	−0.20%
decile 6	0.09%	0.35%	0.37%	0.37%	0.37%	−0.18%	−0.72%	−0.22%	−0.23%	−0.20%
decile 7	0.09%	0.35%	0.37%	0.37%	0.37%	−0.18%	−0.75%	−0.13%	−0.15%	−0.10%
decile 8	0.08%	0.35%	0.37%	0.36%	0.37%	−0.18%	−0.74%	−0.06%	−0.06%	−0.05%
decile 9	0.09%	0.37%	0.39%	0.39%	0.39%	−0.19%	−0.79%	−0.06%	−0.05%	−0.10%
decile 10	0.08%	0.32%	0.34%	0.34%	0.33%	−0.17%	−0.68%	0.02%	0.03%	−0.17%
farmer_small	0.11%	0.44%	0.46%	0.46%	0.46%	−0.22%	−0.90%	−0.21%	−0.15%	−0.10%
farmer_large	0.23%	0.97%	0.99%	0.99%	1.00%	−0.25%	−1.04%	−0.39%	−0.32%	−0.31%
retired_small	0.10%	0.41%	0.41%	0.41%	0.41%	−0.18%	−0.73%	−0.71%	−0.71%	−0.73%
retired_medium	0.05%	0.22%	0.22%	0.22%	0.22%	−0.16%	−0.66%	−0.64%	−0.64%	−0.66%
retired_large	0.06%	0.25%	0.25%	0.25%	0.25%	−0.11%	−0.45%	−0.43%	−0.43%	−0.45%
EA1_small	0.09%	0.37%	0.39%	0.38%	0.40%	−0.22%	−0.91%	−0.25%	−0.30%	−0.07%
EA1+_small	0.11%	0.45%	0.48%	0.48%	0.48%	−0.25%	−1.02%	−0.48%	−0.50%	−0.47%
EA2_small	0.11%	0.44%	0.46%	0.46%	0.46%	−0.24%	−0.98%	−0.20%	−0.16%	−0.14%
EA2+_small	0.12%	0.50%	0.53%	0.53%	0.54%	−0.26%	−1.05%	−0.34%	−0.33%	−0.26%
EA1_large	0.13%	0.52%	0.55%	0.54%	0.55%	−0.27%	−1.13%	−0.19%	−0.20%	−0.18%
EA1+_large	0.06%	0.26%	0.28%	0.27%	0.27%	−0.17%	−0.69%	−0.11%	−0.13%	−0.18%
EA2_large	0.09%	0.37%	0.40%	0.40%	0.40%	−0.19%	−0.77%	0.01%	0.05%	0.01%
EA2+_large	0.09%	0.38%	0.40%	0.40%	0.40%	−0.18%	−0.75%	−0.01%	−0.04%	−0.02%
Aggregated for entire population	0.09%	0.35%	0.37%	0.37%	0.37%	−0.18%	−0.74%	−0.15%	−0.15%	−0.15%

next household segment which would lose the most from fuel pricing and at which mitigating measures might be specifically targeted.

10.5 Policy-Relevant Conclusions

Since 1989, the Czech society and economy, in common with other formerly centrally-planned economies in Central and Eastern Europe, has moved toward a market economy. Economic and political transitions brought many changes both in the market and in the behavior of consumers. This was also the case for the Czech Republic. Since 1993, when the Czech Republic was established, we identified several trends in passenger car ownership and motor fuel consumption for Czech households. In the case of car ownership, we found that Czech households own more passenger cars, and the average car is equipped with a stronger engine. However, the age structure of the fleet did not change as much, and that is mainly due to the fact that most of the passenger vehicles were purchased in the second-hand market. As a result of an increasing number of vehicles with larger engines, airborne pollution from mobile sources in the Czech Republic has been increasing.

So far, the only instrument with some effect on individual road transport has been an excise tax on propellants, but as we have documented, the real rate of this tax has not been rising while the real wealth of households has been continuously increasing. Although the marginal effect of an increase in fuel price is three times stronger than of income, both of them are relatively weak in their effects on car ownership. If a policy intends to regulate emission levels, we therefore recommend the introduction of a measure that can motivate a switch toward more environmentally friendly cars. The specific household segments most likely to own a car in the Czech Republic are richer and larger households, households with children and those living in smaller rather than larger cities. Policy makers might, therefore, target specific measures toward these household segments in order to change their consumption patterns and transportation-related preferences.

The average response of Czech households to income change is about +0.7 and to fuel prices –0.5, while the cross price effect is quite weak, especially for railways. This indicates that the effect of any increase in fuel prices would diminish if wealth increased more than the increase in the price of fuel. In fact, this was the case during almost the entire post-transition period in the Czech Republic, when household wealth was increasing relatively more than the price of fuel. We also find that households with more economically active members, households of pensioners living in medium-sized towns, and households of farmers, are less responsive to price changes, and policy makers would need to pay special attention to the regulation of their fuel consumption. Households with one economically active member and with children are most responsive to changes in income, indicating their potential for the largest increases in fuel demand when they get richer.

Interestingly enough, expenditures on motor fuel did not change so much during the entire analyzed period of 1993–2009. On average, fuel expenditures of fuel users, expressed in absolute terms as well as in terms of budget shares, remained relatively the same across all income deciles and years and amounted to about 14,000 Czech crowns (2005 prices), 5–6% of total household expenditures or 4.5–5% of their total net incomes respectively. Fuel tax is relatively even across all households; this is supported by the Suits index, whose value ranges between −0.02 and −0.08 until 2006.

The burden of further increases in pricing of fuel might therefore be distributed quite evenly across income deciles. Indeed, utilizing the micro-simulation model being embodied with price and income responses of fuel demand, we predict a relatively even distribution of the burden across Czech households. However, this does not hold if we analyze the incidence for several household segments that are defined according to transportation-relevant household and housing characteristics, instead of solely relying on the income-based definition of household segments. One recommendation drawn from our analysis is therefore to undertake the distributional analysis of fuel or passenger vehicle pricing-based policy not only at the level of income deciles, which may not show important differences in consumption patterns across different household segments, but rather at the level of household segments defined according to properly identified transportation-specific household and housing characteristics.

Should certain differences due to fuel pricing appear in the Czech Republic, the overall effect of fuel pricing policy as proposed here would be relatively small, and should the effect on distribution vary across household segments, these burdens would be relatively small both in absolute terms and as a percentage of total household income. Relatively small effects of fuel pricing on distribution is in line with the results of the incidence analysis of fuel taxation as performed for several developing, and a few developed, countries in a recently published book by Sterner (2011). Since our simple modelling doesn't account for the effect of fuel pricing on car ownership, due to the negative effect of a lagged fuel price on car ownership, the predictions provided here might be considered conservative.

A household segment in the Czech Republic that would be affected relatively more would be a household of pensioners and a family with children and living in a smaller municipality. In brief, households with non-active persons such as households of pensioners, households with a person on maternity or paternity leave, or unemployed households would be affected most because of an inability to benefit from the recycling of additional revenues from fuel taxation to labor taxation cuts since all these groups are absent from the labor market. Each of these household segments might be targeted by specific mitigating measures if a policy maker wishes to minimize the adverse social effects. However, such social compensation would result in a weakening of the expected benefits on the side of fuel consumption and of pollution and environmental quality. How environmental improvements from a reduction in individual transportation would be distributed remains to be analyzed in further research.

Acknowledgment This research was supported by the Ministry of Education, Youth and Sports of the Czech Republic, Grant No. 2D06029 *'Distributional and social effects of structural policies'* funded within the National Research Programme II. The support is gratefully acknowledged. Responsibility for any errors remains with the author.

References

Brůha J, Ščasný M (2006) Distributional effects of environmental regulation in the Czech Republic. Paper presented at the 3rd annual congress of association of environmental and resource economics – AERE, Kyoto, Japan, 4–7 July 2006

Brůha J, Ščasný M (2008a) Distributional effects of environmentally-related taxes: empirical applications for the Czech Republic. Paper presented at the 16th annual conference of the European association of environmental and resource economists, Gothenburg, 25–28 June 2008

Brůha J, Ščasný M (2008b) Tax progressivity measurement: empirical applications for the Czech Republic. In: Ščasný M, Braun Kohlová M et al. (eds) Modelling of consumer behaviour and wealth distribution. Matfyzpress, Prague. ISBN: 978-80-7378-039-5, pp. 157–179

CDV (2009) Study on transport trends from environmental viewpoints in the Czech Republic 2008. Transport Research Centre, Brno, August 2009

CHMI (2010) Registr zdrojů znečisťujících látek. Database of Czech Hydrometeorological Institute [online]; www.chmi.cz. Accessed 12 Dec 2010

Dargay J (2006) Household behaviour and environmental policy: review of empirical studies on personal transport choice. Paper presented at workshop, household behaviour and environmental policy: empirical evidence and policy issues' organized by OECD Environment Directorate, 15–16 June 2006, Paris

Deaton A, Muelbauer J (1980) An almost ideal demand system. Am Econ Rev 70:312–326

EUROPEAN COMMISSION (1997), COM (1997) 30 final – Proposal for a Council Directive restructuring the Community framework for the taxation on energy products (97/C 139/07). Brusel

Halvorsen B (2006) Review of the paper 'empirics of residential energy demand' by Bengt Kristrőm. Discussion paper prepared for the OECD workshop on household behaviour and environmental policy: empirical evidence and policy issues. OECD, 15–16 June 2006

Hausman JA (1981) Exact consumer's surplus and dead-weight loss. Am Econ Rev 71:662–676

Johnsotone N, Serret Y, Dargay J (2009), Household Behaviour and Environmental Policy: Personal transport: Car Ownership and Use. Paper prepared for the OECD Conference on 'Household Behaviour and Environmental Policy', 3–4 June 2009, OECD Headquarters, Paris

Ščasný M (2006) Distributional aspects of environmental regulation: theory and empirical evidence in the Czech Republic. PhD thesis. Institute of Economic Studies, Faculty of Social Sciences, Charles University, Prague

Ščasný M (2011) Who pays taxes on fuels and public transport services in the Czech Republic? *Ex post and ex ante* measurement, Chapter 17. In: Sterner T (ed) Fuel taxes and the poor: the distributional effects of gasoline taxation and their implications for climate change. Resource for the Future, Washington, DC, pp. 269–298

Sterner T (eds) (2011) Fuel taxes and the poor: the distributional effects of gasoline taxation and their implications for climate change. Resource For the Future, Washington DC,320 pages, ISBN 9781617260926

Suits D (1977) Measurements of tax progressivity. Am Econ Rev 67:747–752

Vartia YO (1983) Efficient methods of measuring welfare change and compensating income in terms of ordinary demand functions. Econometrica 51:79–98

Verhoef ET (1994) External effects and social costs of road transport. Transportation Research, 28A:273–287

Verhoef ET (2002) Externalities. In: van der Bergh JCJM (ed) Handbook of *Environmental and Resource Economics*, Edward Elgar, Cheltenham, pp. 197–214

Chapter 11
Accelerated Introduction of 'Clean' Cars in Sweden

Muriel Beser Hugosson and Staffan Algers

Abstract The increased focus in Sweden on greenhouse gas emissions, oil dependency and energy efficiency has lead to the implementation of different policy measures in the transport sector. In Sweden there has been a long tradition of buying large, powerful and heavy cars with high fuel consumption and CO_2 emissions. The Swedish car fleet is the heaviest car fleet in all Europe. We describe and discuss effects of major measures that have been implemented to accelerate the introduction of clean cars in the Swedish car fleet. We also briefly describe a decision support tool to evaluate policies affecting the composition of the car fleet. We find that the result of the implemented measures is a high share of clean cars in new car sales and that these policies have lead to a dominance of low emission diesel cars and E85 cars in this share. We also find that the share of biogas cars is still very small and that the use of E85 fuel for E85 cars is quite price sensitive.

11.1 Introduction

The increased focus in Sweden on greenhouse gas emissions, oil dependency and energy efficiency has lead to the implementation of different policy measures in the transport sector. In Sweden there has been a long tradition of buying large powerful and heavy cars with high fuel consumption and CO_2 emissions. The Swedish car fleet is the heaviest car fleet in all Europe. Due to a former high tax on diesel, the Swedish car fleet has just recently started to contain diesel cars. Many of the policy measures implemented in Sweden have focused on transforming the Swedish vehicle fleet to become more CO_2 and energy efficient by accelerating the introduction of Clean Cars.

New car sales statistics also show that the shares of different fuel types have changed significantly toward higher use of alternative fuels since these measures were implemented. But these measures also have costs, and it is important to find

---

M.B. Hugosson (✉)
Department of Transport Sciences, Centre for Transport Studies, Royal Institute of Technology, 100 44 Stockholm, Sweden
e-mail: muriel@kth.se

the most efficient mix of different measures. To establish an efficient policy, effects of separate and combined measures need to be known, as well as effects of other important factors, such as vehicle supply changes.

To be able to understand the mechanism driving the observed changes, a quantitative model for the composition of the vehicle fleet has been developed. This model has been used to help identifying efficient measures, and also to describe the consequences of a combined strategy used in the national transport planning process. In doing so, it needs to be recognized that the composition of the Swedish car fleet affects CO_2 emissions not only by technical properties like fuel type and fuel consumption, but also by the cost of driving which in turn affects car use. If the car fleet becomes more fuel efficient, it may also be cheaper to use, which will increase car use. Such rebound effects have also been considered in the modelling process.

In this chapter we describe different policy measures used in Sweden to accelerate the introduction of clean cars as well as its effects on new car sales and fuel shares. A brief presentation of the Swedish car fleet model will be made and some recent applications and results will be demonstrated.

11.2 Major Policy Measures Implemented in Sweden

In Sweden several measures have been used to transform the car fleet to be more energy and CO_2 efficient. The measures have been applied at both national and local levels. It is important to distinguish between policies to prepare the market for changes and policies making the market change. Examples of policies used to prepare the market are green procurement, laws and regulations. Examples of policies to force the market to change are clean vehicle purchase subsidies, circulation taxes etc. In this section we give a brief description of some of the policy measures, both to prepare the market and to make the market change, that have been used in Sweden.

11.2.1 Swedish Definition of Clean Cars

Before discussing different measures, we first need to make the reader more familiar with definitions used in Sweden related to the concept of 'clean cars' which makes vehicles eligible for different policies.

The definition of an environmental friendly car has changed over time. In the mid-1990s the definition of environmental friendly cars was that such a car should be able to run on renewable fuels or be a hybrid. In 2005 the Swedish Government focused more on energy consumption and adopted a new definition of clean vehicles. In the new definition it is fuel consumption, the fuel type and emission levels that define the clean vehicles. The definition of clean vehicles is (source Swedish Transport Administration, www.trafikverket.se):

- Petrol vehicles, diesel vehicles and electric hybrids that emit less than 120 g of CO_2 per km. Petrol and diesel cars have to meet the Euro 4 standard, the European emission standards of emissions of NO_X, HC, CO and particulate matter (PM10) that have been in place since 2005 in the EU. Diesel vehicles also need to have a particle filter or other effective cleaning device that emits a maximum of 5 mg of particles per kilometer.
- Vehicles that can use ethanol E85. E85 is a fuel containing a mixture of 85% ethanol and 15% unleaded petrol. A clean vehicle which can use E85 may consume a maximum of 9.2 l petrol per 100 km and meet the Euro 4 standard.
- Vehicles that consume biogas or natural gas (CNG). Biogas is methane, produced by anaerobic digestion of organic waste matter. The maximum gas consumption for these vehicles is 9.7 m^3 gas per 100 km and they have to meet the Euro 4 standard.

In this chapter when we refer to the concept 'clean vehicles' or 'clean cars', the above definition holds. When we refer to the concept 'alternatively fuelled cars' then it is the old definition that holds, where only alternatively fuelled cars and hybrids are included, not petrol- or diesel fuelled low-CO_2 emission cars.

11.2.2 Measures to Prepare the Market

At the national level several policy measures have been used to prepare the market for the use of clean cars. The aim with these measures is to avoid problems for the future clean car users and to stimulate the supply side to promote clean car models and distribution of alternative fuels.

11.2.2.1 Increased Supply of Refuelling Stations with Alternative Fuel, National Level

Since 2006, all larger refuelling stations have been obliged to provide at least one renewable fuel. According to a follow up study from the Swedish government, (Trafikutskottet 2009), the number of fuel stations supplying biofuel has increased fourfold from December 2005 to September 2009. About half of all fuel stations supplied biofuel in 2009, as compared to 10% in 2005. Ninety percent of the 1610 fuel stations that supply biofuel have chosen to supply ethanol E85. Only 90 refuelling stations supply biogas or natural gas.

11.2.2.2 Green Procurement, National and Local Level

Green procurement can be very effective both in the sense of introducing new car models on the market but also to force the public sector to buy clean cars. As an example can be mentioned that seven cities in the EU project ZEUS – Zero and Low Emission Vehicles in Urban Society – joined together in a common public procurement effort. The project ran from 1996 to 2000. The cities pooled a large order and were able to negotiate competitive prices. A European wide tender invitation to vehicle manufacturers was published in the Official Journal in February 1997. The cities together bought some 200 electric vehicles and an additional 150 went to buyers outside the project at the same competitive price (BEST 2009).

Another example is the procurement process of buying flexifuel cars in 1999. Flexifuel cars can run on ethanol (E85) and petrol or any mix of those two. Since there was only a single flexifuel car (Ford Taurus), available at that time, in Sweden, the city of Stockholm worked to improve the supply of vehicle models running on ethanol and petrol. The project was supported by the Swedish Delegation for Sustainable Technology. Before the procurement process there were no refuelling facilities because there were no cars and no cars because there were no refuelling facilities. This activity to buy a certain number of a flexifuel car managed to undo this catch 22, and by 2003 8000 Ford Focus had been delivered to Sweden. Even Ford points out that without the technology procurement, the Focus Flexifuel version would not have reached Sweden.

Today the Stockholm City Council requires that Stockholm's administrations and its self owned companies shall, except in special cases, always procure clean vehicles and be fully compensated for the additional costs of acquiring clean vehicles.

11.2.2.3 Tax Exemption for Alternative Fuels, National Level

The alternative fuels, ethanol, biogas and FAME (biodiesel), are exempt from fuel tax.

11.2.3 Measures to Make the Market Change

Measures to make the market change can address the demand side as well as the supply side. The measures can be at different levels – European, national, regional or even local. As the supply side is global, policies aimed at influencing the supply side need to be at an international level, like the EU regulation on setting emission performance standards for new passenger cars (REGULATION (EC) No 443/2009).

As Sweden is a fairly small market for most car manufacturers, mostly demand oriented measures have been employed in Sweden, at the national as well as the local levels. To understand these policies, it needs to be recognized that the demand side consists not only of private buyers, but also of companies providing cars to their employees for private use. About half of the new car sales concern private buyers (Naturvårdsverket 2007).

11.2.3.1 CO_2 Based Circulation Tax, National Level

The tax consists of two parts – one part is a base tax per vehicle (36 Euros per year) and the second part is based on the CO_2 emission. The CO_2 component is 1.5 Euros per gram CO_2 emission (over 100 g/km) for conventional cars and 1 Euro per gram for alternatively fuelled cars. For diesel cars a supplementary environmental tax is added by multiplying the tax for a petrol driven car by 3.15. The reason for this factor is that the diesel fuel tax is lower for diesel than for petrol, and that the environmental requirements are less demanding for diesel cars. From July 2009, clean cars are exempt from this tax the first 5 years.

11.2.3.2 Subsidy for Purchase of New Clean Car (Privately Bought), National Level

A clean vehicle subsidy was introduced in 2007. A subsidy of 1000 Euros was available from the government if a new clean car was purchased by households (not companies). The subsidy ended in July 2009.

11.2.3.3 Company Car Benefit Tax Reduction, National Level

Employees who are being supplied a car by the employer for private travel are taxed for this benefit according to the purchase price of the vehicle. The benefit value has one fixed component and one component based on the purchase price. For a car with a purchase price of 10,000 Euros the benefit value is 2400 Euros, if the purchase price is 40,000 Euros then the benefit value is about 6700 Euros. For clean cars these benefit values are lowered by 20% for ethanol fuelled cars and by 40% for hybrids and gas cars. The cost to the employee is then the marginal tax on the net benefit value. There is an upper limit of the benefit reduction of 1600 Euros. Depending on the marginal taxation level, the maximum reduction is worth between 700 and 1100 Euros per year net after tax.

11.2.3.4 Congestion Charge Exemption for Alternatively Fuelled Cars, Local Level

Alternatively fuelled cars are exempt from congestion charges in Stockholm. The charge is differentiated during the day and varies between 1 and 2 Euros. The maximum fee is 6 Euros per day. The charge affects vehicles entering as well as leaving Stockholm city. The exemption may be worth up to 900 Euros per year for regular car commuters.

11.2.3.5 Free City Residential Parking for Clean Vehicles, Local Level

Some municipalities have introduced reduced or even free parking for residential clean vehicles. The residential parking fee in central Stockholm is about 70 Euros per month or 5 Euros per day. In Stockholm, this policy ended in 2009.

11.2.4 Clean Vehicle Supply Changes as Well

It is not only policies that change over time – also the supply of clean vehicles changes. In the European countries voluntary commitment on CO_2 emission reduction from passenger cars with the car industry have been reached. This may have had

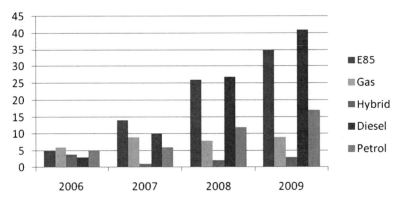

Fig. 11.1 Clean car model supply 2006–2009
Source: Bil Sweden, www.bilsweden.se

some impact on the supply of cars on the European market. On the Swedish market supply changes are not only related to car manufacturers launching new developments, but also to car dealers importing already existing smaller cars with low CO_2 emissions as a consequence of economic incentives for buyers of clean vehicles. Supply changes in Sweden may therefore have been more rapid that supply changes seen in a global context.

In 1994 there were only three clean car models available in Sweden. One of the models was an ethanol car and the other two were electric cars. Only in 1996 biogas vehicles were introduced, and the first hybrid car was introduced in 2000. In 2005, several diesel and petrol cars with low CO_2 emission were introduced on the Swedish market.

After 2005, the number of clean car models available on the Swedish market has increased rapidly, as shown in Fig. 11.1 (models having a sale over 25 vehicles per year).

As can be seen from the figure, E85 models dominated in 2007 and continued to increase in 2008 and 2009. The supply of gas and hybrid vehicles has however been rather constant (the number of hybrid vehicles actually decreased in 2007 as a consequence of sharpening the definition of clean hybrid vehicles). Low CO_2 fossil fuel cars (in particular diesel cars) are now dominating the supply of clean vehicles.

11.3 Changes of the Swedish Car Fleet and the Car Market

In this section statistics are used to illustrate the actual development of sales of clean vehicles and the fuel consumption in the Swedish vehicle fleet. In Fig. 11.2 the development of fuel type share among new cars is shown. The main trends are that diesel cars and E85 capable cars have increased their shares significantly from 2004 to 2009, at the expense of conventional petrol cars.

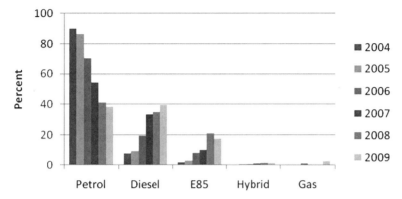

Fig. 11.2 Fuel type market shares for new cars bought in Sweden from 2004 to 2009
Source: Bil Sweden, www.bilsweden.se

Table 11.1 Fuel type market shares (%) for new clean vehicles

Fuel type	Jan–Sept 2009	Jan–Sept 2010
Petrol	16	16
Diesel	21	43
E85	53	31
Hybrids	3	3
Gas	7	7

Source: Bil Sweden, www.bilsweden.se

As can be seen from the figure, the share of E85 vehicles decreases somewhat in 2009. This tendency has been continuing in 2010. In Table 11.1 the distribution of clean vehicles with regard to fuel type is shown for the period January to September for the years 2009 and 2010.

The table reveals a dramatic change between market shares for diesel and E85 cars. The market share for clean cars remains however at 39% of total car sales in both periods. There was no major change in supply, so the change must have other reasons. There are at least three reasons that may be relevant. The first reason is that the 1000 Euros subsidy for privately bought vehicles has been replaced by a vehicle circulation tax exemption for clean vehicles the first 5 years. As will be explained later, this will affect diesel cars much more than other cars. The second reason is the relationship between the price of E85 and petrol/diesel. There is little evidence regarding preference changes, but we know the development of fuel prices. This development is shown in Fig. 11.3.

From the figure it can be seen that the price development has favored E85 up to the beginning of 2009 (with a short exception in the beginning of 2007). This condition has favored E85 cars during that period. If we look at the top five selling models in the year 2008, when fossil fuels (especially diesel), suffered from high prices relative to ethanol, and 2010 (Jan–Sept) when this disadvantage had disappeared, we note that the fuel type mix has changed accordingly (Table 11.2).

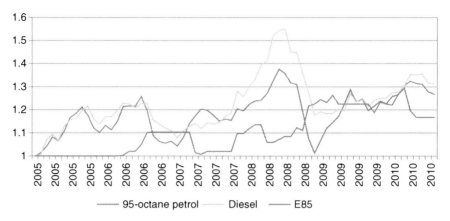

Fig. 11.3 Fuel price development 2005–Sept 2010
Source: Statoil, www.statoil.se

Table 11.2 Top five selling models of clean vehicles in 2008 and 2010

2008		Jan–Sept 2010	
Clean car model	Type	Clean car model	Type
Volvo V70 flexifuel	Ethanol	KIA cee'd Eco	Diesel
Saab 9-3 Biopower	Ethanol	Volvo V50 D	Diesel
Volvo V50 flexi fuel	Ethanol	VW Passat Ecofuel	Biogas (CNG)
Saab 9-5 Biopower	Ethanol	Volvo V70 flexifuel	Ethanol
Ford Focus flexifuel	Ethanol	VW Golf Mutlifuel	Ethanol

Source: Bil Sweden, www.bilsweden.se

As can be seen the compact low-CO_2 fossil fuel cars have become very popular. It is interesting to note that Kia cee'd was introduced on the Swedish market already in 2007, but reached its top position only in 2010.

A third reason would be preference changes, maybe at least partially caused by the debate concerning the real climate impact of ethanol and the food versus fuel debate. This debate concerns the risk of higher food prices caused by increased cultivation demand of crops for biofuel production. Another aspect of preference change may be that the introduction of low emission diesel cars has made the diesel alternative more accepted in Sweden (traditionally diesel cars have been heavily taxed for environmental reasons).

11.3.1 Local Market Changes

A local measure that has been introduced is the congestion charges in Stockholm. The objectives of the charges were to decrease traffic in the city centre as well as decrease environmental impacts from traffic. When the congestion charges were

11 Accelerated Introduction of 'Clean' Cars in Sweden

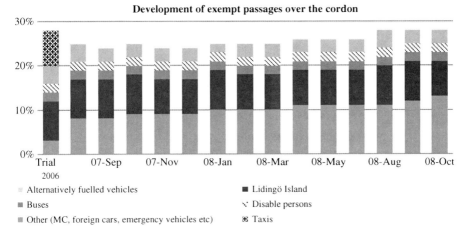

Fig. 11.4 The development of exempt passages over the congestion charges cordon in Stockholm (City of Stockholm Traffic Administration 2009)

initiated, alternatively fuelled cars were exempt from the charges (note the definition of alternatively fuelled car explained in Section 11.2.1). Before the charges passages into or out from the Stockholm city centre made by alternatively fuelled cars were about 3%. In December 2008 the share of passages made by alternatively fuelled cars had increased to 14% (City of Stockholm Traffic Administration 2009). The first year with charges, taxis were also exempt from congestion charges, but after 2007 taxis were no longer exempt and a large share of the alternatively fuelled cars passing the cordon is taxis. In Fig. 11.4 the development of the share of exempt passages over the cordon is shown.

As can be shown in the figure the passages of alternatively fuelled car increased rapidly. The exemption from congestion charges affected the sales of clean cars in Stockholm a great deal (City of Stockholm Environment and Health Administration 2009). Still the charges main objective was to avoid congestion in the city centre and the exemption of alternatively fuelled cars was abolished from January 2009 (but kept to 2012 for cars registered before January 2009). Recent statistics show that the share of alternatively fuelled car exemptions is now decreasing. As clean cars are not anymore exempted from congestion charges, this is not necessarily an indicator of a reduction in the share of clean cars.

Another measure used in Stockholm is free residential parking for alternatively fuelled cars. This has also affected the sales of clean cars in Stockholm, but not to the extent as the congestion charges (City of Stockholm Environment and Health Administration 2009).

In 2008 sales of clean cars grew at a record pace, in comparison to other European countries. One third of all cars sold in Stockholm and a quarter of all cars sold in Sweden were clean cars. Figure 11.5 shows the development of clean car sales from 2001 to 2009.

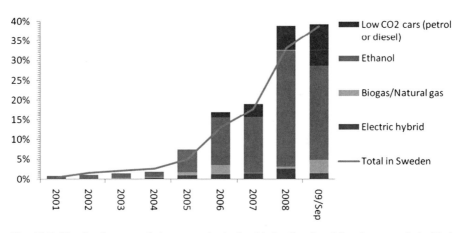

Fig. 11.5 The development of clean car sales in Stockholm (*bars*) and Sweden as a whole (*line*) (City of Stockholm Environment and Health Administration 2009)

In the figure it can be observed when different policies were introduced. Note that Stockholm is included in the line for total in Sweden. In 2006 congestion charges were introduced in Stockholm. This leads to a large increase in sold alternatively fuelled cars. In 2006 it was not yet decided if the charges would be permanent. In 2007 charges were reintroduced, at the same time as the purchase subsidy of 10,000 SEK. This affects both the sales of alternatively fuelled cars (exempt from congestion charges) and the low-CO_2 emitting cars (higher reduction of purchase price and lower consumption of fuels).

11.4 Evaluation of Implemented Measures

A number of different policy measure used to accelerate the introduction of clean cars have been introduced on the Swedish market. We have also observed large changes in the Swedish car market. But what measures have been most efficient? And to what extent have other factors contributed, such as fuel price changes and supply changes? To really understand this, we need to understand and quantitatively describe the mechanism of car buyer behavior. A first attempt to model this mechanism was undertaken in Sweden in 2006. This tool has been used as a decision support for the Swedish transport authorities and the Swedish environmental protection agency to describe future policies, but not to evaluate all currently implemented measures. We will therefore discuss current policies here without a common model evaluation. After that, we will briefly describe the modelling tool and some examples of policy evaluation based on this tool.

Another dimension that cannot be ignored is the general public debate concerning climate change in media and elsewhere. The awareness of the environmental impact of traffic emissions and the wish to do something may also have had an effect. This is however difficult to quantify and this issue is not discussed here.

11.4.1 Current Policy Evaluation

Some of the measures are more general, aiming at preparing the market, other are more aimed at providing economic incentives to the individual car buyer. We first discuss the measures to prepare the market, then the incentives to make the market change.

11.4.1.1 Evaluation of the Measures to Prepare the Market

Increased Supply of Refuelling Stations with Alternative Fuel

The importance of this policy can be judged by comparing the shares of gas vehicles with E85 vehicles. The policy has been successful in providing accessibility to E85, but not with respect to gas. The shortage of gas fuelling stations (and also supply problems for existing gas stations) has seriously hampered the introduction of gas vehicles.

Green Procurement

Green procurement has been particularly useful in early stages of the introduction of clean cars. Public authorities and publicly owned companies own only all small part of the vehicle fleet, so it is difficult to extend the influence to larger parts of the fleet.

11.4.1.2 Evaluation of the Measures to Make the Market Change

Tax Exemptions for Alternative Fuels

Ethanol and gas are exempt from energy and CO_2 taxes (but not from value added tax – VAT). The energy tax on petrol is currently close to 0.4 Euro per liter (including VAT). If a corresponding tax would be levied on E85, it would amount to about 0.3 Euro per liter E85. This would amount to about 400 Euros per year for average car owners.

The tax exemption may affect not only the car purchase but also what fuel that is actually being used for flexifuel vehicles. An illustration of the sensitivity to fuel prices is given in Fig. 11.6 (based on data from the Swedish Petrol Institute and Statoil), where the price advantage of E85 with respect to petrol is plotted together with the E85 consumption per flexifuel vehicle from January 2006 to September 2010.

The price advantage has been calculated as the petrol price per liter minus the E85 price per liter multiplied by 1.35 to reflect the consumption rate difference between petrol and E85. The monthly consumption per flexifuel vehicle has been calculated as the monthly quantity E85 sold divided by the stock of flexifuel vehicles each month.

From the figure it can be seen that the monthly consumption lies around 200 liters when the price advantage is positive, and that it reduced by roughly half when

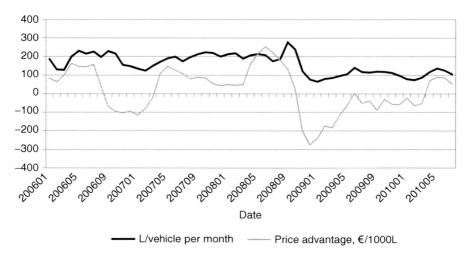

Fig. 11.6 E85 price advantage and monthly consumption, Jan 2006–July 2010 (season adjusted)

the price advantage is negative. The last part of the time series indicates that an increased price advantage does not have the same effect as in the earlier part. This may be explained by media reports of E85 being contaminated with harmful ingredients and advice to use petrol more often in flexifuel cars. If an energy tax is added to the E85 price, it seems very likely that E85 consumption for existing flexifuel vehicles will go down a lot. Most of this reduction is likely to be replaced by petrol. No quantitative analysis has been made for this scenario.

Subsidy for Purchase of New Clean Car (Privately Bought)

The 1000 Euros subsidy for clean vehicles has been claimed to be a great success. It is true that it coincides with a big increase in the clean car sales, but it also coincides with a big increase in the supply of clean cars. It may be that the subsidy has prompted car dealers to supply clean cars already available outside Sweden, but this is difficult to assess. Modelling exercises using the model described later suggest that the pure price effect of the subsidy would be about 5–10% increase.

CO_2 Based Circulation Tax

The CO_2 based circulation tax provides incentives to own an alternatively fuelled car as well as a low emission car. In Table 11.3 examples of the size of tax under different conditions are shown.

The owner will earn about 150 Euros by owning a 120 g CO_2 petrol car as compared to a 220 g CO_2 petrol car. By changing to an alternatively fuelled car, he will earn 10 Euros in the 120 g CO_2 case and 60 Euros in the 220 g CO_2 case. As is obvious form the table, diesel car owners face higher circulation taxes. The incentive to choose a low emission vehicle is much larger in the diesel case, where there

11 Accelerated Introduction of 'Clean' Cars in Sweden

Table 11.3 Yearly circulation tax example

Yearly circulation tax (Euros)	120 g CO_2	220 g CO_2
Petrol	66	216
Alternative fuels	56	156
Diesel	208	680

are 470 Euros to be earned each year by choosing a 120 g CO_2 vehicle instead of a 220 g CO_2 vehicle.

In 2009, the 1000 Euros subsidy for private clean car buyers was replaced by a 5 year exemption from the vehicle circulation tax for clean cars. This means that low emission diesel cars are now more subsidized than low emission petrol cars or alternative fuelled cars, which may be seen as doubtful as the larger circulation tax was to compensate for environmental effects and fuel tax differences. It actually counteracts the introduction of biofuel cars on the market, as can be seen from the registration statistics as presented above.

Company Car Benefit Tax Reduction

Company car benefit tax reduction implies an incentive of maximum 700–1100 Euros per year for the individual car beneficiary. This is the largest single incentive being used in Sweden. It concerns only the company car buyer segment, but this segment accounts for about half of the new car market. The segment consists of larger cars, and as low emission fossil cars are not eligible for tax reductions, this policy will solely promote alternatively fuelled vehicles. In Table 11.4 the fuel shares for private buyers and companies can be compared.

It appears that the shares of alternatively fuelled cars are much higher for companies than for private buyers. This is at the expense of petrol cars, as the diesel share is also higher for companies than for private buyers.

Congestion Charge Exemption for Alternatively Fuelled Cars (Stockholm)

This is the second largest single economic incentive. It may be worth up to 900 Euros per year for regular car commuters. The measure has shown to be efficient (as reported above), but it concerns only a small segment.

Table 11.4 Fuel shares for private buyers and companies

Buyer	Petrol	Diesel	E85	Hybrid	Gas
Private	51	34	14	1	1
Company	24	46	23	2	5

Free City Residential Parking for Alternatively Fuelled Cars (Stockholm)

This measure also represents a large economic incentive, about 800 Euros per year. Also this measure concerns a very small segment, BEST (2009) reports that about 6000 residents enjoyed free residential parking.

11.4.2 Summary of Effects of Current Policies

A number of different measures have been employed to accelerate the introduction of clean cars in Sweden. In addition to the specific measures taken, the Swedish market has also benefitted from the introduction of already existing cars models. The result has been that the share of clean vehicles is now about 40% of the new car sales, mainly consisting of low emission diesel cars and E85 capable cars. In spite of the significant share of clean cars, there is a problem associated with this situation. The problem is that E85 cars to a large extent seem to be driven on petrol when the price relation between E85 and petrol is not favorable. It seems that an accelerated introduction of clean cars may not be enough, and that long run policies need to be implemented to assure that flexifuel cars are not only capable of using E85, but also do so. The development has still lead to reduced fuel consumption for new cars as shown in Table 11.5 (using petrol equivalents for E85 and gas cars).

The increased supply of clean cars and diesel cars has also made it possible to choose a car with higher performance (more powerful engines) at the same time decreasing fuel consumption or fuel cost. A consequence of this is that the average fuel consumption remained reasonably stable for each fuel type at the same time as the change of fuel type shares has lead to a decrease in the average fuel consumption (and hence to a corresponding decrease in CO_2 emissions). Thus, the choice of cars with higher performance (higher than the petrol equivalent) reduces some of the potential savings of CO_2 emissions (Sprei 2009). This shows that even though it is important to prepare the market by introducing incentives to promote clean cars, more can be gained by providing further incentives to support less consuming vehicles when the market is ready.

11.5 Evaluation of Future Measures

Changing the car fleet is a slow process. It is likely that future climate and energy policies need to include further measures with respect to the car fleet composition and use. To be able to evaluate different policies in these respects, a quantitative

Table 11.5 Average fuel consumption development

Average fuel consumption development				
Year	2005	2006	2007	2008
l/100 km	8,0	7,8	7,3	7,1

Source: The Swedish Transport Administration, www.transportstyrelsen.se

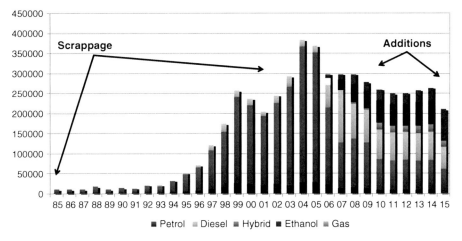

Fig. 11.7 Example of the Swedish car fleet composition in year 2015

model for the composition of the vehicle fleet has been developed. This model has been used to help identifying efficient measures, and also to describe the consequences of a combined strategy used in the national transport planning process. In doing so, it needs to be recognized that the composition of the Swedish car fleet affects not only CO_2 emissions by technical properties like fuel type and fuel consumption, but also by the cost of driving which in turn affects car use. If the car fleet becomes more fuel efficient, it may also be cheaper to use, which will increase car use. Such rebound effects have also been considered in the modelling process.

Even if a new policy can have a large effect on purchase behavior of new cars, new cars introduced to the car fleet only amount to a small part of the total fleet. A vehicle serves in the fleet between 10 and 15 years, so if we want to change the fleet in twenty years the purchase behavior must be affected starting today. In Fig. 11.7 an example of the Swedish car fleet composition in year 2015 is shown.

The figure shows the distribution of the car fleet on vintage and fuel type in the year 2015. The car fleet has transformed from the current year by scrappage of older vintages and addition of future vintages. Most of the younger vintages still remain in the car fleet. The car fleet is thus a slow giant; changes made today will have a significant effect on the total car fleet composition first after several years.

As shown above, there has been a large change in purchase pattern in the last years. It is not possible to understand from statistics only which policy measure has had the largest effect, neither how the market would react on new policy measures. That implies that a model reflecting car buyers' behavior is needed to explain the vehicle fleet development.

11.5.1 The Swedish Car Fleet Modelling Tool

In 2006 a vehicle fleet model was developed for the Swedish Road Administration (Transek 2006, Beser Hugosson and Algers 2010). In this section we give a brief description of this model system.

The vehicle fleet model system annually updates the stock of vehicles by subtracting scrapped vehicles and adding new vehicles (a cohort model). The output of the model is not only the numbers of vehicles of different types, but also average fuel consumption rates and average fuel costs. These are calculated as an average over all vehicle types and vintages in the vehicle fleet.

The Swedish car fleet model system consists of three sub models:

- A scrapping model
- A total fleet size model
- A vehicle type choice model (for new vehicles)

To make a forecast with the model system the car fleet composition is thus calculated yearly from the base year to the forecast year of interest. The year by year calculation is made by subtracting all scrapped cars by type and vintage, and adding all new purchased cars by type to the existing car fleet. To update the car fleet for the next year, a new calculation is made. The result from the model system is a distribution of car types of different vintages and types, having different fuel and different fuel consumption. From this distribution the average fuel consumption, average fuel price and average CO_2 emissions are calculated for different forecast years.

The model allows the user to evaluate how different polices will affect the development of the car fleet, under assumptions related to the future supply of cars including their characteristics (explained below), to the distribution on different market segments (private and company buyers) and to external factors like fuel costs.

11.5.1.1 The Scrapping Model

The scrapping model is a simple model, based on data from the car register. It calculates the share of cars that will be scrapped during a year. The share is depending on two variables only: vehicle age and make.

11.5.1.2 Car Ownership Model

The total fleet size model is a car ownership model used in the national planning process developed by VTI (2002). This model is also a cohort model, based on individual car ownership entry and exit probabilities. The number of car owners is annually updated using models for these probabilities. The models are mainly driven by socio economic variables, and the only policy variable is petrol cost. This variable is meant to represent costs of driving, and might be justified at the time when the model was developed more than ten year ago. At that time the share of petrol cars was over 90% in Sweden. When the car ownership model is used as part of the vehicle fleet model, this implies some inconsistency as the average cost of driving will depend on the mix of car types. This is a subject for future model improvement.

11.5.1.3 New Vehicle Type Choice Model

To model what vehicles will be added to the vehicle fleet, we need to model consumer behavior. We therefore need to model how the consumer chooses a vehicle from a set of available vehicles. A convenient theory for this problem is the theory of discrete choice (Ben-Akiva and Lerman 1985).

The model for car type choice for purchases of new vehicles is therefore formulated as a discrete choice model. Such models have been estimated in different applications (Bunch et al. 1993, Train and Winston 2007), in the EU TREMOVE model system (TREMOVE 2006), and integrated with car use (Bhat and Sen 2006).

The model for new car purchase is of course the one that is most responsive to car promotion policies, and this model is also the one that is most elaborated. The modelling work is based on two major types of sources describing new purchase behavior. The largest source is the Swedish Car Register, which contains data for the complete Swedish vehicle stock. From this data source the current fleet, new purchases and scrapping was retrieved for a few years. The other data source is a Stated Choice survey directed to persons who bought a new car in the beginning of the autumn 2005.

The scope of the vehicle fleet model is to be able to model effects of different policies directed toward purchase of new cars, and thus influencing the vehicles fleet composition. In order to do this, it is necessary to understand what influences car type choice. Therefore, in addition to the data on car type sales, data describing attributes of the brands and models on the market are needed. In addition to the data sources already mentioned, data describing a number of attributes of different car makes and models was also collected.

Modelling car type choice on car register data has to be restricted to attributes already existing. Using only existing data can also be difficult from a statistical point of view, as there may be strong correlation between variables such as price, size, horsepower etc. These two reasons were the main reasons to conduct the stated choice study, in which attributes contained also in the larger data set could be varied in a statistically more efficient way, and where additional attributes such as fuel infrastructure and fuel type could be better analyzed.

Different consumers may trade car attributes differently. In Sweden, cars bought by companies and to a large extent provided by the employer to the employee (as a fringe benefit) make up a substantial part of the new car market (about half) (Naturvårdsverket 2007). This part of the market is more oriented toward large cars. Therefore, the new car purchase model is segmented, comprising three segments. One segment is private persons, and the other two are company cars, without and with a leasing contract respectively. The last segment is typically the fringe benefit car segment (although such cars can also be used as a shared car by the company).

For each of these segments, a choice set of more than 300 different car models was established. The estimated choice model is of the nested logit type, which in this case implies that elasticities are larger between model of the same model family (like the BMW 3xx series) that between cars of different brands. If a new car model – perhaps an electric vehicle – is introduced, this will affect the market share of close car models relatively more than for other car models.

The model estimation results reflect consumer preferences for different attributes associated with different car types for each segment. Car buyers consider a number of different factors when choosing a new car and it turned out that the following factors were found to significantly affect what kind of new cars being bought:

- Price/benefit tax
- Running cost (fuel and vehicle tax)
- Size class
- Fuel type
- Tank volume
- Rust protection guarantee
- Safety (NCAP/Folksam insurance company classification)
- Engine power (hp)
- Share of fuel stations with alternative fuel
- Make

11.6 Application of the Model System and Results

The model system has been used to describe policy effects in several projects (Naturvårdsverket 2007, Swedish Authorities 2007, WSP 2008). The aim has been to predict the impact different climate policies would have on the Swedish car fleet. To use the vehicle type choice model in forecasting, a choice set for the forecasting year needs to be defined. This means that changes in supply and policies need to be formulated in one or more scenarios in terms of the variables used in the models. Also, the shares of the different consumer segments need to be defined, as they are exogenous to the model. The result of the model is market shares for all vehicle models in each segment. When multiplied with the number of car buyers in each segment, the addition of new vehicles to the vehicle fleet is quantified. In the following section a brief presentation of an application is made.

Research aimed at improving the model in a number or aspects is currently ongoing. These are related to the overall model structure, the data on which the model was estimated, and the particular sub-models. Experience from using the model also shows that assumptions on supply are crucial. Such assumptions are quite uncertain in the long run, which has to be addressed by testing the robustness of different policies with respect to possible supply scenarios.

11.6.1 Using the Model System for Predicting the Effects of Different Policy Measures

The model was used to describe effects of different climate policies in a report to the Swedish Environmental Protection Agency (Naturvårdsverket 2007). Consequences of national climate policy changes were analyzed by defining the following scenarios.

- S1: Stronger CO_2-based vehicle tax
- S2: CO_2-based company car benefit taxation rules
- S3: Instant fuel tax increase
- S4: GDP adjusted fuel tax

The first scenario implied a steeper increase in the vehicle circulation tax, making the difference between a 120 g CO_2/km and a 200 g CO/km increase by about 200 Euros per year. The second scenario implied a dynamic development of the car benefit taxation rules. In this scenario, the benefit value was kept about the same as today for vehicles emitting less that 135 g CO_2/km, whereas it was increased by about 3000 Euros per year for a 200 g CO_2/km car, and increasing up to 4000 Euros over a 5 year period (for a 30,000 Euros car). This would mean about the half, or 1500–2000 Euros net after tax extra cost per year for an employee having such a 'benefit'. This scenario does not affect privately bought vehicles.

The third scenario implied an instant fuel cost increase by 0.4 Euro per liter for petrol and diesel. The fourth scenario implied adjusting the fuel tax by the GDP growth, which would correspond to an increased petrol price by about 0.17 Euro/liter after ten years.

The scenarios have different impacts and differ consequently with respect to CO_2 effects and fuel consumption. In Fig. 11.8 the total effects on CO_2 emissions of the different measures are shown, including rebound effects. Except for emissions caused by the combustion, emissions are also incurred in production and distribution of the fuel. For petrol and diesel this is less important than for ethanol, which can be produced under very different conditions. The latter can be discussed at length, and to avoid that in this chapter only emissions from the combustion are considered here. Hence, the CO_2 emissions are taken as the tail-pipe emissions i.e. not considering the life cycle emissions including renewable or fossil origin of the fuels. The model provides CO_2 emission results for fossil as well as renewable fuels.

The total mileage is increasing over time. This is due to population growth, increased car ownership and economic growth in general. The two first scenarios are mainly affecting the CO_2 efficiency of the car fleet. The two following scenarios affect fuel prices and hence driving costs to a larger extent, and have therefore a larger effect on car fleet use. The mileage increases by about 19% in the base alternative between 2006 and 2020, while the increase in mileage is only about a third

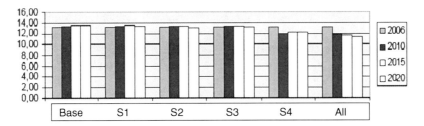

Fig. 11.8 Total CO_2 emission 2006–2020 (million tons per year)

of this in the combined scenario. The effects on CO_2 emissions are net effects of the change in mileage and the CO_2 efficiency in each scenario. The total amount of CO_2 emissions is calculated to increase somewhat over time, as the mileage increases somewhat faster than the CO_2 efficiency of the vehicle fleet. The combined scenario gives of course the largest reduction of the total CO_2 emissions, about 17% compared to the base scenario for the year 2020.

11.6.2 Rebound Effects

As mentioned above, efficient policies with respect to the vehicle fleet may also imply reduced running costs, which to some extent may counteract policy intentions. This is modelled by first calculating the average fuel cost of the future vehicle fleet, and then feeding this number into the transport usage model used for the national Swedish planning process. This model, called Sampers, comprises all personal travel in Sweden, including local, regional, long distance and international trips.

The magnitude of the rebound effect corresponds to an elasticity of about –0.3. This elasticity results from calculations using the Sampers model. This means that if a policy results in an average fuel consumption decrease of 10%, then there will be an increase of the number of kilometers travelled by 3% (as the marginal cost of travel will be reduced). The policy may of course have an impact of the total number of cars, but this effect is not yet in the model.

11.7 Conclusions

The challenge to increase energy efficiency and to reduce greenhouse gas emissions has led to the implementation of policies to accelerate the introduction of clean cars in Sweden to meet these goals. These policies have been shown to have a significant effect on the new car sales and some effect on the fuel type shares.

The main trends are that diesel cars and E85 capable cars have increased their shares significantly from 2005 to 2010, at the expense of conventional petrol cars.

Several measures have been implemented, of which some can be seen as more important. One of these is the policy to support availability of alternative fuel over the country. The success of this policy is however limited to the supply of E85 fuel. Other important policies are those that provide major economic incentives, like the car benefit reduction policies for company cars and the tax exemption for biofuels. At the local level, exemption from congestion charges has also shown to have a large impact.

The sales of new clean cars in Sweden have increased from being about less than 5% in 2005 to be over 40% in 2010. The top five selling clean car model types have also changed from 2005 to 2010, due to combinations of policies and supply. Even though the measures taken have been successful from the point of share of clean vehicles, there is a problem in that E85 vehicles are run on petrol to a large extent when the price relation is unfavorable.

The demand for transport has however increased, and can be assumed to continue to do so. The vehicle fleet is a slow giant, and it will take time to make it sufficiently energy efficient and low emitting. It is therefore necessary to find efficient policies to support this process. Experience from practice and planning has shown the necessity of a decision tool able to assist in this process.

To be able to evaluate different policy options, a quantitative model for the vehicle fleet composition is needed. For such a model to be useful it has to be consistent with car buyer behavior, reflecting preferences from different buyer segments. Such a decision tool has been developed in Sweden, and applications have shown how it can (and has been) be used for analyzing separate measures and measures combined to strategies. As a successful evaluation of different policies will depend on how well the model reflects reality, it is necessary to assure that the ability of the model is good enough. The car fleet model has the ability to quantify the policy effects on the car fleet (i.e. to have a model at all). It has also the ability to consider a number of different factors when forecasting the demand for new vehicles. This allows testing different policies, at the same time considering other developments in car supply, such as increased energy efficiency, safety and reliability which are important to consumers. Research is going on to improve the model in a number of aspects. These are related to the overall model structure, the data on which the model was estimated, and the particular sub-models.

Many of the measures reported here are based on particular conditions in Sweden, like the large share of fringe benefit cars and exemptions from congestion charging. Some measures may however allow more general conclusions. In particular we believe that findings related to the importance of price relations between fossil and renewable fuels for the consumption of alternative fuels are more general. Similar findings emerge also from the Brazilian experience (Pacini and Silveira 2010). Whether the short term policy to accelerate the introduction of flexifuel or biofuel cars will be successful in terms of CO_2 mitigation depends very much on the long term willingness to assure price relations that make consumers choose the alternative fuel.

References

Ben-Akiva M, Lerman SR (1985) Discrete choice analysis: theory and application to travel demand. London

Beser Hugosson M, Algers S (2010) Transforming the Swedish vehicle fleet – policies and effects. Paper presented at the 2010 international energy workshop in Stockholm

BEST (2009) Promoting clean cars – case study of Stockholm and Sweden. EU-project BEST Deliverable No 5.12

Bhat CR, Sen S (2006) Household vehicle type holdings and usage: an application of the multiple discrete-continuous extreme value (MDCEV) model. Transportation Res Part B, 40(1):35–53

Bunch et al. (1993) Demand for clean-fuel vehicles in California: a discrete-choice stated preference pilot project. Transportation Res Part A: Policy Practice 27(3)

City of Stockholm Environment and Health Administration (2009) Försäljning av miljöbilar och förnybara drivmedel i Stockholm (Sales of clean cars and alternative fuels in Stockholm). Report, Sweden

City of Stockholm Traffic Administration (2009) Analysis of traffic in Stockholm – with special focus on the effects of the congestion tax 2005–2008. Report, Sweden

Naturvårdsverket (2007) Drivkrafter till bilars minskade koldioxidutsläpp (Driving forces behind reduced CO_2 emissions from cars). Summary in English. Report 5755, November 2007, ISBN 978-91-620-5755-8.pdf

Pacini H, Silveira S (2010) Ethanol or gasoline? Consumer choice in face of different fuel pricing systems in Brazil and Sweden. Biofuels 1(5):685–695

Sprei F (2009) Vilka Styrmedel har ökat personbilarnas energieffektivitet i Sverige? (What policy measures have increased the energy efficiency of cars in Sweden?) Fysisk Resursteori, Institutionen för Energi och Miljö, Chalmers Tekniska Högskola, Göteborg, Sweden

Swedish Authorities (2007) Strategi för effektivare energianvändning och transporter, EET Underlag till Miljömålsrådets fördjupade utvärdering av miljökvalitetsmålen (Strategies for more efficient use of energy and transport). Naturvårdsverket report 5777 (in Swedish), ISBN 91-620-5777-0

Trafikutskottet (2009) Pumplagen – uppföljning av lagen om skyldighet att tillhandahålla förnybara drivmedel (Assessment of the obligation to provide renewable fuels at public refuelling stations). Riksdagstryckeriet, Stockholm, Sweden, ISBN 978-91-85943-83-8 (in Swedish)

Train KE, Winston CM (2007) Vehicle choice behaviour and the declining market share of U.S. automakers. Int Econ Rev 48(4)

Transek (2006) Bilparksmodell (Car fleet model). Report 2006:19, Transek AB, Sweden

TREMOVE 2 (2006) FINAL REPORT PART 1: Description of model version 2.44 European Commission DG ENV

VTI (2002) Modeller och prognoser för regionalt bilinnehav i Sverige (Models and forecasts for regional car ownership in Sweden). VTI report 476 2002

WSP – Analys och strategi (2008) Bilparksprognos i åtgärdsplaneringen – EET-scenario och referensscenario (Car fleet forecast in the Swedish transport infrastructure planning). Report (in Swedish)

Chapter 12
Making People Independent from the Car – Multimodality as a Strategic Concept to Reduce CO_2-Emissions

Bastian Chlond

Abstract Carbon dioxide emissions can be reduced not only by technical improvements but also by appropriate planning concepts aimed at individuals, with a view to changing their travel behavior, mainly through the use of other modes of transport. This article illustrates what should be understood as a change in travel behavior, why observed effects are still few, and the conclusions to be drawn from these. The article shows how politics and planning can create framework conditions that allow for decisions favoring multimodal behavior. The general concept of multimodality and a catalogue of measures are introduced, and illustrated by means of the example of the city of Karlsruhe in Germany, where such planning concepts have been successfully implemented and positive effects are becoming measurable.

12.1 Introduction

There seems to be a general belief that transport volumes and their resulting greenhouse-gas emissions are destined to increase as long as economic growth persists. However, there is some evidence that this link can be decoupled, as illustrated in the example of Germany. During the years of the German 'economic miracle,' with its decades of steady economic growth after the war, car demand, energy consumption and greenhouse gas emissions in Germany followed an upward trend, slowing down only after the two petrol crises and the breakdown of the industrial sector in East Germany after reunification. As shown in Fig. 12.1, consumption peaked in 1999.

Nevertheless, during the past decade Germany has managed to reduce energy consumption, and thus carbon dioxide emissions, in the transport sector considerably, and this in spite of a still-increasing demand for travel, a developing economy within a globalized world with massive increases in freight transport volumes, and still-growing car ownership rates: The reduction in energy consumption and emissions results both from technological improvements that will not be the focus here,

B. Chlond (✉)
Institute for Transport Studies, Karlsruhe Institute for Technology, 76128 Karlsruhe, Germany
e-mail: Chlond@kit.edu

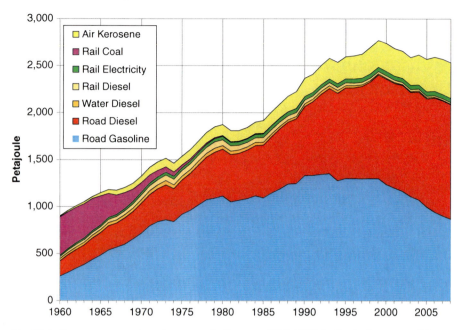

Fig. 12.1 Energy consumption of transport in Germany 1960–2008 in Petajoule per year (Knoerr et al. 2010)

but much more from a change in behavior, mostly in terms of the use of modes of transport. The average German more and more frequently uses public transport instead of the car. Over the past decade car use has ceased to grow, public transport has achieved better acceptance and patronage than ever, and the bicycle has become an accepted mode of transport and not simply an item of sports equipment.

This development is consistent with the objective of reducing carbon dioxide. Nevertheless, it has to be kept in mind that the discussion about climatic change and its importance for life on earth is comparatively recent. However, achievements in the reduction of fossil energy consumption in transport are not a result of more-or-less recent interventions, measures and policies. They are, rather, the outcome of processes and interventions that were initiated and motivated mostly decades earlier.

This chapter documents those processes, changing framework conditions, and interventions that triggered those behavioral changes that today yield fruit in reducing carbon dioxide. Nevertheless, it must be stressed that these processes need a considerable amount of time. The combination of measures used and their effectiveness will be shown by means of the example of the city of Karlsruhe.

12.2 Changing Behavior: Toward Less Car Dependency

12.2.1 A Look at the Past – The Role of the Car and the Car-Oriented Society

For the first 40 years after the war (the 1950s to the 1980s) (West) Germany experienced the so-called *economic miracle*. Among its other consequences the steady increase in wealth resulted in an enormous increase in car availability and car ownership rates. The car-oriented society became the overriding concept. As a result road infrastructure was developed extensively (e.g., the development of high-capacity roads within cities; the development of the federal motorways and the trunk road network). As a consequence of this suburban life became feasible and eventually mainstream in the society.

This process was fostered until the nineties by demographic changes. The generation of baby boomers born after the war fed both economic development and transport demand. Also driving this process has been the reduction in the price of gasoline – at least in work time equivalents. There was no reason not to rely on the car. As a consequence patronage of public means of transport went down. As at about the same time many first-generation rail-oriented public transport systems were wearing out, many tramway systems were abandoned as a consequence. The service quality of public transport declined, caused by the self-reinforcing spiral of supply reduction and decrease in demand. This also affected modal use considerably.

The first interventions designed to reduce the growth in car travel demand were initiated and installed by the late 60s and the beginning of the 70s through, for example, the development of high capacity public transport systems. However, these measures and policies had no immediately visible effect. It is possible that the increase in car demand was slowed down, but it could not be prevented completely. Beyond that, the reunification in 1990 led to another push in car demand as the East German population also wanted to keep pace with development in the west and therefore adopted the life and mobility style of West Germany.

Furthermore, not only Germany's transport system, but also its industry, and thus society itself, is based on the car, with the automotive industry also playing a significant role in terms of its importance for GDP. As a consequence, every intervention in the transport sector is discussed in terms of its negative impact on the car industry and of its relevance for the destruction of jobs. This greatly affects the enforceability of interventions and illustrates the difficulties faced by politicians and planners working toward changing travel behavior.

We can assume that sooner or later Germany would have become totally 'car dependent,' not only in terms of the car manufacturing industry but also in terms of spatial structures, suburban lifestyle behavior and the resulting structures for trade and economy, with massive consequences in terms of the consumption of fossil energy and of carbon dioxide emissions.

12.2.2 From Monomodality to Multimodality

It is worth examining the success of the car in order to understand how measures and interventions seeking to change behavior can become effective: The car has to be regarded as a 'universal' mode. It is obviously not only comfortable and always available but it has its fields of application in urban traffic as well as regional traffic or long distance transport. Compared with this other transportation modes are very specialized: They have their fields of application in certain markets: The bicycle is superior to the car only on short distances in urban areas, where parking might be difficult. The situation of public transport is similar. It is also a 'specialist'. It, also, is superior to the car only in certain situations (e.g., commuting into the city centre, where parking is difficult, or within cities if the service is frequent and excellent).

Therefore, the overall concept of a behavioral change in terms of modal use requires fundamental rethinking: It is easy to understand that resigning from using a car overnight is nearly impossible to achieve, as daily life is orientated and organized according to the private car.

Apart from the required fundamental change in attitudes it becomes clear that a carless life is impossible to achieve, at least within the framework of current spatial conditions: Even people with changed attitudes would not be able to maintain a new way of life within their former environment and neighborhood. At the minimum they would need to change their place of residence to a neighborhood where a car is not needed any more. It is easy to comprehend that such a change is rather unlikely, particularly as in Germany a change of residence or house is rather uncommon. People who buy their own house usually stay there to the end of their lives. To resume: Such a change becomes unlikely.

On the other hand particular measures and interventions are implemented which make people make use alternative modes, such as public transport or a bicycle instead of a car, at least once in a while, or for certain purposes and to certain destinations. The effects of these measures on carbon dioxide reductions will be quite small, as the basic mobility will remain unchanged.

Nevertheless, these occasional changes of modes and behavior should not be underestimated in terms of their relevance. People 'learn' to use other modes and can assess their characteristics and utility. They are becoming 'multimodals' (as using different modes, Kuhnimhof et al. 2006) compared with the 'monomodals' or 'captives' (who are bound to one mode) as has been the typical situation in the past.[1]

If more and more persons behave multimodally and their use of alternatives to the car becomes more frequent, the effects will slowly accumulate. And a variety of different measures can complement each other and may once in a while create conditions where it is possible to live without a car. A set of, or combination of,

---

[1] Captives have been understood in the past as those persons bound to public transport because they had no alternative means of transportation. Today we also include car drivers who do not use any other modes within the category of 'captive' as they are car-dependent (Zumkeller et al. 2006).

alternative modes can compete with the universality of the car. Every alternative mode has its special field of application and these specializations need to complement each other. If modal behavior is to change, therefore, a transport system must be created in which a combination of different modes, each with specialized characteristics and fields of application, which can compete with the 'universality' of the car. That this is a realistic overall concept will be explained in Section 12.4.

12.2.3 Behavioral Change as a Demographic Process

On the other hand, it is clear from the results shown in Fig. 12.1 that changed travel behavior in the German population is progressing, albeit slowly. This seems to be in contradiction to the mentioned fact, that we can expect only a few rare cases and that more general changes in behavior are unlikely to happen. If we try to observe the effects of measures and interventions (e.g., a new direct tramway connection between home and work), we usually only identify a few *individuals* with changed behavior, as, for example, in those changing to another mode for commuting from the year before, and these tend to be those having direct advantages and utilities, such as a shorter travel time.[2]

Nevertheless, for *the population as a whole* we do observe changed behavior: However, this is less a change in the behavior of individuals within that population than an exchange of population. That is, people with different ('changed') behaviors are replacing those with 'conventional' i.e., unchanged, behaviors. The general development and trend is therefore determined by the composition of the population with different life and mobility styles.

To illustrate this: At this time the car availability for senior citizens is still increasing. The retiring persons of today have been accustomed to owning a car during their active work life. They have arranged their lives according to the framework conditions of that 'car-oriented' and thus monomodally and car-dependent time. This car-dependent generation is now going to replace the retired people of the past, who usually lived without a car. This increase in car ownership and use in the age group of the retired, therefore, does not involve a change in behavior, it involves a so-called 'cohort effect'.

On the other hand, recent research results (IFMO 2011, Kuhnimhof et al. 2011) show that car-ownership rates and car use is declining for the younger generation in Germany. A growing number of those below 30 of today organize and arrange their lives to be less car oriented than their predecessors in the same age group one, two or more decades earlier.

---

[2] It has to be mentioned that for the identification of behavioral changes it is necessary to have adequate empirical sources, such as panel surveys or at least panel data, which observe individuals' behavior repeatedly. Results from the German Mobility Panel, a survey which observes individuals in their mobility behavior repeatedly, show that a real change in behavior (e.g., car use and car ownership) is a relatively rare event (Dargay et al. 2003).

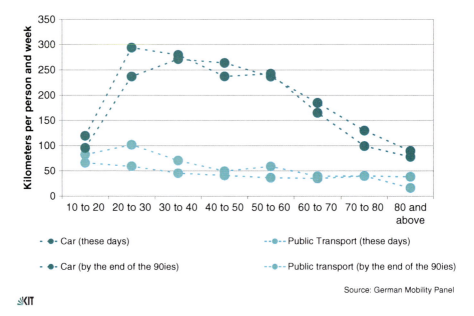

Fig. 12.2 Different developments in modal behavior for young and old people (Chlond et al. 2009)

Figure 12.2 illustrates the slow but steady change in mobility, at least in modal behavior, for different age groups. This aggregates for the average of the German population to show only minimal changes (an increase in public transport/increase in cycling) because of the cancelling-out effect within the population. That is, car use is increasing for those over 60 years of age, whereas the decline in car use is clear for the under-30s. Aggregate effects in terms of modal use and carbon dioxide emissions are obviously determined by which groups and life styles are becoming dominant.

To sum up: A radical change in the behavior of individuals is very unlikely. What becomes relevant over time is the difference in the behavior of new generations as they replace the old. How fast such a process can take place depends on the size of the cohorts, the percentage of the new cohort displaying different behavior and, of course, the framework conditions set.

We have to ask, what are the reasons for the changed behavior of the younger population? Why are the young people of today behaving differently? What is causing their different behavior? Obviously there must be a change in general conditions on the one side, in combination with changed orientations and attitudes toward the car and its use. As most of the youngsters of today live in relative wealth, far away from the fear of social deprivation, there is no financial reason for young people to drive less. They could potentially afford a car. The question is, whether today's general conditions give an advantage to those who own and frequently use a car versus those who use other modes as well. And we have to ask whether, and which, interventions and changing framework conditions have really influenced the utility of the car.

All policies and interventions to change modal use become effective by altering the characteristics of the competing modes. But as people habitually show a lot of inertia, voluntary changes in behavior are unlikely. People do not willingly resign from the suburban life style and daily routines that they have developed over years and by means of dependence on the car. This is basically the case for the older generation of today. They will try to maintain their car-oriented lifestyle for as long as the framework conditions allow. Or, in other words, lifestyle and modal orientation will only be changed if the alteration of the framework conditions is so massive that they are forced to do so. This has not been the case up to now.

A change in mobility behavior becomes more likely when something else is changed in an individual's situation (e.g., a change of job, a change in the family's situation, or a change of house). These events force people to adjust to their new life circumstances whether previously known or unexpected, and to find an optimal solution individually within a changed general framework, a framework set by processes in the globalized world and the (already installed or unexpected) interventions from politics and planning.

And this is probably the case for young people. Young people find themselves more frequently in a completely new situation (e.g., starting university or a new job) and they are informed about imminent global developments like rising energy prices. They perhaps also feel the need to adopt behavior less harmful to the environment.

This opens up avenues of change, such as the adoption of a lifestyle that is less car oriented.

To conclude:

- A change in behavior becomes more likely if there is both a change in an individual's situation and the general framework for change is set appropriately.
- The implications of this for politics and planning are that a 'climate' has to be created to enable people to function with a different mobility style.
- To achieve a significant change in behavior plenty of time is needed. Relevant numbers of people in the population have to experience new life circumstances with altered general conditions.

12.2.4 How can Measures Become Effective?

12.2.4.1 Measures Have To Be Understood Against the General Conditions

A change in behavior toward a less car-oriented lifestyle can only be achieved if general conditions are appropriate and favorable for this to happen. Observed effects in modal behavior are a combination of external developments, policies and conditions from the federal side as well as those decisions and implemented measures made at a local level. That means that these 'locally' implemented measures have to be understood against the background of a relevant framework set by national legislation or by global markets.

12.2.4.2 Push and Pull to Change Modal Behavior

All strategies aiming to influence behavior change, e.g., modal use and thus the modal split, can be achieved by either an increased attractiveness of public transport or improved conditions for non motorized travel (i.e., 'Pull'), or by decreasing the attractiveness of car use ('Push'). The silver bullet is to do both.

12.2.4.3 Sequence and the Importance of Role Models

First the advantages of the alternatives and the changed behavior have to become obvious and clearly visible, and then they have to be adopted by a minority of first adopters who will serve as role models for those who follow. Later, a soft but increasing pressure can be exerted in order to change the behavior of the majority. People have to learn the advantages first, then they can softly adapt their behavior. This means that for interventions to be successful they must be introduced stepwise and slowly.

The following sections provide only an exemplary and incomplete overview toward understanding the way in which these elements complement each other in order to bring about changed behavior. Like elsewhere in this chapter, the focus will be on Germany.

A more general overview about potential measures and policies worldwide is shown in Chapter 14 of this book.

12.3 Prerequisites and Strategies to Change Modal Behavior

12.3.1 External Developments: Awareness About the Future

Measures and interventions are not sufficient and will not be effective as long as people do not understand why a behavioral change is necessary. Indeed, politicians and planners may implement measures, but their efficacy and acceptance will depend on an assessment of the relevant external global developments, which come mainly from outside the sphere of influence of politics and planning. These developments will change attitudes and increase people's willingness to accept the measures and to change behavior.

The threat of climatic change can be regarded as broadly accepted among the German population and across all political parties. Beyond that, it is regarded as an accepted fact that energy will become more expensive in the future as a result of peak oil, which also is today accepted as an external development which cannot be avoided.

The same holds true for expectations of future income growth. The German population is aging and the total population is likely to shrink within the next years, with financial implications for an increase in social insurance dues, and an expected decrease in retirement pensions. The majority of the population has no long-run expectation of an increase in available income. This also affects the decisions they make now, such as their choice of a place of residence, and their long-run expectations of the cost of mobility. This knowledge probably influences travel behavior.

Obviously, the dissemination of these inconvenient truths has taken some years, and this is why changes in behavior are only recently becoming obvious. And to conclude: The assessment of future framework conditions is a central prerequisite for changing behavior.

12.3.2 Shaping the Future Through Legislation

The awareness of external developments has to be supported by appropriate policies. In the following paragraphs some issues will be discussed. These are chosen as among the most relevant from the author's perspective, but with no claims of completeness.

12.3.2.1 Creating a Framework for Funding the Development of the Necessary Infrastructure

It has become clear that infrastructure development needs appropriate funding. By the mid-60s there was recognition of the need for national legislation to fund the development of (mainly) a public transport infrastructure by means of an additional tax on motor fuel. This law has had a lot of impact, as it motivated local authorities to add their own funds to improve the infrastructure. Competition between municipalities for this generous funding has been tight and an incentive has been developed in the last four decades for the building of a tramway and light-rail system. It would not have been possible for local authorities to cover the complete funding of these usually rather large investments. The funding can be understood as a 'pull' measure by the German national, i.e., federal, government to motivate the local authorities.

12.3.2.2 The Role of Fuel Prices and Fuel Taxation

Over decades the price of gasoline in Germany decreased more or less steadily, not only in absolute figures but also in relation to its work-time equivalent. In spite of the nominal increases in gasoline prices at filling stations, the price in purchase power decreased up to 1990 (Esso 2010). This has changed massively within the past two decades. Gasoline after 1990 became more expensive both nominally and in relation to purchasing power. This development has partly been driven by world markets, as shown above, but also by a more-or-less steady increase in taxation on motor fuel. Originally, the tax increases were motivated by the lack of available funds (both, as mentioned above, for funding public transport infrastructure, but also later for the funding of German reunification). From 1999 onward, the insight that conventional carbon-dioxide-emitting energy is too cheap led to the introduction of eco-taxes on fuel, in several steps, for the funding of the development of alternative energy sources.

Thus, taxes were raised by 113% between 1990 and 2010[3] (German Ministry of Finance 2010). This results in a tax share on gasoline in 2010 of about 67% of the selling price at the filling stations of 1.40 €/l.

These additional price rises, set to the external energy price development, indicated the preciousness of energy and made people aware of the likelihood of future price rises. People are 'pushed' to change behavior (or at least to choose patterns of living that make change possible in the future). And as a by-product the car manufacturing industry has been indirectly pushed to foster technological advances.

12.3.2.3 The Role of Spatial Planning and Land Use Planning

Germany as a federal republic is based on 'the principle of subsidiarity'. This means that the responsibility for implementation of decisions can be met locally ('principle of the independence of local authorities'), but within the principles and the framework set by the legal framework of superior authorities.

In parallel to the increase in car ownership, a suburban lifestyle with commuting by car became the relevant role model in society, and it became obvious by the 70s that spatial planning was needed to prevent urban sprawl. As a consequence, regional planning authorities were installed, to develop state or regional land use plans according to accepted, appropriate and promising principles. These principles included the development of systems around central areas, and the setting up of development axes along the rail infrastructure. Understandably these plans are often in conflict with the interests of local municipalities (competition for workplaces, tax-payers etc.) and as these institutions ('Regionalverbände'/'Raumordnungsverbände', the councils for spatial planning) have nearly no legal power they have often been regarded as 'paper tigers' and thus, in the public and political discussion, as dispensable. But besides the development of the master plans these institutions also mediate between the diverging interests. The problem is that the success of the development plans and the effectiveness of both the institutions and the plans could never be measured and evaluated.

Nevertheless, planning principles, at least in the more urbanized, densely-populated regions, today show positive effects as land use has taken place along the axes for development, and the relevant infrastructure (not only for transport but also for provision of goods and services to the local population) could be developed, at least where implemented, according to principles that are *not* car oriented. These principles, which are now usually applied, allow for relatively compact forms of settlement, and additionally provide the majority of the population with

---

[3] It has to be mentioned that there exists an additional VAT (at the moment in 2010 19%).

a comprehensive public transportation system, at least in regions above a certain minimum density. It is exactly these density standards that may be achieved by the prevention of further urban sprawl.

Another example is in the prevention of urban sprawl by strengthening the positioning of urban shopping facilities in central places, in contrast to the trend in favor of greenfield development. The usability of private cars has, by necessity, been made less attractive by policies such as parking restrictions within cities. This, however, has resulted in an unwanted trend toward shopping malls and shopping centers outside cities, pushed also by competition between municipalities for tax revenues. Car traffic, and thus emissions, not only were not reduced by these policies but even grew, the emissions output moving to other destinations at a higher demand and emission level. The role of the regional planning authorities in mediating between (competing) municipalities or different levels of political authorities was sometimes successful in preventing further urban sprawl.

It must be admitted, that this was not the case from the beginning, nor did it happen everywhere. Real success can be seen only in those regions where the typical suburbanization processes took place later, and where it was possible to learn from the bad examples of other regions.

Unfortunately, these successes are not directly measurable or quantifiable. It can only be surmised what would have happened if the legal framework, institutions and master plans had not existed, or by comparison with other countries where spatial planning is less restrictive and institutionalized.

To sum up, it can be concluded that the role of integrated spatial planning has to be regarded as important for making the German population less car dependent.

12.3.3 Strategies to Create a Multimodal Transport System

12.3.3.1 General Principles

To reduce carbon dioxide emissions from car travel and trigger a change in behavior, the implementation of measures has to be done locally. Following those principles mentioned above, it can be achieved by avoiding unnecessary (car) travel or by replacing the car by other modes. More and more people should be motivated to use public transport and the bicycle more frequently, i.e., people should be enabled to behave multimodally.

Avoiding Travel at All – Making Living in the Cities More Attractive

To *avoid* any travel at all the best strategy is to make people live as centrally as possible. This involves both spatial planning (creating appropriate housing areas) and transport planning. Living in the city must be made attractive. This has not been the case in the past, and that has been one of the reasons for suburbanization. Fortunately, this has changed in Germany: City living has become more and more attractive within the last decade. Polluting and noise-emitting processing and manufacturing businesses have moved out and have been replaced by service-oriented businesses.

But the rising attractiveness of the cities gives them power, and cities can act and make decisions with increasing self-confidence. They can afford to reduce commuting by car into the city from outside, and the less car traffic coming from outside, the higher the quality and attractiveness of life in the city, a self-reinforcing positive spiral that city administrators must become aware of. Of course, traffic and travel cannot completely be avoided, but the opportunity for cities to grow again but with more concentrated populations should be seen as very relevant for the reduction in carbon dioxide emissions.

Once again, the competition between municipalities for inhabitants, jobs or taxes has to be mentioned. Municipalities have, of course, to remain aware of what is acceptable to or wished for by inhabitants and enterprises. But as attitudes have changed within the last decades, the suburban car-oriented life has ceased to be the societal ideal. This allows for measures and interventions in accordance with mainstream attitudes in society toward making city living more attractive, even by taking measures against commuters from outside!

Enabling Multimodal Behavior

Enabling people to behave multimodally means creating a system in which alternatives to the car are available in most situations in everyday life. This does not mean the promotion of one mode, instead creating a system in which the combination of alternative modes offers an alternative to more-or-less exclusive car use. This system should provide a competitive alternative to the universality of the car (see Fig. 12.3).

This is obviously easier to achieve within cities, where the use of a car is often more a burden than an advantage. But multimodality can also be fostered for those commuting into the cities. They should have the option to make use of public transport for at least some trip patterns. Here, too, a combination of measures is necessary to enable a change in behavior.

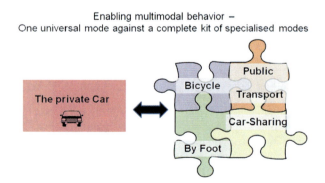

Fig. 12.3 Competing with the car by a kit of complementing modes

12.3.3.2 Pull Strategies

Public Transport

Improving public transport: This is easy to propose, but rather difficult and costly to implement. Furthermore, it takes a lot of time: The example of Karlsruhe, in the case study shown later, illustrates how a system could be developed within approximately three decades. Other possibilities and measures will not be discussed in detail, but some examples will be introduced that seem relevant from the author's subjective point of view in terms of multimodality.

From demand orientation toward supply orientation: According to past understanding of the roles of transport modes, public transport was regarded as a necessity for the socially disadvantaged, that is for those who were not able to drive or could not afford a car. Thus it was regarded as an add-on, relevant for a minority. This has changed completely. For about 25 years public transport has defined itself as supply oriented at least in the densely urbanized regions. Supply orientation means offering quality and frequent services in order to convince those customers with alternatives to use public transport. This can be achieved by offering quality in terms of vehicles, modernity, a positive image, frequent services in accord with the rhythms of a modern society (that is, for night-time leisure activities and not, as in the past, only to fulfil basic needs).

Understanding public transport as an integrated system: In the past public transport was traditionally locally oriented. People who wanted to use public transport for one trip across different providers needed to buy different tickets. This made the use of public transport (e.g., for commuting from outside into a city) rather unattractive. Both the interest in offering users journeys by public transport that required just one ticket and changed federal and state legislation allowed and pushed the installation of the so-called 'Verkehrsverbünde' (public transport associations). These allowed for integrated ticketing and integrated timetables geared mainly toward potential customers living outside the cities. As a result commuting with public transport, for example, became cheaper and easier, with the main result being that public transport has become a *system* from the perspective of the potential customer.

Making public transport cheaper: Apart from the usual subsidies it makes sense to make public transport cheaper for certain target groups. Usually the fixed costs of car ownership heavily impact modal choice.

Job-tickets as season passes for public transport, especially for work commuters: Usually employers offer parking spaces for their staff, which is expensive in the cities. But where subsidized tickets for commuters are introduced both firms and employers can save money since no parking space has to be offered. This move can be triggered by local authorities by taking away the existing obligation to create parking spaces. A subsidized job-ticket is also aimed at commuters from outside, but clearly the effectiveness of a job-ticket is dependent on public transport of a reasonable and acceptable quality (supply orientation). Also here the complementarity of measures becomes obvious.

Student tickets: Since the mid-90s most German universities have cooperated with the transport associations in the introduction of subsidized students passes.

The effects are multifold: Students of today are motivated to organize their lives without a car, and even after having previously lived in a suburban setting, they 'learn' to behave multimodally. That, as mentioned above, for people under 30 car use is declining, can be regarded partly as a result of the more-or-less obligatory student passes for public transport. The relevance of these policies must not be underestimated.

Improvement of the Conditions for Cycling

Often cycling and public transport are considered as competitors in that their clientele are those who do not have a driving license or who cannot afford a car.

But the combination of cycling and public transport allows for synergies that promote life without a car. From the point of view of the traveller both the bicycle and public transport are specialists: They both have their fields of application where they are superior to the car, but for both modes these fields are typically small and specialized. As mentioned above, a private car is universal. Thus, for an individual to cope with the different challenges in daily life without a car a combination of modes is necessary. This, at least within cities, can be achieved by the complementary combination of public transport and the bicycle.

In terms of measures and policies this means the creation of a climate that is positive to, and appropriate for cycling. These are typical pull measures.

These will not be discussed in detail. It should be mentioned that the creation of cycling schemes, the introduction of cycling paths and cycling lanes, the creation of real networks as well as the permanent observation of specialized lanes is today state-of-the-art in presumably every German city. It should be mentioned that all the measures for cycling are comparatively cheap, at least when compared to the cost of the infrastructure for public transport and, of course, the car. And beyond this, the system's perspective should be mentioned as well. All the measures can only be effective when introduced area-wide. The relevance of the measures has to be seen in their totality and in their cumulative effects on demand. Only a combination of measures will be help to create a climate in which cycling is regarded as an accepted mode for a major part of the population.

Cycles for rent: This measure does not aim at the typical cyclist but much more at those who basically use public transport but need the bicycle as a complementary mode once in a while. Therefore, the implementation of such schemes occurs in cities where public transport customers from outside are commuting into the city or are coming in as visitors.

Installation of Car Sharing as a Backup System

A multimodal system has to allow for mobility in every situation and without the need for car ownership. But in a modern society, which developed over decades within framework conditions favorable to the car, many people who rely mainly on public transport and the bicycle will still need a car at least once in a while. Car-sharing closes this gap and has to be understood as an additional pull measure to encourage multimodal behavior with reduced car use. Up to today formal, commercially-organized, car-sharing systems have been introduced in many German cities where the system of public transportation is reasonably good.

Car ownership is usually not given up immediately; individuals learn to make use of the car less and less frequently until one day the car becomes more and more a burden (e.g., the finding of a parking space) rather than an advantage and, if the situation is right, they will get rid of it (Chlond and Waßmuth 1997). Also, for car sharing role models are important. What was regarded as a niche market for environmentalists some years ago has in the last few years become a status symbol for a modern life style. The use of modern technologies for booking a vehicle eases access and opens up the possibility of achieving a lifestyle in which other modes form the basic form of travel and where the (shared) car is regarded only as a backup system.

12.3.3.3 Push Strategies

Parking Policies

Parking management: The most effective measures for reducing car use and thus influencing the modal split are parking policies. The reduction of available parking space, as well as appropriate pricing schemes, can control mode choice effectively, insofar as the local administration has control of demand. This can be difficult, for example where parking spaces are on private ground. Beyond that, both the interests of residents as well as of local businesses have to be taken into consideration. Here also some relevant aspects have to be pointed out.

- Residents should be able to own a car. Commuting from outside by car and parking in residential areas should be prevented. Both can be achieved by a prioritization of residents.
- Sufficient parking space should always be available for the more-or-less necessary[4] commercial traffic. This can be achieved by a pricing scheme that makes long-term parking expensive and that always provides a sufficient number of available parking spaces.[5]
- Reduction in total parking space availability: This can come along with a reconfiguration of road space in combination with measures to improve the living quality in a city.

It should be taken for granted that all parking policies have to be accompanied by an appropriate enforcement!

---

[4] For the definition of necessary car traffic see (Haag 1996).

[5] Experience shows that mainly for business and commercial traffic it is more important to find a parking space at all than it is to find one for free. In other words, business and trade have learned that parking management and parking charges offer an advantage for those who really need a space.

Abandonment of the Obligation to Offer Parking Space

Until recently there has been an obligation in Germany – mainly in the case for new buildings or a changed designation – to offer parking space in accordance with demand and the laws of the federal states. This made building within the cities expensive and, as a consequence of the provision of parking space, encouraged car use. Legislation now allows the municipalities to abandon this obligation both for housing and for office buildings. Instead of this investors have to pay the municipality a fee which can be used for the improvement of alternative modes. The effect of this measure is twofold: car mobility is impeded due to a lack of parking space, and funding for public transport is available from another source, which additionally increases its attractiveness.

Rededicating Road Space and Capacity in the Cities

An improvement in public transport and cycling conditions does not mean at the same time a decline in the conditions for car use. But the preferential treatment of or prioritization of public transport at traffic signals, for example, means that capacity in terms of the number of passengers able to pass through an intersection will be redistributed from car drivers to public transport users. The effects are obvious: car drivers can see that public transport is the better alternative (pull), and, as the patronage of public transport is obviously high this prioritization is accepted. The improvement of public transport is then directly connected to a decline in the convenience of using a car (push). This example shows once again that a concerted and balanced combination of measures can change behavior.

Road Pricing

Additional charges on the utilization of the road infrastructure are certainly highly efficient for pushing car drivers into public transport. Nevertheless, this has not yet been implemented in German cities. Although it is a topic of discussion, this is less from the perspective of mobility management than for funding the road infrastructure (Ammoser 2010). As the legal and regulative framework for parking and spatial policies is comparatively effective the need has not yet been seen to foster road pricing schemes. Furthermore, there are obstacles in the form of different levels of responsibility for the road infrastructure (federal, state, local) and, once again, competition between municipalities.

12.3.3.4 Conclusions

This catalogue of measures is neither complete nor explained in detail,[6] and it must be emphasized that the relevance and contribution of each of the mentioned

---

[6] Chapter 14 of this book gives a rather comprehensive overview of measures that may be introduced to influence travel behavior.

framework condition developments, factors, policies and measures, cannot be separated from each other. Each of the mentioned factors has some influence but all are complementary to each other (or with other more mathematical words, certain are necessary but no one will be sufficient).

12.4 The Example of Karlsruhe – The Development Toward a Multimodal City

12.4.1 Why Karlsruhe?

The policies as mentioned above will be illustrated, taking the city of Karlsruhe as an example in which the framework conditions for using a car and the possibilities for use of other alternative modes have changed massively. Other cities are pursuing similar policies and overall concepts; nevertheless, the example of Karlsruhe is comparatively well developed.

Karlsruhe is located in Southern Germany on the Rhine Plain near the border with France. The city was founded in 1715 as the capital of Badenia and today has 285,000 inhabitants and can be regarded as a rather typical German city. Karlsruhe and its surroundings belong to one of the wealthier regions in Germany (administration, research and industry), most of the households have no financial reason to prevent ownership and operation of their private cars, and the level of car ownership is consequently high. The city is surrounded by important car-manufacturing plants and other firms connected to the automotive industry. Karlsruhe can therefore be regarded as a 'car town', and, incidentally, Carl Benz, who is regarded as the inventor of the automobile, was born in Karlsruhe and studied at the university. Following general trends, Karlsruhe followed the usual path with the building of road and parking infrastructure into the 70s.

Nevertheless, Karlsruhe has managed to reduce the role of the car considerably during the past decades. The measures and interventions and their effects can be regarded as an example of the possibility of changing behavior and creating framework conditions allowing for mobility with less car usage and a reduction in carbon dioxide emissions without a loss of life quality. As a consequence more and more people have become aware of the possibilities and the use of alternative modes. This has caused a change in perspective. Additional measures to support multimodal behavior and alternatives to the car are better accepted than ever. Beyond that, the early adopters form a role model for those who follow both now and in the future. The process has developed a kind of self-reinforcing process as the adaption to changed behavior gains speed, at least through the resulting demand.

It has to be clarified that there was a development process, that massive investments were necessary, and that this development was not possible within only a few years. The results, however, give scope for optimism. The example shows that with development of the supply side ('Pull') it is possible to slowly develop also 'Push' measures that encourage appropriate effects.

12.4.2 The Pull Measures

12.4.2.1 Development and Improvements of Public Transport

During the 60s and 70s many German cities decided to abandon their tramway systems. Karlsruhe did not follow this trend, but rather extended the existing dense tramway network into newly-built urban districts. A previously abandoned railway was reconstructed in 1979 as a light rail line connecting the outskirts directly with the city centre, which provided seamless travel without the need of a transfer for the passengers. The result of the success of this integration was the birth of the idea of also connecting other existing railway lines around Karlsruhe with the tramway network in the form of a light railway system, in order to create direct public transport connections from the region right into the city centre of Karlsruhe. In addition to the necessity of building some new infrastructure (e.g., 'transition points' as links between the tramway and the railway tracks), new light rail vehicles had to be developed that were compatible with the different power supplies, signal systems and technical standards of both the railway and the tramway system. This kind of extension began no earlier than 1992.

The *Karlsruhe Model* (Karlsruher Modell 2010) incorporates the seamless transition from an inner-city tram trip to a regional train journey. Karlsruhe has a dense public transport network both with tramways and regional light rail which provides commuters with fast and convenient direct travel without additional transfers from outside to their destinations in the city centre. The network development incorporates massively upgraded existing infrastructure, revitalized railway lines, which had been abandoned decades earlier, as well as completely newly built infrastructure.

This process of the development of the network and services was accompanied by the foundation of the Karlsruhe regional public transport association in 1994. This allowed for integrated ticketing, integrated timetables and other improvements in the services, with increased utility, mainly for the customers living outside Karlsruhe. Commuting with public transport became cheaper, easier and, by reasons of the infrastructure enlargements and direct connections into the centre of Karlsruhe, faster and more convenient. It is a matter of course that all other potential policies for making public transport more attractive have been introduced (e.g., cheap student passes, electronic ticketing, information systems, etc.).

With few exceptions all tramway/light rail infrastructure within Karlsruhe is also separated from car traffic, allowing for fast and punctual travelling that avoids traffic jams.[7] Another relevant element is the prioritization of public transport modes at signalized intersections, which allows for fast and comfortable public transport rides without any stops.

Altogether the authorities in Karlsruhe and surroundings were able to develop a public transport system that is a role model in the world. Most of the improvements in infrastructure outside the city into the surrounding region have been built within a period of only two decades. At the beginning there was a lot of skepticism, as the

---

[7] In 2010 the construction of a tram tunnel was started in order to avoid traffic jams caused by tramways.

more car-oriented population in the region did not perceive the potential advantages. This changed rapidly leading to competition amongst the municipalities as to who will be connected to the public transport network next.

Unfortunately, the success of the public transport improvements in terms of demand increases has not been documented in detail. A prominent example is the conversion of the former railway line to a light rail line between Karlsruhe and Bretten in 1992, which caused a more-or-less immediate increase in ridership of roughly 400%. Up to 2001 the demand on workdays had risen sixfold over former demand. 40% of the total demand is said to be former car traffic, resulting in a reduction in car use within this corridor of roughly 10% (Bahn-Ville 2002). The success of this first new line resulted in some consequences, with all the other municipalities in the region wanting to follow this example of a transfer-free connection into Karlsruhe's city centre. New transit stops in Bretten have improved the availability of transit to housing areas, schools, and companies. Based on the great success of the initial tram line, political officials were encouraged to extend the light-rail system. Further development was financially supported by federal funding schemes.

It has to be mentioned that the administration of Karlsruhe and of the neighboring municipalities as a consequence made use of the existing funding schemes of the federal government to develop this infrastructure, and that there was growing consensus in the municipal or regional councils, even among different political parties, to also fund the necessary 'own contribution'. This has grown as a result of the scheme's obvious success, as illustrated by the increase in ridership within the regional public transport association of 73% between 1995 and 2009 (Karlsruher Verkehrsverbund 2010).

12.4.2.2 Improving the Conditions for Cyclists

The bicycle development plan is rather recent. Before certain measures could be implemented reductions in car traffic load had to be achieved in order to improve and enlarge the bicycle infrastructure. From 2008 onward former car lanes (or lanes which had been used for parking) have been rededicated for cyclists. Acceptance is high. What was impossible because of protests by car drivers some years ago is now widely accepted. Central elements of the bicycle development plans are (amongst others):

- Signposting and development of dedicated bicycle networks, mainly to increase the accessibility of the city for cyclists, e.g., by opening of one-way streets for cyclists against the main direction of traffic
- Dedicated places for parking bicycles
- Increasing traffic safety by creating own lanes/elimination of black accident spots
- Rededicating car lanes into cycle lanes
- Installing a bicycle renting system in 2007 in cooperation with Deutsche Bahn (the German national rail company). This additionally illustrates the complementarity between public transport and the bicycle.
- A lottery for new bicycles among new residents in Karlsruhe (mainly students) to offer an incentive to change behavior.

12.4.3 Push Measures

It has already been mentioned that any restrictions against car driving are difficult to achieve. Therefore, the idea of 'pushing' has to be understood as a comparatively smooth process. The push measures must never be regarded as harassment or chicanery.

Success in modal terms can more or less be traced back to the improvements in the public transport system. However, some improvements for public transport and cycling mean at the same time a worsening in the conditions for car use. But these *push*-measures were accepted as they could be comprehended as reasonable. At all intersections, where public transport and car traffic is competing for the scarce time slots, public transport has full priority. Car drivers can see that public transport is obviously the better alternative (pull), and as a result the patronage of public transport is high. The improvement of public transport is directly connected with a worsening of car transport comfort (push).

The same holds true for the rededication of former car lanes for cyclists: With the increasing demand for cycling the attractiveness of the intervention becomes visible. As a consequence more car drivers will use a car less frequently and use a bicycle instead.

One of the most relevant push factors has been the situation in terms of parking. As the rate of car ownership is high in Karlsruhe, parking space is a scarce resource, and especially in the districts around the city centre, demand has always been higher than supply.

Parking policies have been implemented according to the principles already mentioned:

- Residents of a city quarter are favored.
- Availability of sufficient parking space for the relatively necessary commercial traffic.
- Traffic calming since the 1980s with a reconfiguration of road space (effects: reduction in parking space, improvement of living quality through the greening of urban streets).

The results are:

- Short-term parking is available everywhere at a comparatively high cost: Those who really need to use a car may do so.
- Parking space became available for the residents, but only for residents and only in the district of the resident's own apartment.
- Long-term parking space is only available at a cost (with the exception of parking space in enterprises for commuters and for shops).

12.4.4 Complementary Measures

It has become clear that the combination of pull and push measures have been proved successful. As a result, many inhabitants, mainly in the central districts of the city, do not really need a car any more. They are able to perform their basic mobility needs with public transport and the bicycle. But even in an optimally public-transport equipped city like Karlsruhe a car will be needed at least once in a while, for example, to go to destinations that are inaccessible even by means of the excellent public transport system.

To close the gap between having a car and not having a car and being totally reliant on public transport and the bicycle, a car-sharing system was installed in 1995, following the example of the city of Zurich in Switzerland. Interesting enough the car-sharing system was not set up by the administration but by private initiative and capital which identified the obvious gap. Of course, the administration supports the car-sharing system, which also collaborates with the public transport association.

Figure 12.4 shows the development of participation up to today. It can be assumed that the increasing number of participants means at the same time that most of those either do not own a car any more, or did not buy a car when otherwise, in the absence of the system, they would presumably have done so.

Interesting enough the increase in participants within the last two, to three years is higher than before. This diagram illustrates that the development toward a less car-oriented multimodal city takes time, but that as soon as the conditions are set appropriately, the process can become self reinforcing.

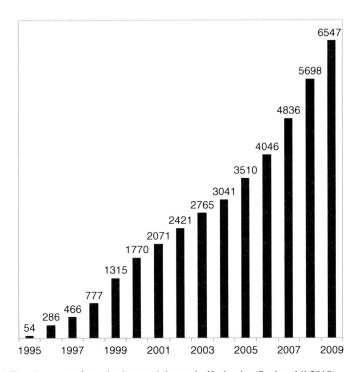

Fig. 12.4 Development of car-sharing participants in Karlsruhe (Stadtmobil 2010)

Of course, it is obvious that the car-sharing participants live mainly in those neighborhoods where there is excellent public transport and where, on the other side, the usability of a private car is comparatively poor. Up to now this is mainly the case around the city centre or neighboring districts.

12.4.4.1 Spatial Policies

Through the process of de-industrialization in Karlsruhe some former industrial sites as well as former military and railway sites became redundant. These brownfields are today built to a high density with expensive, high quality apartments with a restricted number of available car parks. Nevertheless, accessibility to and integration into the public transport system is taken for granted as is the integration of car sharing. As a result the number of inhabitants in the city centre could be increased, with these new residents usually showing a multimodal behavior, as the system enabled them to do so.

Would it have been possible to follow an alternative path? Of course the process of suburbanization is still continuing within the region of Karlsruhe, which by reason of its attractiveness has still increasing numbers of residents. Nevertheless, with the existence of the regional light rail system, which seamlessly connects the suburban space with the city centre, multimodal behavior is also feasible for residents living outside the city. The regional planning authority successfully channelled urban sprawl by controlling the development of new housing areas in the suburban parts of the regions alongside the development axes of the public transport system. Commuters may use public transport directly by means of the light rail system or at least take advantage of intermodal combinations such as park and ride as well as bike and ride.

As an illustration of the positive effects of such planning principles: the price of real estate with good accessibility to public transport is usually comparable to that at the edge of the city (Bahn-Ville 2002).

Here the combination of infrastructure measures (light rail) and planning principles (alongside the axes of public transport) slowly shows effects.

Another relevant development worthy of mention is that as a result of the excellent accessibility and increased attractiveness of the city centre, new shopping malls have developed within recent years which are performing extremely successfully. Fifty six percent of all shopping customers in the city of Karlsruhe use public transport.

12.4.5 Effects

The effects of the integrated strategies in Karlsruhe are difficult to quantify. As has been shown above, the public transport system is gaining ridership and the number of car-sharing participants is growing. Unfortunately, no overview exists to illustrate what has been achieved so far in terms of car-kilometers saved or carbon dioxide emissions prevented. Travel behavior surveys that could illustrate the behavior of

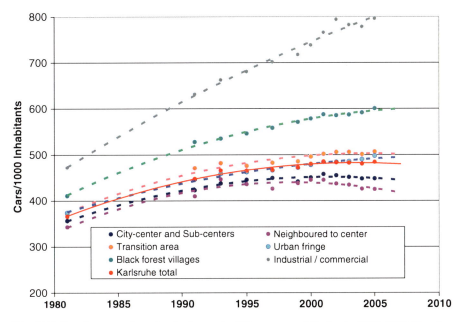

Fig. 12.5 Development of car-density figures in different parts of Karlsruhe 1980–2005 (based on data of the Office of statistics of the City of Karlsruhe)

the population today compared to the situation decades ago are not available (at least not for the current situation). Nevertheless, there are indicators that show that the development and the measures are successful.

Figure 12.5 illustrates the success in terms of car ownership. In spite of the economic success of the city, total car density in Karlsruhe is stagnating and decreasing within the city centre and the neighboring zones. Also this process corresponds with the earlier described developments. In the suburbs where public transport is not as well developed compared with the centre car densities are still growing (e.g., Black Forest villages). These districts were settled decades earlier and as a result the population today has a higher mean age and more conventional habits. Nevertheless, it has to be mentioned that these are also mostly multimodals who are likely to use public transport to destinations in the city centre. But the younger population, which has to make decisions on location and mobility style, more often decides to live in the central areas where a car is frequently not really helpful.

It is worth mentioning also that the stagnation/decrease in car ownership is a relatively recent development. This illustrates, as well, that a lot of effort has been necessary, and which is only today yielding fruit. Also, the numbers given in Fig. 12.4 (development of car sharing customers) show that the development and the success of the measures implemented is still under way.

Beyond that, according to the *Downs-Thomson-Paradoxon* (Mogridge 1990), the conditions for car traffic that remains have improved, which also affects emissions positively through a reduction in congestion. Note that because of the reduced traffic load investment in alternative modes did not negatively affect travel speeds for cars.

The accessibility of Karlsruhe and its permeability by car is high. Anyone who really needs to go by car can and is allowed to do so.

12.5 Appraisal

Karlsruhe is probably the city in Germany with the most advanced public transport system for its size. Also, the implementation of the other measures is well advanced, so it can be regarded as an early bird. Nevertheless, most cities in Germany do, in principle, follow the same kind of policies, at least in recent years, by changing the framework for the competing modes. The objective of reducing greenhouse gas emissions has become within recent years a central element in the transport master plans of nearly every city in Germany. The illustration in Fig. 12.1 suggests that similar processes are likely to take place all over Germany, at least in urban areas.

Nevertheless, the effects and results of the policies are becoming visible comparatively 'late', or delayed in relation to their first implementation. That is, instantaneous success or immediate reaction to implemented pull measures can usually not be expected. It took decades for the development of the public transport system and the implementation of measures to reduce car use. Nevertheless, the combination of pull- and push-measures has slowly shown results, not so much because large numbers of individuals have changed their behavior but much more because new generations are enabled to implement a different modal behavior than their predecessors of one or two generations earlier. Perhaps the success became visible earlier in Karlsruhe because the exchange of population is comparably high. By reason of the large number of jobs in research institutions and the different universities, there are a lot of dynamics in the population. The number of people who are confronted with the need to decide where to live and how to be mobile is high. Sooner or later more and more people find themselves in a situation where they have to make similar decisions. This gives an opportunity for politics and planning. Global framework conditions are likely to change by themselves, as the processes are easy to understand, are predictable and calculable.

And therefore administrators and politicians have to develop appropriate frameworks on the one side but also must send signals through appropriate policies that enable people to adapt their behavior accordingly. The example as shown illustrates that this challenge can be managed.

If such a turnover is feasible in an environment that has been developed for decades in terms of economic, societal and spatial structures favoring the car, this can be achieved anywhere else, at least in regional contexts with a certain degree of urbanization. And beyond that, a development toward a more sustainable transport system would be even easier to manage and to fund in a society that had not yet become car-dependent. The latter case roughly describes the situation in many societies worldwide. In those cases only development has to be managed and funded, and is likely to be more cost efficient compared with the redevelopment of structures that were formerly car oriented.

This gives hope for optimism.

References

Ammoser H (2010) Vom Tabu zur Entscheidung – Perspektiven der Mauterhebung in Deutschlands Metropolen und Ballungsräumen (From a tabu to a decision – perpectives of road pricing in Germany's metropolitan regions). Internationales Verkehrswesen 62:23–24

Bahn-Ville (2002) Bahn.Ville AP3/Karlsruhe (Report of workpackage 3 about the effects of the Stadtbahnlinie Karlsruhe – Bretten). http://www.bahn-ville.net/fr/2_etapes/phase_3/ligne_de/WP3_DE-2.3.Karlsruhe_de.pdf. Accessed 20 Nov 2010

Chlond B, Waßmuth V (1997) Can car sharing substantially reduce the problems arising from car ownership – some empirical findings from Germany. In: Publications and conference proceedings of the 24th European transport conference 1997, Seminar C, P413, pp 291–309

Chlond B, Kagerbauer M, Kuhnimhof T, Ottmann P, Wirtz M, Zumkeller D (2009) Erhebungswellen zur Alltagsmobilität (Herbst 2008) sowie zu Fahrleistungen und Treibstoffverbräuchen (Frühjahr 2009) (Report of the annual survey of the German Mobility Panel 2008 and 2009). Institut für Verkehrswesen, Karlsruhe. http://mobilitaetspanel.ifv.uni-karlsruhe.de/de/downloads/panelberichte/index.html. Accessed 17 Oct 2010

Dargay J, Chlond B, Hanly M, Madre J-L, Hivert L (2003) Demotorisation seen through panel surveys: a comparison of France, Britain and Germany. Conference paper at: 10th International Conference on Travel Behaviour Research, Lucerne, 10–15. August 2003

Esso (2010) Time for an average German employee to work for affording 1 liter of gasoline. http://www.esso.de/auftanken/rund_ums_zahlen/statistiken/minuten.html, Accessed 11 Jan 2011

German Ministry of Finance (2010) Entwicklung der Energie-(vormals Mineralöl-) und Stromsteuersätze in der Bundesrepublik Deutschland (development of the taxation on energy (formerly fuel) and electricity in the Federal Republic of Germany). http://www.bundesfinanzministerium.de/nn_4192/DE/BMF__Startseite/Service/Downloads/Abt__IV/060,templateId=raw,property=publicationFile.pdf. Accessed 17 Oct 2010

Haag M (1996) Notwendiger Autoverkehr in der Stadt (Necessary car traffic in urban environements). Fachgebiet Verkehrswesen, Universität Kaiserslautern

IFMO (2011) Mobilität junger Menschen im Wandel. Institut für Mobilitätsforschung (BMW group), München

Karlsruher Modell (2010) Das Karlsruher Modell (Karlsruhe model). http://www.karlsruher-modell.de. Accessed 20 Nov 2010

Karlsruher Verkehrsverbund (2010) Verbundbericht 2009 (Annual report of the Karlsruhe public transport association). http://www.kvv.de/kvv/documentpool/KVV_Verbundbericht_2009.pdf. Accessed 15 Oct 2010

Knoerr W, Kutzner F, Lambrecht U, Schacht A (2010) Fortschreibung und Erweiterung 'Daten- und Rechenmodell: Energieverbrauch und Schadstoffemissionen des motorisierten Verkehrs in Deutschland 1960–2030. Bericht im Auftrag des Umweltbundesamtes (Update of the data and calculation model for energy consumption and pollutants of motorized transport, report on behalf of the German Federal Environment Agency). IFEU, Heidelberg

Kuhnimhof T, Buehler R, Dargay J (2011) A New Generation: Travel Trends among young Germans and Britons. Paper presented at the 90th Transportation Research Board Meeting. Accepted for publication in the Transportation Research Record

Kuhnimhof T, Chlond B, von der Ruhren S (2006) Users of transport modes and multimodal travel behaviour: steps toward understanding travelers' options and choices. Transportation Research Record 1985, J Transp Res Board, Washington 2006, pp 40–48

Mogridge, M (1990) Travel in towns: jam yesterday, jam today and jam tomorrow? Macmillan Press, London

Stadtmobil (2010) Stadtmobil is the largest provider of car-sharing systems in Germany. http://www.stadtmobil.de/karlsruhe/index.html. Accessed 15 Oct 2010

Zumkeller D, Ottmann P, Chlond B (2006) Car dependency and motorization development in Germany. Report PREDIT 0303C0029 on behalf of ADEME (French Agency for Environment and Energy Issues) Institut for Transport Studies, Karlsruhe

Chapter 13
National Road User Charging: Theory and Implementation

Bryan Matthews and John Nellthorp

Abstract The potential benefits of introducing road user charging, in terms of limiting the negative effects of driving such as congestion and harmful emissions, outstrip those that could be achieved with fuel tax alone. In this chapter, we explore the theory and evidence underpinning road user charging, with a particular focus on the role of the policy in addressing climate change damage costs. In practice, countries have been hesitant to adopt national road user charging for cars. With the exception of the city state of Singapore, none has yet done so. Instead, road-related taxes, mainly on fuel and car ownership, are commonplace and a major source of tax revenue. It emerges that there is a diverse patchwork of approaches to charging for road use throughout the world, and varying calls for reform. Greenhouse gas reduction is not always at the centre of the road user charging debate. However, national road user charging appears to offer a holistic solution: it is therefore interesting to explore what role within it climate change costs should take, and what impact on climate change such a solution might have.

> *If road pricing was implemented nationwide, we would all face different prices for the trips we make. When we travelled on uncongested roads we would generally pay less, but on congested roads we would generally pay more. Paying the family road bill would probably be like paying the phone bill (DfT 2004a:2).*

> *The benefits of road pricing come not so much from the overall cost, but from the differentiation in cost that it makes possible. Major benefits could be obtained without road users overall paying more than they otherwise would in fuel duty. But additional revenue could fund more transport infrastructure or services, as well as providing higher environmental benefits (DfT 2004a:3).*

B. Matthews (✉)
Institute for Transport Studies, University of Leeds, Leeds LS2 9JT, UK
e-mail: b.matthews@its.leeds.ac.uk

13.1 Introduction

Road user charges have been the subject of considerable analysis, discussion and debate over the past 50 years. For much of this period, interest has focused on their potential as a policy tool to combat traffic congestion (Vickrey 1969, Walters 1961, 1968) and to raise revenue in an efficient manner (Mohring and Harwitz 1962, Newbery 1989). Over time, the context of the debate has broadened and attention has focused on the notion of establishing efficient prices designed to internalize external costs, said to encompass road damage costs, congestion costs, accident costs and environmental costs (European Commission 1995, 1998, Nash and Matthews 2005). In this chapter, we explore the theory and evidence underpinning this modern interpretation of road user charging, with a particular focus on the role of the policy in addressing climate change damage costs.

National road user charging is the logical extension of the policy to all roads, which has become technically feasible due to a combination of recent technologies, in particular:

- automatic number plate recognition (ANPR), which is used to enforce the area permit system in central London;
- microwave systems, which are used to collect a charge from an onboard unit (OBU) in passing vehicles in Singapore and Trondheim, and on specific links in Melbourne, Toronto, Italy, California, London and the West Midlands (UK); and
- cellular phone networks (GSM/GPRS) and global satellite navigation systems (GNSS), which are used to deliver the national Heavy Goods Vehicle (HGV) charge in Germany, and have the great advantage of wide national coverage (DfT 2004b).

In practice, countries have been hesitant to adopt national road user charging for cars. With the exception of the city state of Singapore, none has yet done so – although the UK has carried out a full national feasibility study (DfT 2004a). Instead, road-related taxes, mainly on fuel and car ownership, are commonplace and a major source of tax revenue.

Car ownership taxes are beginning to be differentiated by CO_2 emissions performance, and this is now the case in 17 of the 27 European Union (EU) countries (ACEA 2010). For example, in the UK the emissions-based annual tax rates shown in Table 13.1 apply.

To get a sense of how these tax levels relate to the annual climate change cost of car driving, we can take an average UK car emitting 208.3 grams of CO_2 equivalent per kilometer (gCO_2e/km) (Defra/DECC 2010) over 13,550 km per annum (DfT 2010a). Using a carbon reduction value of £188 per tonne of carbon equivalent (TCE) in 2010 on the central estimate (DfT 2010b) gives an annual greenhouse gas (GHG) cost of £155. For comparison with the table above, such a vehicle would be labelled in Band I – the official estimates of 'real world' GHG emissions factors are 15% higher than the labelled value applied to new vehicles (Defra/DECC 2010). It

13 National Road User Charging: Theory and Implementation

Table 13.1 CO_2 differentiated car ownership tax in the UK

Band	CO_2 emissions, g/km	Annual tax for petrol or diesel car, 2010–11 (£1≈1.2 Euros)
A	up to 100	£0
B	100–110	£20
C	111–120	£30
D	121–130	£90
E	131–140	£110
F	141–150	£125
G	151–165	£155
H	166–175	£180
I	176–185	£200
J	186–200	£235
K	201–225	£245
L	226–255	£425
M	over 255	£435

Source: DVLA (2010)

therefore appears that the annual car ownership tax is higher, at £200, than the GHG cost – at least on the central estimate.

Taking into account the GHG emissions associated with car-based mobility as a whole changes the picture somewhat. Using estimates by Chester and Horvath (2009) in a US context to give broad-brush adjustments for fuel production (not only fuel combustion) and for vehicle production and maintenance, suggests emissions per car kilometer should be increased by +17% and +19% respectively, giving a total of £210 per annum.

On the other hand, focusing only on the emissions differential between the 'average car' and a Band A vehicle labelled at 100 gCO_2e/km, for which vehicle tax is zero, the incremental GHG costs are potentially £85 based on fuel combustion alone, or £116 on the wider measure. Of course all these calculations are very rough.

So in the absence of national road user charging in most countries, vehicle ownership taxes appear to be playing an increasingly important role in confronting the consumer with the costs of GHG emissions, and incentivising us to purchase more carbon-efficient vehicles. However, these arrangements fail to provide a *marginal* signal to efficient car use, when individuals are deciding whether to make a particular trip by car or how far to drive. A marginal signal would need to be linked to the cost of making a particular trip so as to influence individuals' decisions on whether, how, when and by what route to make that trip. Currently, the 'out of pocket' costs to the individual of making a particular trip, as we will illustrate, represent only a part of the costs to society of making that trip, and hence currently serve as a poor marginal signal.

Both fuel tax and road user charging, in its various forms, have the potential to provide a much more efficient marginal signal, both being closely linked to actual car use. In general, road user charging has much greater potential to be linked to the particular costs of particular trips than the relatively 'blunt' instrument of fuel tax.

Consequently, the potential benefits of introducing road user charging, in terms of limiting the negative effects of driving such as congestion and harmful emissions, outstrip those that could be achieved with fuel tax alone. We will consider whether this is true for climate change costs alone in Section 13.2. On the other hand, the cases for road user charging and for fuel tax are also influenced by considerations of the ease and costs of implementation, and by their ability to reflect the other negative impacts of driving. We will consider these too within this chapter.

There is a limited but growing body of experience of direct charges for the use of roads being implemented, in the form of 'congestion charges' for access to cities and tolls on major highways. We will outline this experience, and describe important ways in which it lays the groundwork for national road user charging.

At the same time, a number of European countries have introduced national distance-based road pricing systems for heavy goods vehicles using their national motorway networks: Austria in 2004, Germany in 2005 and the Czech Republic in 2007. While these are based on infrastructure capital, maintenance and operational costs and might be seen as an alternative means of roads finance, they do incorporate some price differentiation according to environmental factors. The one European country to have introduced charges for HGVs on all roads which explicitly include environmental costs is Switzerland (not a member of the EU), part of whose charges are earmarked for investment in new rail infrastructure (Nash et al. 2008b).

The European Commission has concentrated on establishing a common framework throughout the EU for the charging of road goods vehicles, leaving charges for private car use to the member states in line with the principle of subsidiarity. The Commission's framework is based broadly on the notion of internalizing external costs of road transport (European Parliament 2006). Meanwhile, an Interoperability Directive (European Parliament 2004) lays down the conditions for interoperability of electronic road toll systems in Europe, which will help to ensure that car travel across national borders remains straightforward if and when member states do go forward with national road user charging schemes.

As we describe it in this chapter, there is a diverse patchwork of approaches to charging for road use throughout the world, and varying calls for reform. Greenhouse gas reduction is not always at the centre of the road user charging debate. However, national road user charging appears to offer a holistic solution: it is therefore interesting to explore what role within it climate change costs should take, and what impact on climate change such a solution might have.

In the rest of this chapter, we briefly review the theory on which the case for national road user charging is based (Section 13.2), present some estimates of the expected impacts of the policy were it to be implemented (Section 13.3), give a global overview of attempts at implementation in a number of countries (Section 13.4), focus on areas of progress and conversely on the sticking points with this policy (Section 13.5), and finally draw conclusions on the extent to which national road user charging might be a valuable policy tool with which to tackle climate change in future (Section 13.6).

13.2 Theory

13.2.1 Road User Charging as a Pigouvian Tax

The use of a car imposes various costs on the user him/herself (internal marginal costs) and also a set of costs on other users of the transport system and on society more widely (external marginal costs) including climate change (Table 13.2).

The main economic argument for road user charging is that it allows these marginal external costs to be internalized, i.e. the driver to bear them, and so to take them into account when making decisions related to transport. For example: whether to drive to work or take public transport; how far to travel to shop or for leisure activities; how to respond to rising levels of congestion on the road network; or whether/how to react to increasing estimates of the cost of climate change damage. By providing efficient incentives to individual behavior, the policy leads toward an optimal level of resource use – in this case the use of scarce road space and environmental resources in particular. Road user charging is therefore a classic application of a Pigouvian tax (Pigou 1920, Knight 1924).

Table 13.2 Marginal external costs of car use in Great Britain, at peak/off peak times, low estimates, 1998

Cost item	Peak times (0700–1000 and 1600–1900 weekdays)	Off-peak times (other)
Infrastructure operating cost and depreciation	0.0005	0.0005
Congestion	0.1322	0.0701
External accident costs	0.0078	0.0080
Air pollution	0.0018	0.0018
Noise	0.0001	0.0001
Climate change		
• original 'low' estimate, based on £7.3/tCO_2	0.0012	0.0012
• revised 'low' estimate, based on £19.7/tCO_2 (extrapolation from Defra/DECC, 2010, table 3)	0.0032	0.0032
• revised 'low' estimate, based on £19.7/tCO_2 plus 36% for fuel production and car production/maintenance	0.0044	0.0044
Total	0.144/0.146/0.147	0.082/0.084/0.085

Sources: Sansom et al. (2001); additional analysis

13.2.2 Alternative Policy Options

Environmental policy theory (e.g. Baumol and Oates 1988) indicates that Pigouvian taxes are an efficient way to internalize externalities when damage cost is measurable and the damage cost function is linear, not exhibiting a threshold in the expected range. Stern (2006) finds that the damage cost function for climate change is *non-linear* with respect to temperature change, although the damage function is likely to be convex rather than having a distinct step or threshold (Fig. 13.1), therefore we need to ask what policy alternatives exist.

In broad terms, as also explained in previous chapters of this book, the leading alternatives are:

- Standards - i.e. CO_2 emissions standards on vehicles. This can be done, for example, by the EU setting targets for average emissions across the new vehicle fleet (120 gCO_2/km by 2012 (European Commission 2007)), although such a target is vulnerable to an increase in total CO_2 emissions via the quantity of vehicles and/or kilometers travelled. In practice, a standards approach is being pursued in Europe alongside the Pigouvian tax policy. Table 13.3 shows the progress made in relation to the EU targets, based on the Commission's progress report (European Commission 2010).
- Tradable permits. In theory this option is attractive because it guarantees the quantity of emissions – subject to enforcement – while the market price varies to control emissions, clear the market and provide incentives to consumers and suppliers. Under the EU Emissions Trading System, the aviation sector will participate from 2012, however there is no plan to include road transport even in phase 3, from 2013 to 2020 (European Commission 2008, 2009). In the international context, a key disadvantage of permit trading has been the difficulty of negotiating national GHG allowances, which form the basis for a quantity

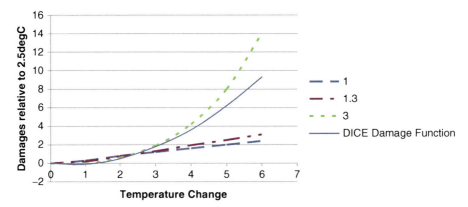

Fig. 13.1 Range of climate change damage functions considered by Stern (2006)
Source: Stern (2006), Fig. PA.1

Table 13.3 Progress relative to EU targets for CO_2 emissions from newly-registered cars

Average CO_2 emissions from new passenger cars registered in the EU, gCO_2/km												
	2000	2001	2002	2003	2004	2005	2006	2007	2008	2009	2012	2015
Actual	172.2	169.7	167.2	165.5	163.4	162.4	161.3	158.7	153.6	145.7		
2007 Target (unlikely to be met)											120	
Target (Regulation (EC) No 443/2009)												130

Source: European Commission (2010), table 2

cap – and an extension of the policy to road transport seems at odds with the goal of reaching agreement based on a narrow set of sectors in the short term.

At least within the next decade, Pigouvian taxation backed-up by vehicle standards appears to be a feasible combination of policies. Tradable permits, making maximum use of the market mechanism to achieve GHG targets, have some appeal in the longer term despite current political and operational barriers.

13.2.3 Detail of a Pigouvian Tax for Road Use Externalities

On a given part of the road network, the driver faces increasing *internal costs*, c_{int}, as the traffic level rises (Fig. 13.2). These internal costs comprise a mix of money and non-money costs, the latter – travel time in particular – being convertible to money using established valuation techniques (Mackie et al. 2003; Bickel et al. 2006). These costs are borne by the driver without any intervention taking place. Table 13.4 lists a typical set of internal costs of car use.

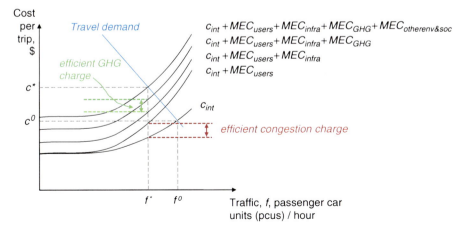

Fig. 13.2 Efficient greenhouse gas and congestion charges

Table 13.4 Marginal costs of road use

To the user, c_{int}	Time
	Fuel
	Consumables (oil, etc)
	Vehicle wear and tear (including tires)
	Internal accident risk
To other users, MEC_{users}	Congestion delay and related costs
	External accident risk
To the infrastructure provider, MEC_{infra}	Wear and tear, depreciation
	Marginal infrastructure operating costs
To the rest of society and the environment, $MEC_{other\ env\&soc}$	Contribution to climate change
	Air pollution
	Noise

Source: Adapted from DfT (2004c)

When a driver decides to join the road and traffic increases by one unit, the impact is not simply a small increase in c_{int} for that driver, but a small increase in c_{int} for all drivers – the road being a congestible resource. The marginal social cost across users is then

$$MSC_{users} = \frac{d(c_{int} \cdot f)}{df} = c_{int} + \frac{d(c_{int})}{df} \cdot f \qquad (13.1)$$

In other words, the impact is the internal cost to the new driver plus a marginal external cost to other users, MEC_{users}, given by the final term in (13.1). By levying a 'congestion charge' equal to MEC_{users}, any congestion-related costs on other users will be internalized.

To incorporate other externalities, including climate change (MEC_{GHG}), additional MEC terms can be added:

$$MEC = MEC_{users} + MEC_{GHG} + MEC_{infra} + MEC_{otherenv\&soc} \qquad (13.2)$$

Each additional term could be a 'flat rate' if the cost item varies primarily with distance ($dMEC/df = 0$), or variable with traffic if it is sensitive to travel time (and hence congestion). Data tables in Sansom et al. (2001) suggest that GHG emissions vary only modestly with travel time: for example, their low estimates of MEC_{GHG} in Table 13.2 are not sensitive to Peak/Off-Peak travel conditions, to two significant figures.

In general, GHG emissions can be expected to vary in proportion with fuel consumption. Defra/DECC (2010, table 2b) gives the following marginal emissions factors for petrol and diesel fuels in 2009:

- petrol 2.272 kgCO$_2$/l;
- diesel 2.563 kgCO$_2$/l.

13 National Road User Charging: Theory and Implementation

The same marginal cost of carbon, £51/tCO₂e, is applied to each, therefore *MEC_GHG varies directly with fuel consumed* for each fuel type:

$$MEC_{GHG} = £0.116/l \text{ (petrol)}$$
$$MEC_{GHG} = £0.131/l \text{ (diesel)}$$

It is easy to appreciate how a Fuel Tax, differentiated between petrol and diesel, can be used to internalize MEC_{GHG} given this data.

Would a road user charge for GHG based on kilometers travelled match the cost data so closely? Figure 13.3 shows how MEC_{GHG} per kilometer varies with speed, based on UK fuel consumption functions (DfT 2010c). The implication is that at low speeds in particular, i.e. in highly congested conditions with an average speed of less than 20 km/h, a constant kilometre-based GHG charge could substantially underestimate the external costs, and under-incentivize demand restraint.

Alternatively, how would a travel time-based road user charge for GHG match the cost data? Figure 13.3 also shows how MEC_{GHG} per minute varies with speed. This time MEC_{GHG} is increasing throughout the relevant range: a charge based on a constant MEC_{GHG} per minute travel time would be likely to substantially overstate the MEC_{GHG} in congested conditions, and substantially understate it at high speeds.

In conclusion, an optimized road user charging system with access to data such as traffic speeds, journey times and distance travelled, but *not* fuel consumption data, would require:

- two out of three variables (speed, *v*, journey time, *t*, and distance travelled, *m*)
- in order to estimate the fuel consumption function per vehicle type

$$\text{Fuel, litres} = \frac{a + bv + cv^2 + dv^3}{v} \cdot m \tag{13.3}$$

where *v* can be substituted by *m/t*, or *m* substituted by *vt* as necessary, and *a*, *b*, *c* & *d* are parameters

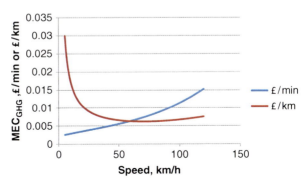

Fig. 13.3 Marginal external cost of GHG emissions, per kilometre and per minute, by traffic speed
Note: Basis: Euro V standard Petrol Car, <2.5t, 1400–2000cc
Sources: TRL (2009), Defra/DECC (2010)

- emission factors for CO_2 from fuel consumed;
- marginal costs for CO_2 emissions in order to set efficient charges to internalize MEC_{GHG}.

13.3 Modelling the Impacts of Road User Charging

13.3.1 At the EU Level

A number of studies have sought to model the expected impacts of road user charging, either on its own or in combination with a package of other policy measures, so as to help demonstrate and quantify the potential benefits of the policy. We review here two such exercises undertaken at the European level: firstly work undertaken as part of the 'Mid-Term Assessment of the EU Transport White Paper' (TML 2006); and secondly work undertaken for the GRACE project (Nash et al. 2008a). The EU Transport White Paper (European Commission 2001) proposed a package of over 50 policy measures aimed at 'shifting the balance' between modes of transport, 'eliminating bottlenecks', 'placing users at the heart of transport policy' and 'managing the globalization of transport'. The detail of these measures are set out in the Annex to the White Paper (European Commission 2001).

The ASSESS Project used a combination of EU-scale models – principally the SCENES Transport Model – to assess possible EU policy implementation scenarios to the year 2010, and the longer term prospects to the year 2020. Four scenarios were developed, in increasing level of ambition:

(i) Null scenario (N-scenario): assumes that none of the White Paper measures has been implemented, neither at the European level nor in the Member States. The N-scenario is the autonomous trend development and acts as the reference case.
(ii) Partial implementation scenario (P-scenario): includes only measures that will most likely be implemented before 2010. This means that the measure is already implemented or that there are clear indications that implementation will take place soon. The latter is the case when approved EU-directives include deadlines for Member States to adapt national legislation accordingly. This scenario is derived from the results of the policy review up to 2005 undertaken as part of the ASSESS project.
(iii) Full implementation scenario (F-scenario): includes all measures introduced in the White Paper.
(iv) Extended scenario (E-scenario): for most measures the extended scenario follows the full scenario while for some measures the partial scenario is followed because there is no indication that the full implementation is feasible. An example of the latter case is kerosene tax. Since global implementation seems infeasible a compromise that applies the tax only to intra-European flights is included in the extended scenario. Additional to this two policy changes

13 National Road User Charging: Theory and Implementation

were introduced. Firstly, the extended scenario includes more pricing measures, most importantly higher prices for freight haulage and introduction of road pricing for passengers. Secondly, it includes a faster uptake by market parties of the opportunities that are enabled by the new EU legislation on liberalization by providing the financial incentives and technological means. This means a faster implementation of the RIS, EMRTS and SESAME technological projects in respectively inland waterways, rail and air transport, a faster introduction of Galileo applications (the European GNSS) and more effort on competitive tendering and market opening in the rail sector to accelerate reform in the passenger sector.

Table 13.5 presents a sample of the results of this modelling work, isolating the traffic and carbon emissions impacts. It can be seen that the N scenario *leads to CO_2 emissions rising although less quickly than transport use*, due mainly to growth in transport activity being compensated for by increases in the fuel efficiency of the road vehicle fleet. It was estimated that emissions would increase in the new EU Member States[1] due to their much stronger growth in transport activity. Compared to N, the F scenario would lead to a very small decrease in CO_2 emissions, associated with the increased use of biofuel. In the E scenario, in which there is full implementation of marginal social cost pricing in the freight sector and partial marginal social cost pricing for passenger car and air transport, CO_2 emissions are significantly lower than those in the other scenarios. The P-scenario is not included in the table because it does not include road user charging measures.

In another EU-scale modelling exercise, undertaken as part of the GRACE project (Nash et al. 2008a), the relative merits of fuel tax and road user charging were explored. This work used the TREMOVE model to analyze the impacts of three pricing scenarios each of varying complexity. TREMOVE allows for the estimation of the demand reactions and modal shifts which follow on from the initial pricing reforms, for the variation of some external costs (e.g. congestion) as a function of the volume of transport, and for the estimation of welfare effects depending on how the transport revenues are used. Furthermore, the pricing scenarios used

Table 13.5 Transport performance in EU25 countries for N, F and E scenarios, relative to 2000 (=100)

EU25		2000	2005	2010			2020		
				N	F	E	N	F	E
passenger km	pkm/year	100	108	117	118	118	135	136	127
tonne km	tkm/year	100	108	117	116	116	139	133	131
CO_2	tonnes/year	100	103	102	103	103	107	107	101

Source: TML (2006)

---

[1] New Member states' are countries that became EU members in 2004, mostly from Central and Eastern Europe.

the most recent (by that time) estimates of marginal external cost generated in the GRACE project. All scenarios were based on the abolition of all existing taxes, charges and subsidies on transport and on non road modes covering their variable costs and marginal external environmental and noise cost. The three scenarios were:

- Scenario 1 – fuel taxes plus a flat rate kilometer charge for heavy goods vehicles;
- Scenario 2 – country and vehicle specific kilometer charges for all vehicles; and
- Scenario 3 – differentiation of the kilometer charge more finely in time and space.

For each of the 3 scenarios two variants are defined that help to understand the role of the use of the net change in transport revenues that result from the policy change. Firstly, it is assumed that all net changes in transport tax revenues are used to decrease general taxes outside the transport sector, in which case 1 € of extra tax revenues collected from non-commuting transport and used to decrease general taxes is given a value slightly higher than 1. This means that this general tax decrease generates a small extra beneficial welfare effect. In the second variant, it is assumed that the change in transport tax revenues is used to decrease existing labor taxes, leading to a much stronger beneficial effect on the labor market; the value of the extra € ranges between 1.26 and 2.52 depending on the national labor taxes. The reason is that taxes are shifted away from labor, directly alleviating the implied distortion of the labor market.

The aggregate results (EU27+4 non-EU European countries) from this work are summarized in Table 13.6.

Whilst these results do not specifically highlight impacts on CO_2 emissions, four key lessons can be drawn relating to the relative merits of fuel tax and national road user charging. Firstly, it is clearly very difficult to use the fuel tax as the only instrument to address all the externalities of cars and motorcycles. Scenario 1 shows that this requires enormous increases in fuel taxes, large increases in tax revenues (by a factor of 3) but only a tiny efficiency gain (if we rule out the pure effect of recycling the revenues to alleviate labor market distortions). Secondly, when a km charge for cars and trucks takes over as the main pricing instrument (scenario 2), revenues are double those in the reference scenario and welfare improves strongly – overall

Table 13.6 Revenues, welfare change and transport use (EU27+4) from GRACE project

In % of GDP	Total revenues	Welfare change when general taxes are decreased	Welfare change when labor taxes are decreased	Change in tonne/km in % of reference	Change in pass/km in % of reference
Reference	2.298	0	0	0	0
Scenario 1	6.224	0.034	1.706	−10.7	−17.4
Scenario 2	5.402	1.191	2.725	−11.0	−11.5
Scenario 3	5.391	1.181	2.702	−10.8	−11.2

Source: Nash et al. (2008a)

transport volumes decrease by some 11%. Thirdly, the benefits of finer spatial and temporal differentiation (Scenario 3 compared to Scenario 2) give higher congestion relief benefits but generate less revenues – because of the large weight given to the increase in tax revenues, the result is that scenario 3 generates a smaller welfare gain than scenario 2 if taxes are equal to marginal external costs – if taxes could be optimized in both scenarios scenario 3 would produce clearly better results than scenario 2. Finally, it is well known that the introduction of a more refined (area and time based) charging and taxing regime increases a scheme's transaction costs (billing, enforcement etc.); this is not yet taken into account in the welfare computation and this needs to be checked region by region as a more refined pricing regime may only make sense in heavily congested areas.

13.3.2 Modelling at the National Level

Having set optimal road user charges equal to *MEC* in equation (13.2) above, on all parts of the network, and allowed demand and supply to equilibrate – a substantial modelling challenge (DfT 2004d) – an assessment can be made of the policy impacts. Table 13.7 shows a selection of findings from the UK feasibility study.

These different pricing policy tests reveal that:

- MEC-based charging would yield the greatest benefits overall, and for congestion reduction in particular;
- greater differentiation of charges, by time of day, area type, etc, increases the total benefits further;

Table 13.7 Key findings from the UK national road user charging feasibility study

	Impacts		Annual benefits, £bn (2010)		
Pricing policy tests	ΔCO_2 emissions	Δ Congestion	Time savings	Environment and safety	Total benefits
Charging MEC: 75 charges	–4%	–48%	+11.8	+0.5	+10.2
Charging MEC: 10 charges, capped at £0.8/km	–5%	–46%	+11.3	+0.5	+9.9
Charging MEC: 9 charges, capped at £0.5/km	–4%	–42%	+10.2	+0.5	+9.0
Revenue neutral: 10 charges	–1%	–41%	+10.1	+0.1	+7.8
Increase Fuel Duty, to raise same revenue as MEC: 10 charges	–4%	–7%	+2.1	+0.5	+2.8

Source: DfT (2004d), tables B3-5

- CO_2 impacts would be relatively modest, mostly –4% or –5%, with the exception of the Revenue Neutral policy which reduced Fuel Duty to offset the aggregate revenue yielded by the MEC-based tariff with ten charges – this policy would produce a much smaller CO_2 reduction and correspondingly smaller environmental benefits;
- increasing Fuel Duty could achieve the same CO_2 reduction as the more complex MEC-based charging systems, but would have little beneficial impact on congestion for an equivalent revenue take from road users.

It is clear from these findings that the optimum CO_2 abatement can be achieved through either fuel tax or national road user charging – provided that the data on speed, distance and vehicle type identified in Section 13.2.3 is collected. In choosing between these two 'good options', it is the congestion-reduction potential of national road user charging versus the additional implementation costs of national road user charging that are likely to be decisive. The ability to differentiate charges by local environmental conditions, not possible with fuel tax, is also a part of the case.

If a country was considering switching over to national road user charging as the best mechanism to manage congestion costs and to generate revenue to cover road network costs, it might be efficient to bundle climate change costs in with that – to eliminate transactions costs associated with collecting Fuel Duty, to simplify the charging structure for the benefit of consumers, and thus to increase acceptability of the policy. In doing so, it would be desirable to differentiate the charge component for climate change (MEC_{GHG}) by vehicle fuel consumption category (per Section 13.2.3), and to vary the charge with both travel time and distance for maximum efficiency. We now turn to experience of implementation, and the lessons learned.

13.4 Attempts at Implementation in a Number of Countries

13.4.1 A Brief Survey

Several countries have reviewed the ways in which they charge for road use, and climate change has figured in a number of these reviews. Hence, in January 2011 the authors undertook a snapshot survey to ascertain and confirm the status of policy toward national road user charging in a selection of 14 countries. The countries were selected not on the basis that they are somehow representative of the average global picture, but rather on the basis that they were known by the authors to have advanced the case of road user charging, in some form or another. The survey was administered via email and was directed to key informants – academics and/or government officials – to whom we are grateful for their responses. The countries covered by the survey were as follows:

- Australia;
- Austria;

13 National Road User Charging: Theory and Implementation

- Canada;
- Czech Republic;
- Denmark;
- Germany;
- Italy;
- Netherlands;
- New Zealand;
- Norway;
- Singapore;
- Sweden;
- Switzerland; and
- United States of America.

For each of these countries, the survey asked two key questions:

1. Does the country have (or has it had in the past) a policy on national road user charging?
2. If not, has there been a feasibility study (or similar)?

Table 13.8 summarizes the results of the survey. Some of the survey responses also volunteered the information that several countries have policies under review at the present time, so where this is the case we included the information in an additional column, otherwise we simply indicate 'not stated' for this item.

It can be seen that several countries from our sample do currently have a policy on national road user charging; most of these countries have also implemented charging in one form or another. In some cases, policies are focused on charges for heavy goods vehicles (HGVs) – Australia; Czech Republic; New Zealand; and Switzerland. In other cases, the policy encompasses all motor vehicles but is limited in some other way, e.g. in Austria where charging policy only relates to the use of motorways, and Denmark, Norway and Singapore where policy is more focused on urban roads. In Australia, Czech Republic, New Zealand, Switzerland and Austria policy is associated with a national implemented scheme. In Norway, the national policy is associated with three implemented urban schemes, whilst in Denmark, there is, as yet, no implemented scheme associated with the national policy. Singapore is the clear exception, where there is a long-standing national policy and a fully operational system of Electronic Road Pricing (ERP) for all vehicles, though Singapore is also exceptional in that it is a city state. Where there are policies relating to inter-urban roads, these tend to be focused on raising revenue to cover infrastructure costs; only in Switzerland is there explicit consideration of environmental costs. Where policy relates to urban pricing, the emphasis tends to be firmly placed on tackling congestion costs.

It can also be seen from Table 13.8 that national road user charging policy is actively under review in a number of countries. Australia's wide-ranging independent review of tax policy, undertaken by former Australian Treasurer Ken Henry, includes proposals to abolish fuel and vehicle registration taxes so long as they

Table 13.8 Results of a snapshot survey

	A policy now?	A policy under review	Implemented scheme	Feasibility study
Australia	Yes; but only in relation to heavy goods vehicles	Yes; as part of the wide-ranging Henry Tax Review	Yes; nationwide scheme but only for HGVs	Yes; but only to examine changes to the HGV system
Austria	Yes; but only in relation to the motorway network	Not stated	Yes; tolls for HGVs and vignettes for cars and motorbikes using the motorway network	No
Canada	No	No	No	No
Czech Republic	Yes; but only in relation to heavy goods vehicles	No	Yes; nationwide scheme but only for HGVs on Freeways and Expressways	Not stated
Denmark	Yes; part of the Green Transport Policy, 2009	Not stated	No	Yes; e.g. FORTRIN and AKTA
Germany	No	No	Yes; nationwide scheme but only for HGVs	Yes; the Pällmann Commission which proposed nationwide road user charging
Italy	No	Not stated	No	None at the national level, though some at city level
Netherlands	No	Yes; exploring moves to transfer fixed car costs to fuel tax, following the rejection of a well-developed national road user charging policy for all vehicles	No	Yes; in connection with the 2009 proposals to introduce national road user charging for all vehicles by 2012

13 National Road User Charging: Theory and Implementation

Table 13.8 (continued)

	A policy now?	A policy under review	Implemented scheme	Feasibility study
New Zealand	Yes; but only in relation to diesel-powered vehicles and HGVs	Not stated	Yes; nationwide scheme of distance-based vignettes for diesel-powered vehicles and HGVs	Yes; e.g. NZ Land Transport Pricing Study, 1996 and Surface Transport Costs and charges, 2005
Norway	Yes; part of the National Transport Plan 2010–2019, but more focused on urban charging schemes than a nationwide scheme	Not stated	No nationwide scheme, but 3 urban schemes	No nationwide study, but several project-specific urban studies
Singapore	Yes; part of the Land Transport Masterplan, 1998	Implemented	Yes; state-wide ERP for all vehicles	Implemented
Sweden	No	Yes, in relation to charges for HGVs and as a means of raising finance	No nationwide scheme, but one urban scheme in Stockholm and another proposed for Göteborg, plus a number of tolled bridges	Yes, but only in relation to HGVs (the ARENA trials) and particular urban areas
Switzerland	Yes; but only in relation to the HVF for heavy goods vehicles	Yes, with a long term view of moving to 'Mobility Pricing'	Yes; nationwide scheme but only for HGVs	Yes; in relation to urban charging for private cars and in relation to 'Mobility Pricing'
USA	No	Not stated	No nationwide scheme, but several local demonstration projects	No nationwide study, but several project-specific urban studies

could be replaced by efficient road user charging (Henry Tax Review 2009). In Switzerland, proposals were put forward in 2007 to start with trials of urban road pricing but were met with parliamentary opposition and were subsequently shelved. However, in light of predicted reductions in fuel tax revenues, new work is now underway to explore mechanisms for shifting away from relying on revenue from fuel tax for the purposes of future financing, toward some other form of infrastructure charging, being referred to as 'Mobility Pricing'. In the Netherlands from 2007 onward, the government had prepared and were making progress toward implementing a national system of road user charging for all vehicles on all roads (see Steen, 2009). However, a change of government in 2010 resulted in those plans being dropped and now, somewhat in contrast with Switzerland, the government is looking at increased emphasis being placed on fuel tax. Policy is also under review in Sweden, where the technical and operational feasibility of a system of charging for HGVs is being examined and, separately, an enquiry to identify the potential for charging to serve as an infrastructure co-financing mechanism is underway.

From Table 13.8 we can identify a third group of countries – those with no stated national policy but which do, nevertheless, have one or more forms of implemented road user charging scheme. Germany has no explicitly-stated policy on national road user charging but has, since 2005, had a nationwide system of charges for HGVs using the country's network of over 12,000 km of motorways (Autobahnen). Furthermore, this scheme is listed as one of the 29 measures included in a 2007 federal package of future climate and energy policies. Sweden has no policy at the national level (though, as noted above it is trialling charges for HGVs) but Stockholm introduced its congestion charging scheme, first on a trial basis from January to July 2006, before it was made permanent in August 2007 following a close-run public referendum. In the USA there is no national policy on road user charging but there has been a number of government-fostered local demonstration projects, generally under the auspices of the 'Value Pricing Pilot Program', to explore the use of pricing to tackle road congestion and financing.

This leaves a fourth group of countries where there is reported to be no policy, no implementation and no national feasibility work. Interestingly, Canada did come close to undertaking a national feasibility study, as it is reported that, in 2008, the federal government issued a Request for Proposals for 'A Comprehensive Study of Pricing as a Tool for Inducing Greater Efficiency and Sustainability in Urban Transportation in Canada', only for the request to be withdrawn just 5 days later after a federal election was called. This points toward the political sensitivity surrounding national road user charging policy, even when simply exploring its feasibility.

13.4.2 The Case of the UK

In the UK, road user charging has been in and out of the political and analytical spotlight numerous times over the past five decades. In 1964, the Smeed committee was charged with studying the technical feasibility of 'improving the pricing system

13 National Road User Charging: Theory and Implementation

relating to the use of roads, and on relevant economic considerations' (Ministry of Transport 1964). In the 1970s the focus turned to London, with the Greater London Council exploring the potential for a supplementary licensing scheme, similar to that introduced in 1975 in Singapore. Later, in the mid-1990s, further studies of the potential for road user charging in London were commissioned by the government, demonstrating a strong economic case. However, on publication, the then minister for transport indicated that the government had no plans to introduce road user charging and stated doubts regarding the technology required as being a key reason.

With a change of government in 1997 came a renewed impetus toward 'integrated transport policy'. By 2000 this had led to the passage of legislation enabling local authorities to introduce urban road user charging, with a previously unheard of acceptance that they could retain and ring-fence the revenues for local transport investment. The government also published its Ten Year Transport Plan, which set out ambitious plans for investment and demand management, and anticipated the introduction of eight urban road user charging schemes by the year 2010. Somewhat separate to these developments, the post of a London Mayor was created and the successful candidate, Ken Livingstone, was swept to office on a program that included, as one of its central policies, the introduction of the London Congestion Charge. The introduction of the charge in February 2003, and its rapid success and acceptance then served to change the national debate again, sewing the seeds for the government minister to commission a full-scale feasibility study in 2004.

The National Road User Charging Feasibility study was charged with advising 'the Secretary of State on practical options for the design and implementation of a new system for charging for road use in the UK' (DfT 2004a). The study was overseen by a Steering Group of stakeholders and civil servants, and published its report in July 2004. The Steering Group's overall conclusion was that 'road pricing would help unblock roads to the overall benefit of the economy and the environment. The time savings and reliability benefits that we would get in return for the prices we pay are potentially large' (DfT 2004a). It was envisaged that the national charging system would need to relate charges to location, the time of day and the distance travelled, but that appropriate satellite-based systems would not be affordable until 2014 and that fitting the required on-board units to the entire vehicle fleet would be a major challenge. Furthermore, the costs of implementing and operating a national charge system were estimated to comprise total start-up costs of at least £23 billion, and annual operating costs of between £2 billion and £3 billion. In acknowledging these substantial costs, the Steering Group noted that they are 'lower than the potential value of the benefits' and that they should be set against the £60 billion a year spent on private motoring (DfT 2004a). Interestingly, reservations were explicitly raised about a mixed system in which direct charges apply to the congested parts of the network, with fuel duty providing the base charge.

Then, in May 2006, the Prime Minister wrote to the incoming minister for Transport urging him to 'advance the debate on the introduction of a national road-user charging scheme' and asking him to 'identify the other key steps for the successful introduction of road-user charging within the next decade' (cited in House of Commons Select Committee on Transport 2009). Also in 2006, Sir Rod

Eddington was asked to advise the Government on the long-term links between transport and the UK's economic productivity, growth and stability, within the context of the Government's commitment to sustainable development (Eddington 2006). A key part of his conclusion, delivered at the end of 2006, was that unless the government introduces a national road user charging scheme by 2015 the UK will require 'very significantly' more transport infrastructure.

In early 2007, a petition on the official Number 10 Downing Street website calling on Prime Minister Tony Blair 'to scrap the planned vehicle tracking and road pricing policy' gained momentum and major headlines, with some 1.8 million signatures. The subsequent Government response to the Eddington Transport Study, published later in 2007, stated that no decision on the introduction of road pricing on inter-urban roads would be made yet, and described it as 'a decision for the future' (DfT 2007). Enthusiasm at the local level has also subsided, with a high profile set of proposals to introduce congestion charging in the city of Manchester, alongside a series of major investment projects, having been clearly rejected in a public referendum in 2008.

So whilst studies have demonstrated that a system of national road user charging in the UK could create substantial benefits to society, and that it could be designed as a sensible way of dealing with carbon emissions and other negative impacts, at the time of writing the policy seems no closer to being implemented than it was ten or more years ago. Still, a number of important voices continue to keep it on the political agenda. For instance the Mayor of London's 2009 draft Transport Strategy states that it will be almost impossible to meet carbon emissions reduction targets or to contain the worsening of road congestion without some form of London-wide road charging. The Royal Automobile Club (RAC) also envisage extensions of charging in London and at the national level. Furthermore, the government's Committee on Climate Change, in their first report to parliament (CCC 2009), finds that, after rapid improvements in vehicle technology, road charging would be the most effective means of achieving the Government's national, long term greenhouse gas reduction targets. They also point out that, in any event, the policy of moving to low-carbon vehicles over time and greater dependency on low-carbon electricity will dictate a change away from duty on petrol and diesel fuel to some alternative.

13.4.3 European Policy

At the level of the European Union, policy regarding infrastructure pricing for road transport largely concerns road freight traffic; the issue of pricing for the use of roads by the private car being an issue left to individual member states. Policy was initially, in the mid-1990s, aimed at limiting competitive problems within the road freight sector caused by the existence of very different methods and levels of pricing for infrastructure use in different member states. For example, vehicles licensed in a country with low annual license duty plus supplementary tolls may have an unfair competitive advantage when competing with a vehicle licensed in a country with high license duty and no supplementary tolls. Thus kilometer charges were permitted to be levied in a non-discriminatory way on heavy goods vehicles wherever they were registered.

Directive 2006/38/EC revised the Eurovignette regime and represents the legal position on European road goods vehicle pricing that was still in force as of this writing (European Commission 2006). The 2006 Directive allows the toll to be applied to all HGVs (vehicles weighing over 3.5 tonnes) as from 2012, replacing the 12 tonnes limit applicable until then. It applies to the trans-European network (TEN) but permits application of pricing to other roads as well. It is also recommended that 'revenues from tolls or user charges should be used for the maintenance of the infrastructure concerned and for the transport sector as a whole, in the interest of the balanced and sustainable development of transport networks' (European Parliament 2006).

In terms of differentiation, the 2006 Directive provides for variations according to a number of factors such as distance travelled, infrastructure type and location, vehicle type and time of day. Thus prices can be differentiated to reflect the key variables determining marginal social cost. However, the legislation still ties the average charge to the average cost of building and maintaining the infrastructure, excluding externalities (except in the case of sensitive areas such as the Alps, where a mark-up can be applied and used to finance alternative transport infrastructure). Although proposals have been brought forward to permit charging for externalities in the level as well as the structure of prices, and a handbook produced on how to measure the relevant external costs (CE Delft 2008), it has been impossible so far to get agreement on their implementation.

13.5 Areas of Progress and Sticking Points

13.5.1 Urban Road User Charging

The first such scheme in the world was introduced in Singapore in 1975. Initially, this was an Area Licensing Scheme to reduce congestion in the city centre. Drivers had to purchase supplementary paper licenses for a day or a month to allow them to enter the defined area between 0730 and 1015. The initial charge was three Singapore dollars (S$3); this was raised to S$4 in 1976. Vehicles with four or more occupants were exempt. Enforcement was by manual inspection. Subsequent modifications involved extensions to the evening peak, the working day and Saturdays, to a set of charging points on expressways, and to all cars however many occupants they had. Different charges were levied for different types of vehicle. The introduction of the scheme led to a reduction in traffic levels of 44%, with subsequent slow traffic growth as the city developed. In 1998 the Area Licensing Scheme was replaced by an Electronic Road Pricing Scheme. Ninety seven percent of the 700,000 vehicles in Singapore were fitted with on board units, in which smart cards were inserted. Gantries at the Area Licensing Scheme entry points and expressway charging points were equipped to identify, interrogate, charge and, if necessary for enforcement, photograph, all vehicles passing. Charges are now levied per crossing rather than per day, and vary by time of day and vehicle type. Charges are revised quarterly to maintain speeds at between 20 km/h and 30 km/h in the city centre, and 45 km/h and 60 km/h on expressways. As a result charges are lower than with the Area Licensing Scheme for much of the day and have been waived on Saturdays but the introduction of the electronic scheme is said to have led to a further reduction in traffic levels of approximately 10–15%. A more detailed assessment of the scheme and its impacts is set out in Chin (2009).

In 2003, London introduced an area licensing scheme known as the London Congestion Charge. At first this covered an 8-square-mile area of Central London. Subsequently, in 2007, the charging zone was approximately doubled by including an area west of the original zone but that western extension was subsequently removed from operation at the beginning of 2011. Drivers wishing to enter the zone between 7:00 am and 6:00 pm, Monday to Friday are required to pay the Congestion Charge. The charge was £5 per day when first implemented but has since risen to £8. However, there are a number of exemptions which apply to motorbikes, mopeds, taxis, buses, emergency vehicles, vehicles using alternative fuels and vehicles whose drivers are disabled, whilst residents of the zone receive a 90 percent discount. It is possible to pay the charge in advance on a daily, weekly, monthly or annual basis, either by phone, mail, internet or at retail outlets. If paid on the following day, the charge is £10. Entry into the charging zone is indicated by a mix of street signs and pavement markings and enforcement is via automatic number plate recognition (ANPR), facilitated by a network of fixed and mobile cameras. A fine of £120 is levied in cases of non-payment, though this is halved if paid within 14 days. Implementation costs in the first 2 years were £190 million – more than twice the amount expected – and annual operating expenses are approximately £130 million. The system has covered its capital and operating expenses every year since its inception, revenues amounting to some £268 million in the year ending June 2008 for example. All proceeds are ring-fenced for spending on improving transport within Greater London.

At the outset, traffic was reduced by 18% and congestion by 30%. More recently, assessments show that traffic levels in London are still reduced but congestion has actually returned to pre-charge levels. The scheme is estimated to have delivered a 19.5% reduction in CO_2 emissions across the charged area (Beevers and Carslaw 2004). This results from a combination of a reduction in vehicle kilometers with the benefit of improved fuel efficiency brought about by increased speed. Even outside the charging area, the benefit of increased speed serves to counteract increases in traffic. For a more wide-ranging review see Santos and Fraser (2006).

The most recent large international city to introduce congestion charging was Stockholm. A system of cordon-based variable pricing was deployed first on a trial basis from January 2006 to July 2006, before it was made permanent in August 2007 following a close-run public referendum in September 2006. The system charges vehicles registered in Sweden when they pass one of 18 'control points' entering or exiting the cordon (based on the CBD) between 6:30 am and 6:30 pm, Monday–Friday. The rates vary from 10 SEK (US$1.50) to 15 SEK (US$2.25) depending on the time of day for crossing a control point, up to a maximum charge per vehicle per day of 60 SEK (US $9). Exemptions from the charge are awarded to motorcycles, buses, taxis, certain alternative fuel vehicles (ECO-cars, LPG, and electric), emergency vehicles, those with disabled drivers and foreign-registered vehicles. There are no resident discounts, except for residents of one land-locked island, from which mainland Sweden is accessible only via the cordoned area. Enforcement is via ANPR, using digital imaging cameras mounted on overhead gantries. The costs of implementation included a 1.3 billion SEK (US$180 million) investment for the tolling system plus a massive 2 billion SEK (US$280 million) investment in related public transport improvements.

In terms of its impacts, the scheme is said to have achieved a 20% reduction in traffic levels. The decline in traffic as a consequence of congestion charging is estimated to have reduced emissions of greenhouse gases from traffic by 14% in the city centre, equating to a 2.7% reduction across Stockholm County (42.5 thousand tonnes). A more detailed assessment of the scheme and its impacts is provided in Eliasson (2009).

13.5.2 Inter-Urban HGV Pricing

Switzerland was the first country to introduce a kilometre-based charge for HGVs, on 1 January 2001. Not being a member of the EU, this charge was not constrained to follow the EU Eurovignette Directive. The Swiss Heavy Vehicle Fee (HVF) is levied on the entire Swiss public road network, applying to both Swiss and foreign vehicles alike, weighing over 3.5 tonnes. It coincided with Switzerland giving way to pressure from the EU to permit heavier goods vehicles, with the weight limit rising from 28 tonnes, first to 34 and then to 40. The charge level of the fee was calculated as the average uncovered cost per tonne-kilometer. The first step was to calculate the uncovered costs of heavy traffic. This included uncovered road infrastructure costs and the monetary valuation of external air pollution, noise and accident costs caused by heavy vehicles; notably, congestion and climate change costs were excluded. This was then divided by tonne-kilometer to obtain the level of charge. The fee varies according to three factors: distance (kilometers travelled on Swiss territory), weight (admissible weight of vehicle and trailer) and the emissions of the vehicle.

Balmer (2003) states that there are three decisive reasons for the political implementation of the HVF. First, before the final implementation project started, it was criticized heavily, but the political deal of introducing the HVF to outbalance the negative effects of the higher weight limit ensured that the project was on safe political grounds again. Second, the way the revenue of a pricing project is used is important. A large majority of people agreed that up to two-thirds of the revenue from the HVF should be used for projects in public transport. This decision fits well with the strategy of shifting goods from road to rail and helps finance the new railway lines. The remaining third goes to the cantons where it is used mainly for road purposes. And finally, one of the strongest arguments in favor of the HVF was its link to the polluter-pays principle.

The charges are relatively high, averaging 1.6 Eurocents per tonne-kilometer (or for a lorry with a payload of 20 tonnes, 32 cents per vehicle-kilometre). Balmer (2003) explains that the combination of the introduction of the HVF in Switzerland with the allowance of heavier vehicles led to remarkable changes within road transport. There was a change in fleet composition because in the year before the introduction of the HVF, sales of HGVs increased by 45 per cent. Truck owners saved money as new vehicles belong to the lowest and therefore cheapest emission class and the admissible weight of the trucks in the fleet could be better matched to the actual needs of the market. The HVF system led to a concentration in the haulier industry, either through mergers or closure of smaller firms. Larger firms were able to manage their vehicles more efficiently and avoid empty runs.

A 4-year study found that there were no significant changes in the modal split, rail retaining its unusually high market share. The study states that the new traffic regime has led to a sustained change in the road haulage sector. The trend toward an ever growing number of lorries on the roads has been broken and the negative effect on the environment shows a significant decrease. The rail sector's share of freight remained steady (Swiss Federal Statistical Office 2004).

McKinnon and McClelland (2005) stated that once the new trans-Alpine rail tunnels, which are largely funded by HVF revenue, were opened in 2007 and 2014/15, rail would capture a much larger share of the Swiss freight market. Balmer (2003) concludes that the system works well overall. Truck traffic has been reduced and there is an incentive to buy cleaner vehicles. The rail market share has been protected despite the advent of heavier goods vehicles.

The German HGV charge was introduced on 1 January 2005, applying to all lorries exceeding 12 tonnes gross weight. The tax is calculated based on the vehicle's environmental status (engine emission levels) and the number of axles.

Rothengatter (2003) explains that the objectives of the study into the HGV charge were to derive fair and efficient user charges for the different vehicle categories using the federal roads and to ensure that charges for infrastructure costs recovered all costs, including capital costs, and took into account future re-investment cycles, new investment and current expenditures. It was necessary that all users should bear exactly the costs that they were responsible for. The Eurovignette required that the toll rate had to be based on actual infrastructure costs: 'The weighted average tolls shall be related to the costs of constructing, operating and developing the infrastructure network concerned'. External costs were not included. The vehicle category charge had to be based on the category's average infrastructure cost. It was possible to differentiate the charge by the time of day (peak/off-peak) and by environmental performance (emission category). The German government decided initially to differentiate only according to environmental performance.

By introducing the HGV toll system, the German government believed that there would be more rigorous application of the user-pays principle to domestic and foreign users. HGVs are responsible for much of the cost of construction, maintenance and operation of motorways, and a distance-based toll will allow HGVs to make a contribution toward infrastructure costs. It was suggested that more efficient use would be made of transport infrastructure capacity due to the tolls (Hahn 2002).

The German government decided to invite bids for a private sector operator to run the system of upgrading, maintenance, operation and financing. The idea was to have a combination of tolling and public-private partnership models and the operator has to pre-finance the system. This allows the private operator to receive a share of the tolls collected on a stretch of motorway. There was additional relief for public budgets by switching from tax- to user-funded infrastructure.

The German system mainly relies on satellite tracking to determine the distance trucks travel on the motorway (Autobahn) network. In mid-2005, around 70% of the trucks on the network were fitted with OBUs which use GPS satellite signals and other positioning sensors to track vehicle movement, calculate the toll charge and communicate information to the agency responsible for collecting the toll. Toll revenue is then collected at the end of each month by direct debit from registered accounts, credit cards or fuel cards. For vehicles without OBUs, payments can be made for particular trips in advance either online or at any of the 3500 toll station terminals. Thus Germany has the most sophisticated pricing system in Europe, which in principle could be extended to cover all roads, and to differentiate in space and time as well as by vehicle type. The German charges are, however, relatively low, averaging 12.4 eurocents per vehicle-kilometre. The scheme was expected to raise around €3 billion a year, which is proposed to be spent on road and rail infrastructure.

13.5.3 What are the Issues for MSC-Based National Road User Charging?

Whilst there is a strong theoretical case for national road user charging, and a body of evidence from modelling and implemented cases demonstrating its potential and actual benefits, there are a number of reasons why a simple 'textbook' approach to marginal cost pricing, as applied to transport, may require further adaptation in practice. The difficulties with a pure MSC pricing, 'textbook' approach are comprehensively identified by Rothengatter (2003), and may be summarized as follows:

(a) measurement is complex;
(b) equity is ignored;
(c) dynamic effects, including investment decisions and technology choice, are ignored;
(d) financing issues are ignored;
(e) institutional issues are ignored;
(f) price distortions elsewhere in the economy are ignored;
(g) the administrative costs associated with implementation may not always be justified by the benefits.

Development of national road user charging proposals will typically address many of these issues: for example, the UK road pricing feasibility study considers different levels of complexity in the charge to reflect MSC at different levels of detail; gives a full financial assessment; considers acceptability and equity; and recommends a scheme for which the administrative costs are more than justified by the benefits (DfT 2004a, Nash 2007). The experiences of Singapore, London, Australia, Austria, Czech Republic, Germany, New Zealand, Norway and Switzerland, also suggest that there are benefits to be had from road user charging in practice even if the charges are not exactly the theoretically 'optimal' ones. The implication is that developing practical national schemes requires balancing the factors identified above with the efficiency gains identified in Tables 13.6 and 13.7.

13.6 Conclusions

We have discussed national road user charging as a policy with which to tackle the climate change impacts of car use, and have compared it with fuel tax in that role. We found that both policies are capable of internalizing climate change costs to a reasonable approximation, and that the selection between them is likely to be based on other, practical criteria. For example, road user charging faces acceptability barriers which are gradually being addressed through feasibility studies and careful scheme design. If national road user charging is adopted, then the administrative cost will be a sunk cost, and there will be a clear case to move charging for CO_2 emissions over from fuel tax (which could be abolished entirely) to become part of the road user charge. If this is what happens, we noted that charges will need to be differentiated by vehicle type and the per-km charge should ideally be differentiated by speed.

We also conducted a brief survey, which found that in addition to the one country which has adopted national road user charging – the city state of Singapore – at least another ten countries have adopted or are considering adopting national road user charges, albeit with some limitations: e.g. sometimes only for heavy freight vehicles, or only in urban areas. Given that the potential benefits identified by modelling work are approximately 0.5% of GDP, or as high as 2.7% in some studies, and given the continuing progress made toward implementation, it seems a reasonable prospect that this means of internalizing climate change costs of car use could become a practicable option for some countries within the next decade.

References

ACEA (2010) Overview of CO_2 based motor vehicle taxes in the EU, 20.04.2010. ACEA, Brussels
Balmer U (2003) Practice and experience with implementing transport pricing reform in heavy goods transport in Switzerland, paper presented at the 4th IMPRINT-EUROPE seminar. Brussels
Baumol WJ, Oates WE (1988) The theory of environmental policy, 2nd edn. Cambridge University Press, Cambridge
Beevers SD, Carslaw DC (2004) The impact of congestion charging on vehicle emissions in London. Atmos Environ 39(1):1–5

Bickel P, Hunt A, Laird J, Lieb C, Lindberg G, Odgaard T, Jong de G, Mackie P, Navrud S, Shires J, Tavasszy L (2006) Proposal for harmonised guidelines. HEATCO (Developing harmonised European approaches for transport costing and project assessment) Deliverable 5. Institut für Energiewirtschaft und Rationelle Energieanwendung, Universität Stuttgart, Stuttgart

CE Delft (2008) Handbook on estimation of external costs in the transport sector, deliverable 1 of the impact project prepared for the European commission (DG TREN) in association with INFRAS, Fraunhofer Gesellschaft, ISI and University of Gdansk, Delft

Chester MV, Horvath A (2009) Environmental assessment of passenger transportation should include infrastructure and supply chains, Environ Res Lett 4(2). doi:10.1088/1748-9326/4/2/024008

Chin KK (2009) The Singapore Experience: the evolution of technologies, costs and benefits, and lessons learnt, discussion paper prepared for the ITF/OECD roundtable of 4–5 February 2010 on Implementing Congestion Charging. Paris.

Committee on Climate Change (CCC) (2009) Meeting the carbon budgets—the need for a step change. Retrieved 15 Mar 2010 from http://hmccc.s3.amazonaws.com/21667%20CCC%20Report%20AW%20WEB.pdf

Department for Transport (DfT) (2007) Towards a Sustainable Transport System. DfT, London.

European Commission (1995) Toward fair and efficient pricing in transport. Brussels

European Commission (1998) Fair payment for infrastructure use: a phased approach to a common transport infrastructure charging framework in the EU. Brussels

European Commission (2001) White paper: European transport policy for 2010: time to decide. Brussels.

European Commission (2006) IP/06/383: Sustainable transport – Towards fair and efficient infrastructure charging. Brussels.

European Commission (2007) Communication from the commission to the council and the European parliament 6 results of the review of the community strategy to reduce CO_2 emissions from passenger cars and light-commercial vehicles, {SEC(2007) 60} {SEC(2007) 61}, COM/2007/0019 final

European Commission (2008) Directive 2008/101/EC of the European parliament and of the council amending directive 2003/87/EC so as to include aviation activities in the scheme for greenhouse gas emission allowance trading within the community (19 November 2008)

European Commission (2009) Directive 2009/29/EC of the European parliament and of the council of 23 April 2009 amending directive 2003/87/EC so as to improve and extend the greenhouse gas emission allowance trading scheme of the Community

European Commission (2010) Report from the commission to the European parliament, the council, and the European economic and social committee progress report on implementation of the community's integrated approach to reduce CO_2 emissions from light-duty vehicles, COM/2010/0656 final.DECC (2010), Toolkit for guidance on valuation of energy use and greenhouse gas emissions, Tables 1–29, supporting the toolkit and guidance. http://www.decc.gov.uk/en/content/cms/statistics/analysts_group/analysts_group.aspx. Last accessed Mar 2011

Defra/DECC (2010) 2010 Guidelines to Defra/DECC's GHG conversion factors for company reporting: methodology paper for emission factors, October 2010. Defra, London

Department for Transport (DfT) (2004a) Feasibility study of road pricing in the UK – Full report. DfT, London

Department for Transport (DfT) (2004b) Feasibility study of road pricing in the UK, Annex C 'Charging technologies and existing schemes'. DfT, London

Department for Transport (DfT) (2004c) Feasibility study of road pricing in the UK, Annex A 'The economic case for road pricing'. DfT, London

Department for Transport (DfT) (2004d) Feasibility study of road pricing in the UK, Annex B Modelling results and analysis. DfT, London

Department for Transport (DfT) (2010a) Transport statistics Great Britain 2010, Updated 25 November 2010. http://www.dft.gov.uk/pgr/statistics/datatablespublications/tsgb/. Last accessed Mar 2011

Department for Transport (DfT) (2010b) TAG unit 3.3.5 'The greenhouse gases sub-objective'. www.deft.gov.uk/webtag/documents/expert/pdf/unit3.3.5d.pdf. Last accessed Mar 2011

Department for Transport (DfT) (2010c) TAG unit 3.5.6 'Values of time and operating costs'. www.deft.gov.uk/webtag/documents/expert/pdf/unit3.5.6d.pdf. Last accessed Mar 2011

DVLA (2010) Rates of vehicle tax. DVLA, Swansea

Eddington, Sir R. (2006) The Eddington transport study: the case for action. HM Treasury and Department for Transport, London. http://www.dft.gov.uk/about/strategy/transportstrategy/eddingtonstudy. Last accessed Mar 2011

Eliasson J (2009) A cost–benefit analysis of the Stockholm congestion charging system. Transp. Res Part A 43:468–480

European Parliament (2004) Directive 2004/52/EC of the European parliament and of the council of 29 April 2004 on the interoperability of electronic road toll systems in the community

European Parliament (2006) Directive 2006/38/EC of the European parliament of the council of 17 May 2006 amending directive 1999/62/EC on the charging of heavy goods vehicles for the use of certain infrastructures. Brussels

Hahn W (2002) Implementing transport pricing reform in Germany, paper presented at the 3rd IMPRINT-EUROPE seminar. Brussels.

Henry Tax Review (2009) Australia's future tax system, Report to the Treasurer, December 2009

House of Commons Select Committee on Transport (2009) Taxes and charges on road users. HC103, TSO, paragraph 38 and written evidence by several witnesses, London

Knight FH (1924) Some fallacies in interpretation of social costs. Q J Econ 38(August):582–606

Mackie PJ, Wardman MR, Fowkes AS, Whelan GA, Nellthorp J, Bates JJ (2003) Value of travel time savings in the UK, a report to the Department for Transport. Institute for Transport Studies, Leeds

McKinnon A, McClelland D (2005) An Alternative Method of Road-User Charging for Lorries. Logistics Research Centre, Heriot-Watt University, Edinburgh

Ministry of Transport (1964) Road Pricing: The economic and technical possibilities (the Smeed Committee Report). HMSO, London.

Mohring HD, Harwitz I (1962) Highway benefits: an analytical framework. Northwestern University Press, Evanston, IL

Nash CA (2007) Developments in transport policy – road pricing in Britain. J Transport Econ Policy 41:135–147

Nash CA, Matthews B (2005) Transport pricing policy and the research agenda. In: Nash CA, Matthews B (eds) Measuring the marginal social cost of transport. Elsevier, Amsterdam

Nash C, Matthews B, Link H, Bonsall P, Lindberg G, van der Voorde E, Ricci A, Enei R, Proost S (2008a) Policy conclusions. Deliverable 10 of GRACE (Generalization of Research on Accounts and Cost Estimation), Funded by the European Commission Sixth Framework Programme, University of Leeds, Leeds

Nash CA, Menaz B, Matthews B (2008b) Inter-urban road goods vehicle pricing in Europe. In: Richardson H, Christine Bae C-H (eds) Road congestion and pricing in Europe. Implications for the United States. Edward Elgar, Cheltenham, pp 233–251

Newbery DMG (1989) Cost recovery from optimally designed roads. Economica 56(222):165–185

Pigou AC (1920) Wealth and welfare. Macmillan, London

Rothengatter W (2003) How good is first best? Marginal cost and other pricing principles for user charging transport. Transport Policy 10:121–130

Sansom T, Nash CA, Mackie PJ, Shires JD, Watkiss P (2001) Surface transport costs and charges Great Britain 1998. Institute for Transport Studies, Leeds

Santos G, Fraser G (2006) Road pricing: lessons from London. Econ Policy 21(46):264–310

Steen M (2009) Dutch propose full-scale road pricing system. Retrieved 25 Mar 2010 from http://www.ft.com/cms/s/0/9a17120a-d090-11de-af9c-00144feabdc0.html

Stern N (2006) The economics of climate change: the stern review. Cambridge University Press, Cambridge

Swiss Federal Statistical Office (2004) Monitoring Sustainable Development. MONET, Neuchâtel

TML (2006) ASSESS – Assessment of the contribution of the TEN and other transport policy measures to the mid-term implementation of the White Paper on the European transport policy for 2010 (in association with TNO, WSP, TRT, CA Kiel, UG, ITS-Leeds, SWOV, DLR, IUAS Bad Honnef and ITU), Final Report for the European Commission, Brussels

TRL (2009) Road vehicle emission factors 2009. http://www.dft.gov.uk/pgr/roads/environment/emissions/regulated.xls. Last accessed Mar 2011

Vickrey WS (1969) Congestion theory and transport investment. Am Econ Rev 59(2):251–260

Walters AA (1961) The theory and measurement of private and social cost of highway congestion. Econometrica 29:676–699

Walters AA (1968) The economics of road user charges. World Bank Staff Occasional Paper Number Five. IBRD, Washington

Part IV
The International Context

If per capita carbon emissions in both China and India rise to U.S. per capita levels, then global carbon emissions will increase by 139 percent. If their emissions stop at French levels, global emissions will rise by only 30 percent. Driving and urbanization patterns in these countries may well be the most important environmental issues of the twenty-first century.

Edward Glaeser, Triumph of the City: How Our Greatest Invention Makes Us Richer, Smarter, Greener, Healthier and Happier, Macmillan, London, 2011, p. 15

Chapter 14
Mobility Management Solutions to Transport Problems Around the World

Todd Litman

Abstract This chapter investigates the role that *mobility management* plays in an efficient transport system. It describes the basic principles that a transport system must reflect to optimize efficiency and maximize benefits, identifies existing transport market distortions, and describes reforms that can correct these distortions. This analysis indicates that many common transport policies and planning practices result in economically excessive motor vehicle travel, which is particularly harmful to lower income people and economies. Mobility management strategies include improvements to alternative modes, more efficient transport pricing, and more neutral planning practices. These strategies tend to increase transport system efficiency, and help achieve social equity objectives by improving affordable transport options. Mobility management can provide multiple economic, social and environmental benefits, and so helps create truly sustainable transport systems. This has important implications for developing countries which are still establishing their planning policies and practices.

14.1 Introduction

Motor vehicle travel can provide large benefits but also imposes large costs. Because many vehicle costs are external (imposed on somebody other than the user), individuals tend to drive more than optimal from society's perspective. As a result, policies that reduce vehicle travel, generally called *mobility management* or *transportation demand management* (TDM), are often justified.

For example, from the perspective of an individual who can afford to purchase an automobile, driving often seems to be the best travel option since it is generally faster and more prestigious than other modes. However, as more travellers drive, problems such as traffic and parking congestion, traffic accidents and pollution increase, making all travellers worse off. As a result, everybody can benefit from policies that limit automobile travel and encourage use of efficient modes.

T. Litman (✉)
Victoria Transport Policy Institute, Victoria, BC, Canada V8V 3R7
e-mail: litman@vtpi.org

Mobility management includes various strategies that improve efficient transport options (such as walking, cycling, public transport and telecommunications), incentives to choose the most efficient option for each trip (such as more efficient road, parking and fuel pricing), and policies that encourage more accessible land use development which reduces the distances people must travel to reach activities such as work, schools and stores. Mobility management is increasingly being implemented around the world, in both developed and developing countries, to help achieve a variety of planning objectives, including congestion reduction, road and parking facility cost savings, consumer savings, accident reductions, improved mobility for non-drivers, energy conservation, emission reductions, and improved public fitness and health (Broaddus et al. 2009). When all impacts are considered, mobility management strategies, such as those listed in Table 14.1, are often the most cost effective solution to transport problems.

However, to be implemented to the degree that is optimal mobility management requires changing the way we define transport problems and evaluate potential solutions. Conventional planning tends to overlook many mobility management benefits, so it is often undervalued in policy and planning analysis. Mobility management also faces institutional barriers, such as inadequate funding and support within the existing transport planning process.

This chapter investigates these issues. It asks, 'How much and what type of travel is overall optimal, considering all impacts, and what transport policies can help achieve that optimality.' It discusses the principles required for an efficient transport system, identifies existing transport market distortions that violate these principles, describes policy and planning reforms that correct these distortions, estimates how such reforms would affect transport activity, and discusses the economic, social and environmental benefits that would result. Although previous studies have evaluated most of these reforms individually, few consider their cumulative impacts.

Table 14.1 Mobility management strategies (VTPI 2010)

Improves transport options	Efficient incentives	Land use management
Transit improvements	Congestion pricing	Smart growth
Non-motorized improvements	Distance-based fees	Transit oriented development
Rideshare programs	Commuter financial incentives	Location-efficient development
HOV priority	Parking pricing	Parking management
Flextime	Parking regulations	Carfree planning
Carsharing	Fuel tax increases	Traffic calming
Telework	Transit encouragement	
Guaranteed ride home		

14.2 The New Transport Planning Paradigm

A *paradigm shift* (a fundamental change in the way problems are defined and solutions evaluated) is occurring in the transport planning field (Litman 1999, Leather 2009). The old transportation planning paradigm focused on *mobility*, that is, physical movement. With that paradigm, *transport* generally means motor vehicle travel, *transport problems* consist of excessive vehicle delay and cost, and *transport improvement* consists of strategies that increase motor vehicle traffic speeds and reduce driving costs. Mobility is seldom an end in itself, however. The goal of most transport is to achieve *accessibility*, which refers to the ability of people and businesses to reach desired goods, services and activities. Various factors affect accessibility, including the speed and affordability of mobility, land use factors that affect the distances between activities, and mobility substitutes, such as delivery services and telecommunications which can reduce the need for physical travel (Litman 2003, El-Geneidy and Levinson 2006).

Accessibility-based transport planning expands the range of solutions that can be applied to transport problems. For example, with mobility-based planning, the only solution to traffic congestion is to expand roadways. With accessibility-based planning, potential solutions also include improving alternative modes (better walking, cycling, ridesharing, public transport, and telecommunications), incentives to use alternatives (such as more efficient road and parking pricing, and commute trip reduction programs), and land use policies that reduce the need for residents to travel to access services and jobs.

Mobility-based planning tends to create a self-reinforcing cycle of automobile dependency and sprawl, as illustrated in Fig. 14.1. Although many planning decisions that support automobile dependency and sprawl may individually seem justified, their cumulative effect significantly increases economic, social and environmental costs. Residents of automobile dependent communities must spend more on vehicles, fuel, roads and parking facilities, have higher traffic fatality rates, produce more pollution, consume more land, and are less physically fit than if they lived in more multi-modal communities. In addition, non-drivers are worse off, which is unfair and regressive (it burdens lower-income people more than higher-income people). This is not to suggest that motor vehicle travel provides no benefits, but it does indicate that planning which is unintentionally biased in favor of automobiles tends to result in sub-optimal transport patterns.

The new planning paradigm expands the range of impacts and options considered in transport planning. For example, under the old paradigm, transport planners were primarily concerned with reducing traffic delays. Under the new paradigm they also consider indirect and external costs and benefits (sometimes called *co-benefits*) which can help decision-makers identify the most efficient and equitable transport improvements available, taking into account all impacts (Co-Benefits Asia Hub 2011).

There are many justifications for this paradigm shift. More comprehensive analysis and more multi-modal planning tend to increase overall transport system efficiency which provides multiple benefits, including congestion reduction, road and parking facility savings, consumer savings, improved mobility for non-drivers,

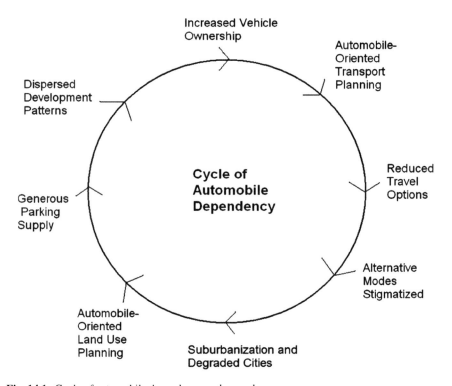

Fig. 14.1 Cycle of automobile dependency and sprawl

traffic safety, energy conservation, emission reductions, more efficient land use development, and improved public fitness and health. Although some people emphasize environmental benefits, mobility management also tends to support economic development. In fact, mobility management tends to reflect basic market principles, and so tends to support economic development, as discussed in the following chapter.

14.3 Transport Market Principles, Distortions and Reforms

According to basic economic and planning theory, an efficient transport system must reflect the following principles (Litman 2006a):

1. *Diverse consumer options* (also called *consumer sovereignty*). Consumers must have access to a variety of travel modes, service quality, and price options so they can choose the bundle that best meets their needs and preferences.
2. *Efficient pricing.* Efficiency requires that prices (what consumers pay for a good) reflect the marginal costs of producing that good unless a subsidy is specifically justified. This means, for example, that users should pay directly for roads and parking facilities, with fees that reflect the congestion, accident risk and pollution

emissions imposed by each trip. This encourages consumers to choose lower-cost options, and insures that people do not impose $10.00 in total costs for travel that users only consider worth $5.00.
3. *Neutral public policy.* Public policies (laws and regulations, pricing and taxes, public infrastructure investments, etc.) should not arbitrarily favor one group or transport mode over others.

Many existing transport policies and planning practices violate these principles:

- Conventional travel surveys and statistics tend to undercount and therefore undervalue non-motorized travel. This skews planning to favor automobile-oriented improvements even when they degrade non-motorized travel conditions.
- A major share of transport funding is dedicated to roads and parking facilities, which encourages communities to favor automobile improvements even if other types of transport improvements provide greater total benefits.
- Roads and parking facilities are generally provided for free, which means they must be funded indirectly rather than through direct user charges.
- Road tolls and parking fees that do apply do not generally vary to reflect marginal costs. For example, road tolls and parking fees do not generally vary by time or location to reflect congestion.
- Vehicle insurance and registration fees are generally fixed – they do not increase with annual mileage. This gives motorists an incentive to maximize their annual mileage, in order to get their money's worth.
- Some tax regulations encourage businesses to subsidize employee parking and vehicles.
- In most countries (except Europe and wealthy Asian countries) fuel taxes are too low to finance total roadway costs, or the full economic and environmental costs of producing and importing that fuel. Many countries subsidize vehicle fuel, either directly, or indirectly through policies such as biofuel subsidies.
- Zoning codes, development policies and infrastructure investments often favor sprawled land use over more compact, infill development.

These distortions tend to reduce transport system efficiency. Problems such as traffic and parking congestion are virtually unsolvable without planning and pricing reforms that encourage urban-peak travellers to use more space-efficient modes. Similarly, underpricing fuel encourages travellers to choose fuel intensive modes, which is economically harmful, particularly for lower-income countries that import petroleum, and increases pollution emissions.

These distortions tend to reduce transport system diversity. For example, planning practices that undervalue non-motorized travel reduce investments in sidewalks and paths; dedicated roadway funding results in wider highways that make walking more difficult; underpricing of roads and parking facilities reduces public transit demand, which over the long-term reduces transit service quality; development policies that limit development density and mix increase the distances between destinations, making them difficult to reach except by automobile. This is particularly harmful to non-drivers, and so tends to be inequitable.

For example, in a typical developing country city only about 10% of households own an automobile. Most travel is by walking, cycling, or public transport. However, because travel surveys undercount short trips, non-commute trips, travel by children, and non-motorized links of public transport and automobile trips, official statistics overlook most non-motorized trips, which exaggerates the importance of motorized travel.

According to standard international practices, transport planners evaluate the transport system quality based on vehicle travel speeds, which directs transport improvement resources to highway and parking facility expansion, and large public transport projects. Few resources are devoted to non-motorized or local public transit improvements. Since expanding urban roadways and increasing urban traffic volumes and speeds tends to degrade walking and cycling conditions, and increased vehicle traffic volumes congest urban streets which reduces bus transit performance, most residents, who rely on walking, cycling and bus transit, experience declining transport performance. This further encourages automobile dependency and sprawl, increasing economic, social and environmental costs.

Described more positively, various policy and planning reforms can correct these distortions. These mobility management strategies favor higher value trips and more efficient modes, increasing overall transport system efficiency. This can lead to reduced traffic congestion, road and parking facility cost savings, consumer savings, increased safety, improved mobility options for non-drivers, energy conservation, emission reductions, more efficient land use development, and improved public fitness and health. Transport planners generally classify these as mobility management strategies. Thus, mobility management is the general term for various reforms that economists and planners recommend for improving transport system efficiency and equity.

14.4 Mobility Management Strategies

Examples of mobility management strategies are described below. For more information see Cambridge Systematics (2009), Cairns et al. (2004), Litman (2007) and VTPI (2010).

14.4.1 Least Cost Transportation Planning (WSDOT 2009)

Least-cost transportation planning is a term for more comprehensive and neutral planning that:

- Considers all significant impacts (costs and benefits), including indirect effects.
- Considers demand management equally with facility capacity solutions.

For example, least cost planning means that funding for roads and parking facilities could be used to improve alternative modes or support mobility management programs if they are more cost effective at achieving transportation planning objectives, such as providing mobility and reducing congestion, considering all benefits and costs.

14.4.2 Commuter Financial Incentives (ICF and CUTR 2005)

Commuter Financial Incentives include several types of incentives that encourage alternative commute modes:

- *Parking Cash Out* means that commuters who are offered subsidized parking are also offered the cash equivalent if they use alternative travel modes. For example, an employee can choose between a free parking space or $75 per month if they use an alternative commute mode.
- *Travel allowances* are a financial payment provided to employees instead of parking subsidies. Commuters can use this money to pay for parking or for another travel mode.
- *Transit and rideshare benefits* are free or discounted transit fares provided to employees.
- *Reduced employee parking subsidies* means that commuters who drive must pay some or all of their parking costs.
- *Company travel reimbursement policies* that reimburse bicycle or transit mileage for business trips when these modes are comparable in speed to driving, rather than only reimbursing automobile mileage.

These strategies are more efficient and equitable than the common practice by businesses of subsidizing parking but offering no comparable benefit to employees who use alternative modes.

Commuter financial incentives can be prorated according to how much employees use alternative modes. For example, employees who drive twice a week would receive 60% of the full Parking Cash Out allowance.

14.4.3 Fuel Taxes – Tax Shifting (Clarke and Prentice 2009, Metschies 2005)

Since governments must tax something to raise revenue, many economists recommend shifting taxes from desirable activities to those that are harmful or risky, for example, reducing taxes on employment and commercial transactions, and increasing taxes on the consumption of polluting, non-renewable resources such as petroleum. Current fuel taxes are relatively low, particularly in the U.S. and many developing countries. There are several specific justifications for increasing taxes on petroleum products in general and motor vehicle fuel in particular:

- To reflect inflation. Fuel taxes are generally unit based (cents per gallon or liter), as opposed to a percentage of the retail price, and so their real value declines with inflation. The real, inflation adjusted value of fuel taxes has declined significantly in many jurisdictions. Increasing taxes and indexing them to inflation is justified to maintain constant revenue.
- As a road user fee. Special fuel taxes are generally considered a road user fee, which should at least pay the costs of building and maintaining roadways, and perhaps more to recover other associated costs, such as traffic services. In many jurisdictions fuel taxes are too low to finance roadway costs, so increases are justified.
- To encourage energy conservation in order to reduce dependence on imported resources, increase economic efficiency, reduce pollution emissions (including climate change emissions) and to leave more petroleum for future generations (Litman 2007).
- To internalize petroleum production subsidies, external costs and tax exemptions.

14.4.4 Pay-As-You-Drive Pricing (USDOT 2010, Litman 1997)

Pay-As-You-Drive (PAYD) pricing (also called *Distance-Based* and *Mileage-Based pricing*) bases vehicle insurance premiums and other fees on the amount a vehicle is driven. This can be done by changing the pricing unit (i.e., how fees are calculated) from the vehicle-year to the vehicle-mile, vehicle-kilometer or vehicle-minute. Existing pricing factors are incorporated so higher-risk motorists pay more per unit than lower-risk drivers. For example, a $375 annual insurance premium becomes 3¢ per mile, and a $1,250 annual premium becomes 10¢ per mile. An average U.S. motorist would pay about 7¢ per mile for PAYD insurance. Similarly, currently fixed vehicle taxes, registration, licensing and lease fees, and taxes can be converted to distance-based fees by dividing existing fees by average annual mileage for each vehicle class. For example, if a vehicle's annual registration fees are $300 and its class averages 12,000 annual miles, the distance-based fee is 2.5¢ per mile.

Pay-As-You-Drive pricing helps achieve several public policy goals including fairness, affordability, road safety, consumer savings and choice, and reduced traffic problems such as traffic congestion, road and parking facility costs, pollution emissions and sprawl. PAYD should reduce average annual mileage of affected vehicles by 10–15%, reduce crash rates by a greater amount, increase equity, and save consumers money. It reduces the need for cross-subsidies currently required to provide 'affordable' unlimited-mileage coverage to high-risk drivers. It can particularly benefit lower-income communities that currently pay excessive premiums. Some insurance companies now offer versions of PAYD pricing, but implementation is limited.

14.4.5 Efficient Road Pricing (FHWA 2009, Schwaab and Thielmann 2001)

Road Pricing means that motorists pay directly for driving on a particular roadway or in a particular area. *Congestion Pricing* (also called *Value Pricing*) refers to road pricing with variable fees designed to reduce traffic congestion. Transportation economists have long advocated road pricing as a way to fund transportation improvements and to reduce congestion problems. Road tolls are justified since many road and bridge projects would otherwise be funded through general taxes, or by taxes paid by motorists who seldom or never use costly new facilities. Some roads include both priced and unpriced lanes, allowing motorists to choose between financial and time savings. Experience with road tolls and various types of congestion pricing indicate that motorists respond to such fees, shifting travel time, route, destination and mode, increasing overall transport system efficiency.

14.4.6 Parking Management (Litman 2006b, Shoup 2005, USEPA 2006)

Parking Management includes a variety of strategies that encourage more efficient use of existing parking facilities, as summarized in Table 14.2. In addition to reducing parking costs, some of these strategies also reduce total automobile travel and therefore costs such as congestion, accidents and pollution.

14.4.7 Transit Service Improvements (EDF 2009, Wright 2007)

There are many ways to improve public transit services, and encourage transit use, including increased service area and frequency, increased transit speed and reliability (including use of transit priority systems that allow transit vehicles to bypass congestion), reduced crowding, more comfortable vehicles, nicer waiting areas (stations and stops), reduced and more convenient fares, improved rider information and marketing programs, transit oriented land use development, pedestrian and cycling improvements around transit stops, bike and transit integration (bike racks on buses, bicycle parking at stations, etc.), park-and-ride facilities, improved security for transit users and pedestrians, and transit services targeting particular needs such as express commuter buses and special event services. Marketing programs that raise the social status of transit travel can also be considered a type of service improvement.

14.4.8 Ridesharing (Ennis 2010, Evans and Pratt 2005)

Ridesharing refers to carpooling and vanpooling, in which vehicles carry multiple passengers. *Carpooling* uses participants' own automobiles, while *vanpools*

Table 14.2 Parking management strategies (VTPI 2010)

Strategy	Description	Typical reduction in parking required	Traffic reduction
Shared Parking	Parking spaces serve multiple users and destinations.	10–30%	
Parking Regulations	Regulations to prioritize use of the most desirable parking spaces.	10–30%	
More Accurate and Flexible Standards	Adjust parking standards to more accurately reflect demand in a particular situation.	10–30%	
Parking Maximums	Establish maximum parking standards.	10–30%	
Remote Parking	Provide off-site or urban fringe parking facilities.	10–30%	
Smart Growth	Encourage more compact, mixed, multi-modal development to allow more parking sharing and use of alternative modes.	10–30%	✓
Walking and Cycling Improvements	Improve walking and cycling conditions to expand the range of destinations serviced by a parking facility.	5–15%	✓
Mobility Management	Use resources that would otherwise be devoted to parking facilities to encourage use of alternative modes.	10–30%	✓
Parking Pricing	Charge motorists directly and efficiently for using parking facilities.	10–30%	✓
Improve Pricing Methods	Use better charging techniques to make pricing more convenient and cost effective.	Varies	
Financial Incentives	Provide financial incentives to shift mode, such as parking cash out.	10–30%	✓
Unbundle Parking	Rent or sell parking facilities separately from building space.	10–30%	✓
Parking Tax Reform	Change tax policies to support parking management objectives.	5–15%	✓
Bicycle Facilities	Provide bicycle storage and changing facilities.	5–15%	✓
Improve User Information	Provide convenient and accurate information on parking availability and price.	5–15%	✓
Overflow Parking	Establish plans to manage occasional peak parking demands.	Varies	

use a larger vehicle that is often leased for the purpose. Ridesharing has minimal incremental costs because it makes use of vehicle seats that would otherwise be unoccupied.

14.4.9 HOV Priority (Turnbull et al. 2006)

HOV Priority refers to strategies that give *High Occupancy Vehicles* (buses, vanpools and carpools) priority over general traffic, such as dedicated lanes, queue-jumping intersection design, and priority parking. HOV priority measures can be justified as a more efficient and equitable allocation of road space (travellers who share a vehicle and therefore *impose* less congestion on other road users, are rewarded by *bearing* less congestion delay), an efficient use of road capacity (they can carry more people than a general use lane), and as an incentive to shift to more efficient modes.

14.4.10 Walking and Cycling Improvements (Cairns et al. 2004, Pucher et al. 2010)

Walking and cycling travel can substitute for some motor vehicle trips directly, and support other alternative modes such as public transit and ridesharing. Residents of communities with good walking and cycling conditions drive less and use transit and rideshare more. There are many ways to improve these modes:

- Improve sidewalks, crosswalks, paths and bike lanes.
- Increase road and path connectivity, with special shortcuts for nonmotorized modes.
- Pedestrian oriented land use and building design.
- Traffic calming, speed reductions and vehicle restrictions, to reduce conflicts between motorized and nonmotorized traffic.
- Safety education, law enforcement and encouragement programs.
- Convenient and secure bicycle parking.
- Address security concerns of pedestrians and cyclists.

14.4.11 Smart Growth Land Use Policies (Ewing et al. 2007)

Current land use policies limit development density, disperse destinations and favor automobile access over alternative modes. *Smart growth* policies, such as those described below, reduce vehicle travel and provide other benefits.

- Encourage compact development with diverse housing types (single and multi-family).
- Create more complete, self-contained communities. For example, locating schools, parks and shops within neighborhoods.
- Encourage infill development, such as redevelopment of older buildings and neighborhoods.
- Concentrate commercial activities in compact centers or districts. Use access management to prevent arterial strip commercial development.

- Use development fees and utility pricing that reflects the higher costs of providing public services at lower-density sites.
- Develop a dense network of interconnected street. Keep streets as narrow as possible, particularly in residential areas and commercial centers.
- Design streets to accommodate walking and cycling. Create a maximum number of connections for non-motorized travel, such as trails that link dead-end streets.
- Apply parking management and reduce parking requirements.

14.4.12 Location Efficient Development (CNT 2008)

Location Efficient Development refers to building, neighborhood and community development that reflects Smart Growth principles. *Location Efficient Mortgages* recognize the savings that result in credit assessments, giving homebuyers more incentive to choose efficient locations.

14.4.13 Mobility Management Marketing (Sloman et al. 2010)

Mobility Management Marketing involves various activities to improve consumers' knowledge and acceptance of alternative modes, and to provide products that better meet travellers' needs and preferences. Given adequate resources, marketing programs can significantly increase use of alternative modes and reduce automobile travel.

14.4.14 Freight Transport Management (Hendrickson et al. 2006)

Freight Transport Management includes various strategies for increasing the efficiency of freight and commercial transport. This can include decreasing the need for vehicle trips by improving distribution practices, shifting freight to more resource efficient modes (such as from air and truck to rail and marine), improving efficient modes such as marine, rail and bicycle; and by reducing the total volume of goods that need to be transported. Because freight vehicles tend to be large, energy-intensive and high polluting, a relatively small improvement in freight efficiency can provide large total benefits.

14.4.15 School and Campus Trip Management (Cairns et al. 2004, NTHP 2010)

These programs help overcome barriers to the use of alternative modes, and provide positive incentives for reduced driving to schools and college or university campuses. School trip management usually involves improving pedestrian and cycling access, providing traffic safety education, promoting ridesharing, and encouraging parents to use alternatives when possible. Campus trip management programs often include discounted transit fares, rideshare promotion, improved pedestrian and cycling facilities, and increased parking fees. These programs give students, parents and staff more travel choices, encourage exercise, and reduce parking and congestion problems.

14.4.16 Institution and Regulatory Reforms (Meakin 2004, Sakamoto 2010)

Mobility management requires institutional reforms to better support and finance demand management policies and programs, and regulatory reforms to allow innovation and competition. Private bus, jitney and taxi services are often restricted to favor existing service providers. Although there are reasons to regulate transportation services to maintain quality, predictability and safety, unnecessary regulations can be changed to address specific problems while encouraging competition, innovation and diversity.

14.4.17 Carsharing (Cairns et al. 2004, Cohen et al. 2008)

Carsharing provides affordable, short-term (hourly and daily rate) motor vehicle rentals in residential areas as an alternative to private ownership. Because it has lower fixed costs and higher variable costs than private vehicle ownership, carsharing tends to significantly reduce annual vehicle mileage by participants.

14.4.18 Streetscaping and Traffic Calming (ITE 2010)

Traffic calming includes various strategies to reduce traffic speeds and volumes on specific roads. Typical strategies include traffic circles at intersections, sidewalk bulbs that reduce intersection crossing distances, raised crosswalks, and partial street closures to discourage short-cut traffic through residential neighborhoods. This increases road safety and community livability, creates a more pedestrian- and bicycle-friendly environment, and can reduce automobile use.

14.5 Summary of Mobility Management Strategies

Table 14.3 summarizes these various Win-Win strategies. This analysis suggests that a well-coordinated program of Win-Win strategies implemented to the degree economically justified would probably reduce total vehicle travel 30–50% compared with current planning and pricing practices (Cambridge Systematics 2009, Ewing et al. 2007, Litman 2010).

This conclusion is supported by comparing the travel behavior and transport costs in different cities and countries with similar levels of economic development but

Table 14.3 Mobility management strategies (Litman 2007)

Name	Description	Transport impacts
Least-Cost Planning	More comprehensive and neutral planning and investment practices.	Increases investment and support for alternative modes and mobility management, improving transport options.
Commute Trip Reduction (CTR)	Programs by employers to encourage alternative commute options.	Reduces automobile commute travel.
Commuter Financial Incentives	Offers commuters financial incentives for using alternative modes.	Encourages use of alternative commute modes.
Fuel Taxes – Tax Shifting	Increases fuel taxes and other vehicle taxes.	Reduces vehicle fuel consumption and mileage.
Pay-As-You-Drive Pricing	Converts fixed vehicle charges into mileage-based fees.	Reduces vehicle mileage.
Road Pricing	Charges users directly for road use, with rates that reflect costs imposed.	Reduces vehicle mileage, particularly under congested conditions.
Parking Management	Various strategies that promote more efficient use of parking facilities.	Reduces parking demand and facility costs, and encourages use of alternative modes.
Parking Pricing	Charges users directly for parking facility use, often with variable rates.	Reduces parking demand and facility costs, and encourages use of alternative modes.
Transit and Rideshare Improvements	Improves transit and rideshare services.	Increases transit use, vanpooling and carpooling.
HOV Priority	Improves transit and rideshare speed and convenience.	Increases transit and rideshare use, particularly in congested conditions.
Walking and Cycling Improvements	Improves walking and cycling conditions.	Encourages use of nonmotorized modes, and supports transit and smart growth.
Smart Growth Policies	More accessible, multi-modal land use development patterns.	Reduces automobile use and trip distances, and increases use of alternative modes.
Location Efficient Housing and Mortgages	Encourage businesses and households to choose more accessible locations.	Reduces automobile use and trip distances, and increases use of alternative modes.
Mobility Management Marketing	Improved information and encouragement for transport options.	Encourages shifts to alternative modes.
Freight Transport Management	Encourage businesses to use more efficient transportation options.	Reduces truck transport.
School and Campus Trip Management	Encourage parents, students and staff to use alternative modes for school commutes.	Reduces driving and increases use of alternative modes by parents, students and staff.

Table 14.3 (continued)

Name	Description	Transport impacts
Regulatory Reforms	Reduced barriers to transportation and land use innovations.	Improves travel options.
Carsharing	Vehicle rental services that substitute for private automobile ownership.	Reduces automobile ownership and use.
Traffic Calming and Traffic Management	Roadway designs that reduce vehicle traffic volumes and speeds.	Reduces driving and improves walking and cycling conditions.

different transport policies and planning practices. For example, residents of wealthy European and Asian countries drive about half as much as in North America, spend much less on vehicles and fuel, and have much lower traffic fatality rates than in North America, due to differences in fuel prices, transport investments and land use development policies (Ewing et al. 2007, Litman 2009).

14.6 Mobility Management Evaluation

Conventional planning tends to undervalue mobility management solutions. These strategies tend to provide multiple economic, social and environmental benefits (Cambridge Systematics 2009, Litman 2007). However, conventional planning is *reductionist*, meaning that each problem is assigned to a particular profession or agency with narrowly defined responsibilities. For example, transport agencies are responsible for traffic congestion and accident reductions, environmental agencies are responsible for emission reductions, social agencies are responsible for helping disadvantaged people, and public health agencies are responsible for encouraging public fitness. This approach tends to undervalue strategies that provide multiple benefits. For example, transport planning agencies tend to evaluate potential transport system improvements based primarily on their impacts on traffic congestion and accidents, but generally ignore impacts on parking costs, mobility for non-drivers, and public fitness and health. Similarly, environmental agencies tend to evaluate transport system improvements based on energy conservation and emission reductions, but generally ignore impacts on congestion and accidents.

Mobility management creates diverse benefits. Therefore it is important to use comprehensive analysis when evaluating these strategies. For example, expanding highways provides only one primary benefit: congestion reductions. By inducing additional vehicle travel over the long run, however, this strategy exacerbates other problems such as traffic accidents, pollution emissions and sprawl (Litman 2005). Similarly, more efficient and alternative fuel vehicles tend to conserve energy and reduce pollution. Reducing vehicle operating costs, however, tends to increase total vehicle travel and therefore congestion, parking and accident problems. Mobility

Table 14.4 Comparing strategies (Litman 2007)

Planning objective	Roadway expansion	Efficient and alt-fuel vehicles	Mobility management
Total Vehicle Travel	*Increased*	*Increased*	*Reduced*
Congestion reduction	✓	✗	✓
Roadway cost savings	✗	✗	✓
Parking costs savings	✗	✗	✓
Consumer costs savings	✗		✓
Traffic safety	✗	✗	✓
Improved mobility options	✗		✓
Energy conservation	✗	✓	✓
Pollution reduction	✗	✓	✓
Efficient land use	✗	✗	✓
Physical fitness & health	✗		✓

Note: Some transport improvement strategies help achieve one or two objectives (✓), but by increasing total vehicle travel contradict others (✗). Mobility management strategies reduce total motor vehicle travel and so can help achieve many planning objectives

management tends to provide a much larger range of benefits, as indicated in Table 14.4. When all impacts are considered, mobility management strategies often turn out to be the most cost effective and beneficial solutions to transport problems.

14.7 Implications for Developing Countries

Mobility management is particularly appropriate in developing countries for the following reasons:

- Most residents do not own automobiles. As a result, improvements to alternative modes provide greater direct user benefits, are more equitable, and do more to increase access to education, employment and services, than do automobile transport improvements.
- Policies that reduce automobile traffic reduce conflicts between motorized and non-motorized travellers, improving access and safety to the majority of travellers who rely on walking and bicycling.
- Developing countries have very limited resources to expand roads and parking facilities, or to provide public infrastructure for sprawled development. Mobility management reduces traffic and parking congestion, and therefore the need to expand roadways.
- By reducing the amount that consumers spend on vehicles and fuel, mobility management reduces the need to import these products. In petroleum producing countries, reduced vehicle use increases the amount of oil that can be exported which improves export exchange and economic competitiveness. This will be increasingly beneficial in the future as international oil prices rise due to peak oil.

14.8 Examples and Case Studies[1]

14.8.1 Innovative Transportation Solutions in Curitiba, Brazil

Curitiba, capital of the Brazilian state Paraná 400 km south east of São Paulo, has over the last 30 years developed a high-quality, cost-effective public transport system. Today it stands as a model recognized internationally. Insightful, long term planning with several innovative solutions has provided the citizens with an effective system that gives priority to public instead of private transport. It has the highest user rates of all Brazilian state capitals, 75% of all weekday commuters. All this during a period of unprecedented city growth.

14.8.2 Bogota, Columbia Transport Initiatives (http://ecoplan.org/votebogota2000/vb2_index.htm)

The city of Bogota, Columbia has a diverse program to improve transportation choices and encourage non-automobile modes. They include:

- *TransMilenio*, a high-capacity public transportation system using articulated buses and convenient, magnetic ticketing.
- *Bikeways.* 120 existing and 180 planned kilometers of cycle paths.
- *Walkways.* Construction of sidewalks and shaded walkways ('alamedas') throughout the city.
- Increased parking fees.
- *Pico y Placa.* Restrictions on private automobile travel, based on each vehicle's license-plate number.
- *Car-Free Day.* An annual Car-Free day.

Because this program includes restrictions on automobile travel it was initially controversial. In October, 2000 a public referendum on the program received more than 62% yes votes indicating a high level of public support.

14.8.3 Africa Safe Routes to School (www.movingtheeconomy.ca/cs_tanzania.html)

The majority of Tanzania's urban dwellers face chronic mobility problems including: high proportions of family income needed for daily travel; long travel distances

---

[1] All Internet links in this section were last accessed in March 2011.

due to fast city growth; a poor route infrastructure network, especially for walking and cycling; and a high number of traffic accidents involving non-motorized transport users.

These problems are even worse for school children, who are sometimes denied access on private buses. Female students are sometimes forced to engage in relationships with male drivers or conductors to facilitate easy entry in the private buses and many children suffer from poor attendance and late arrival at school. The cost of transport also limits access to schools and disrupts education, especially of female pupils.

The Association for Advancing Low Cost Mobility (AALOCOM) was formed to address the mobility needs of Tanzania's urban dwellers, starting with school children. The Safe Routes to School Demonstration Project is in the planning stages at the time of writing, but it is a spectacular example of a community responding to a community problem in a manner that is participatory, broad-based and open. AALOCOM recognizes that the success of the project depends on the participation of the different parties responsible. Using a broad base of stakeholders (parents, teachers, police, NGOs, transportation officials and decision makers), AALOCOM's participatory approach creates a sense of ownership and responsibility around child, pedestrian and cycling safety issues.

The project will be piloted in a medium sized city with significant traffic problems, using schools with a high percentage of children residing 2–3 km away. It will focus on:

- Identifying walking and cycling routes to school where traffic safety is a major concern.
- Educating parents about child pedestrian safety issues and solutions;
- Developing traffic calming and infrastructure plans.
- Working with parents, community leaders and decision makers to reach agreement on what changes to make.
- Facilitating availability of affordable bicycles to teachers and pupils.

14.8.4 Rickshaw Trolley Community Solid Waste Collection (www.movingtheeconomy.ca/cs_rickshaw_trolley.html)

Before the Rickshaw Trolley Community Solid Waste Collection system was introduced, solid waste in most of Mirzapur, India was collected from neighborhood streets in handcarts and then dumped in heaps on bigger streets. From these heaps it was lifted onto bullock carts or tractor trolleys by shovel or a hydraulic loader. While being loaded, tractor trolleys blocked traffic on the narrow streets. This was inefficient, unsanitary and undependable since the city could not afford to keep the loader operating and the staff could not manage to lift more than a little bit of the city's garbage. Eventually garbage blocked many streets and drains, and obstructed maintenance of the drainage and water supply systems. The public had lost confidence in the city services and there was little money available for new equipment.

Solid waste needed to be lifted from the street to tractor trolleys without hydraulic equipment. To do this the municipality designed and introduced a loading platform in 1995 with an access ramp for direct loading into parked tractor trolleys. Now 10 collection depots manage the city's daily solid waste. They use available space along street rights-of-way and do not interfere with traffic movement. To make operation of the depots feasible, the service area had to be increased. This was achieved through the introduction of a three-wheeled rickshaw trolley with a modified frame for easier pedalling, and a tilting bin for easy unloading, designed and built by local workshops. These easy-to-move rickshaw trolleys have twice the capacity of handcarts and double their service area to 400 meters.

This low-cost system eliminated the need for hydraulic lifting throughout the city and dramatically reduced staff physical contact with solid waste. The improvement in city appearance changed public attitudes toward the city. In addition, the municipality donated a rickshaw trolley for replication to the city of Aligarh, provided technical assistance to numerous municipalities from India and Nepal, and is exploring opportunities for private processing of compost.

14.8.5 Malaysian TDM Solutions (www.nctr.usf.edu/jpt/pdf/ JPT11-3Kasipillai.pdf)

Growing motor vehicle ownership and use are imposing significant costs on the Malaysian economy and environment. Kasipillai and Chan (2008) recommend a Transport Development Management-based approach to create more sustainable transportation:

1. Alteration of charges on road taxes and car insurance,
2. Elimination of fuel subsidies,
3. Imposition of fuel taxes and amendments in the bases for car taxation,
4. Congestion charging, particularly in Kuala Lumpur, and
5. National road pricing.

14.8.6 Manila Congestion Pricing (Roth and Villoria 2001)

A study of potential road pricing in Manila, Philippines calculates that an optimal congestion charge of 6–14 pesos per vehicle-km would reduce traffic volumes by 11–24% and increase traffic speeds 44–101%. Estimates of total revenues from congestion fees and a 40 peso per day charge for commercial parking could provide total revenues of 12.6–25.5 billion pesos annually to fund a regional transportation authority.

14.8.7 Ghana (www.ibike.org/ghana-women.htm)

In tackling transport and rural development issues, Ghana faces a host of common challenges: environmental degradation; urban gridlock; the cost of road repair and fuel; health problems for women from carrying heavy loads; and a shortage of foreign exchange for vehicles, parts and construction equipment. Fortunately, Ghana has chosen some innovative projects designed to address these issues in a way that meets the needs of the government, its citizens, and the environment. The goal of this plan is to improve rural quality-of-life to help reduce urbanization trends and the demand for expensive urban infrastructure. Brief descriptions of three current projects follow:

1. *New Road Design*. The Department of Roads and Highways is designing rural roads using new standards that take recognize the needs of the non-motor vehicle using population. One notable outcome is a 'single-blade' (4 m or 13 ft) compacted road. The most significant results of the program, however, are the improved production process, social structure, and resource base and allocation. Due to the changes in design standards, a road can be built more economically by labor intensive methods (costing 10–15% less than with mechanical methods). This new system generates more rural employment as well as growth in the local economy supplying the projects.

    The road program also includes a street-tree component, where citizens plant and maintain trees on both sides of the road, providing shade to non-motorized travellers. This scheme also includes drilling wells for safe drinking water.
2. *Transportation Rehabilitation Project*. One aspect of this project is the development and initial production of 250 bicycle trailers and promotion of bicycles for women. Surveys of women show that the equipment was readily accepted as a substitute for head-portage. Women have embraced the use of bicycles and trailers, exhibiting no cultural resistance to the change. The main problem identified is a lack of money or access to credit to buy the vehicles. This obstacle is being overcome by the purchase of the trailers by local NGO's, who then sell them to community members on instalment payback schemes.
3. *The Ministry Of Local Government's Bicycle Program*. While Ghana pursues a decentralization and democratization process, many of the 7,260 members and staff members of the new district assemblies have difficulty visiting constituents and attending meetings. The problem is most acute in the north where vehicles serve areas with poor road conditions and weak bridges only once a week. There are reports of assembly workers walking 50 km (31 miles) to perform assembly functions. However, the distances involved are moderate for bicycles, the terrain is flat and the weather lends itself to use of bicycles. Starting in 3 districts, the project is making bicycles available through a revolving fund on a hire-to-purchase plan. The program is starting with 200 one speed roadster bikes with the goal of getting 1,000 bicycles.

The Transportation Rehabilitation program is successful and stimulates employment by encouraging local entrepreneurs to produce trailers. The LG bicycle project encourages employment generation by training local youth to assemble, repair and maintain the bicycles. The road design program has so far trained 35 private contractors, employing more than 3,000 people. The target is to hire 70% female employees and to combine employment with nutrition education along with vitamin and mineral supplements. Each employee works for about 3 months, receives food and vitamin/mineral subsidies, earns US$145 and has access to a savings plan to buy a bicycle to use on the new roads. Local NGOs provide education and program coordination, helping to strengthen community organizations and insure their continuity.

14.8.8 Improving Urban Walkability in India (CSE 2009)

Table 14.5 indicates the mode split in Indian cities. The report *Footfalls: Obstacle Course To Livable Cities* (CSE 2009) evaluates walking conditions in Indian cities. Although walking is the dominant mode, representing 16% to 57% of urban trips, it receives little consideration in transport planning and investment: walking conditions are poor, with little investment, insufficient road space, and inadequate design and maintenance standards. The study argues that inadequate support for nonmotorized travel is inefficient and inequitable.

The study developed a *Transport Performance Index* for evaluating urban transportation systems and prioritizing system improvements in Indian cities. It consists of the following factors:

Table 14.5 Indian cities mode split, 2007

City category	City population	Walk (%)	Bicycle (%)	Motor-cycle (%)	Public transport (%)	Car (%)	Auto rickshaw (%)
Category-1a	<500,000, plain terrain	34	3	26	5	27	5
Category-1b	<500,000, hilly terrain	57	1	6	8	28	0
Category-2	500,000–1,000,000	32	20	24	9	12	3
Category-3	1,000,000–2,000,000	24	19	24	13	12	8
Category-4	2,000,000–4,000,000	25	18	29	10	12	6
Category-5	4,000,000–8,000,000	25	11	26	21	10	7
National		28	11	16	27	13	6

Wilbur Smith (2008)

- *Public Transport Accessibility Index* (the inverse of the average distance [in km] to the nearest bus stop/railway station (suburban/metro).
- *Service Accessibility Index* (% of Work trips accessible in 15 min time).
- *Congestion Index* (average peak-period journey speed relative to a target journey speed).
- *Walkability Index* (quantity and quality of walkways relative to roadway lengths).
- *City Bus Transport Supply Index* (bus service supply per capita).
- *Para-Transit Supply Index* (para-transit vehicle supply per capita).
- *Safety Index* (1/traffic fatalities per 100,000 residents).
- *Slow Moving Vehicle (Cycling) Index* (availability of cycling facilities and cycling mode share).
- *On-street Parking Interference Index* (1/(portion of major road length used for on-street parking + on-street parking demand).

14.8.9 Transport Policy Emission Impact Evaluation (www.asiandevbank.org/Documents/Evaluation/Knowledge-Briefs/REG/EKB-REG-2010-16.pdf)

The Asian Development Bank (ADB) helps developing member countries (DMCs) shift their economies onto low-carbon growth paths and reduce their carbon emissions. The Bank has developed models for evaluating how specific transport policy decisions affect energy consumption and pollution emissions. It has expanded its policy and project economic evaluation to consider indirect impacts, including the effects of generated traffic, and co-benefits of demand management. This project has identified many cost effective and beneficial ways to improve overall transport system efficiency and reduce emissions.

14.9 Criticisms

Mobility management may encounter the following criticisms (Litman 2008).

14.9.1 Costly and Dangerous

When evaluated using conventional transport economic analysis, individual mobility management strategies often seem cost ineffective. For example, individual pedestrian, cycling and public transit improvements often do little to reduce traffic congestion, and because per-kilometer crash fatality rates are higher for pedestrians and cyclists than for motor vehicle occupants, efforts to encourage non-motorized travel may seem dangerous.

However, this reflects the inadequacies and biases of current transport economic evaluation, which overlooks many mobility management benefits, and fails to recognize the role that an individual strategy can play as part of a comprehensive mobility management program (Litman 2008). Conventional evaluation overlooks many of the costs of automobile dependency and sprawl, and many of the benefits of a more diverse and efficient transport system. Although an individual strategy may appear to have modest impacts and benefits, an integrated program that includes improvements to alternative modes and incentives to use the most efficient option for each trip can have large impacts and benefits.

14.9.2 Economically Harmful

Critics sometimes argue that, because some economic activities are more efficient with motorized transport, and vehicle travel is associated with economic development and wealth, efforts to reduce vehicle travel reduce productivity and are economically harmful (Pozdena 2009). While it is true that a certain amount of motor vehicle travel can increase productivity, this does not mean that any increase in vehicle travel increases productivity or that any policy that reduces vehicle travel reduces economic development (Litman 2009). As discussed earlier, although motor vehicle travel can provide substantial benefits it also imposes substantial costs, so beyond an optimal level, increased vehicle travel is economically harmful. Mobility management strategies that reflect market and planning principles, such as more efficient pricing and more comprehensive evaluation, tend to increase productivity.

Empirical evidence indicates that, among otherwise similar cities and countries, those that are more automobile dependent are less economically productive, while those that encourage use of alternative modes and have higher vehicle charges are more productive. For example, Fig. 14.2 shows that U.S. states with higher per capita vehicle travel tend to have lower average per capita gross domestic product (GDP).

Figure 14.3 shows that per capita GDP tends to increase with public transit ridership in U.S. cities.

Figure 14.4 shows that per capita economic productivity tends to increase with higher fuel prices, particularly among oil importing countries. This makes sense since higher fuel prices encourage efficient travel behavior and energy conservation, which reduces total transport costs (traffic congestion, road and parking facility costs, accident damages, pollution emissions, etc.), and reduces the export exchange that must be devoted to vehicle and fuel imports. This indicates that substantial increases in vehicle fees can be achieved without reducing overall economic productivity.

14.9.3 Unfair and Intrusive

Critics sometimes complain that a particular mobility management strategy is unfair or intrusive to a particular group. For example, if most roads and parking facilities are unpriced, road tolls and parking fees may seem unfair to urban motorists.

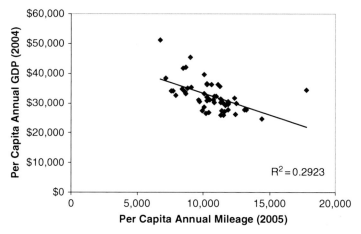

Fig. 14.2 U.S. state per capita GDP and VMT (VTPI 2009)
Note: Information in this and subsequent graphs is contained in the *2009 Urban Transportation Performance Spreadsheet* (www.vtpi.org/Transit2009.xls), based on data from the FHWA's *Highway Statistics*, the TTI's *Urban Mobility Report*, and the Bureau of Economic Account's *Gross Domestic Product By Metropolitan Area* (www.bea.gov/regional/gdpmetro)

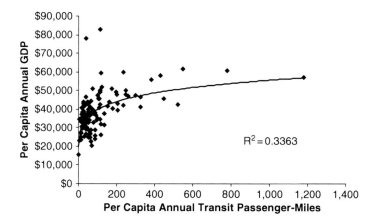

Fig. 14.3 Per capita GDP and transit ridership (VTPI 2009)

Similarly, restrictions on vehicle travel, such as high occupant vehicle lanes or no-drive days may seem unfair. Yet, since automobile travel imposes significant external costs, vehicle travel is unfair to other road users. For example, automobile travel is unfair to bus passengers, who are delayed by traffic congestion caused primarily by automobile travel which requires far more road space per passenger-kilometer, and to pedestrians and cyclists who bear excessive risk and pollution exposure. As a result, significantly higher user fees, and restrictions on vehicle use can be considered fairer overall.

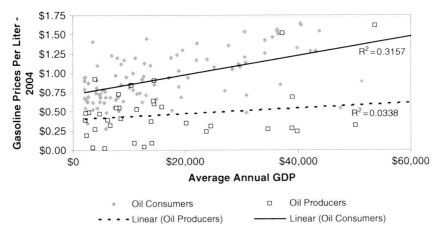

Fig. 14.4 GDP versus fuel prices, countries (Metschies 2005)
Note: Fuel price (www.internationalfuelprices.com), GDP (http://en.wikipedia.org/wiki/List_of_countries_by_GDP_(PPP)_per_capita) Petroleum production (http://en.wikipedia.org/wiki/Petroleum); excluding countries with average annual GDP under $2000

14.10 Conclusions

Motor vehicle travel can provide significant benefits to individuals and society, but it also imposes significant costs. Although a certain amount of motor vehicle travel is efficient and beneficial overall, beyond an optimal level, the incremental costs exceed incremental benefits, resulting in vehicle travel with negative net value.

Many current policy and planning practices tend to favor mobility over accessibility and automobile travel over alternative modes. Many of these violate basic principles of good planning and efficient pricing, resulting in economically excessive motor vehicle travel. Correcting these distortions tends to reduce vehicle travel in ways that increase overall transport system efficiency. A general term for these reforms is *mobility management*, which includes various strategies that increase transport options (better walking, cycling, public transit, etc.), incentives to use more efficient transport options, and more accessible land use development patterns. This favors higher value trips and more efficient modes, increasing overall transport system efficiency.

Mobility management can provide many benefits including congestion reduction, road and parking facility savings, consumer savings, improved mobility for non-drivers, traffic safety, energy conservation, emission reductions, more efficient land use development, and improved public fitness and health. Where these strategies are implemented appropriately people tend to drive less, rely more on alternative modes, and are better off overall as a result. If fully implemented to the degree that is cost effective, mobility management strategies typically reduce motor vehicle travel by 30–50%, and even more in some situations, compared with what results from conventional policies and planning practices.

These reforms are particularly appropriate in developing countries to support economic development, social equity objectives, and to protect the environment for current and future generations.

References

All web-based references were last accessed in March 2011

Broaddus A, Litman T, Menon G (2009) Training document on transportation demand management. Sustainable Urban Transport Project (www.sutp.org) and GTZ (www.gtz.de)

Cairns S et al (2004) Smarter choices – changing the way we travel. UK Department for Transport (www.dft.gov.uk)

Cambridge Systematics (2009) Moving cooler: transportation strategies to reduce greenhouse gas emissions (www.movingcooler.info). Co-sponsored by a variety of organizations; report at http://commerce.uli.org/misc/movingcooler.pdf; summary at http://commerce.uli.org/misc/movingcoolerexecsum.pdf

Clarke H, Prentice D (2009) A conceptual framework for the reform of taxes related to roads and transport. School of Economics and Finance, La Trobe University, for the Australia Treasury Australia's Future Tax System review. http://apo.org.au/research/conceptual-framework-reform-taxes-related-roads-and-transport

CNT (2008) Housing + transportation affordability index. Center for Neighbourhood Technology. http://htaindex.cnt.org

Co-Benefits Asia Hub (2011) Website. www.observatory.ph/co-benefits_asia

Cohen AS, Shaheen S, McKenzie R (2008) Carsharing: a guide for local planners. PAS Memo, May/June. American Planning Association (www.planning.org). www.innovativemobility.org/publications_by/pubs_topic.shtml

CSE (2009) Footfalls: Obstacle Course to Livable Cities. Right to Clean Air Campaign. Centre for Science and Environment (www.cseindia.org)

EDF (2009) Reinventing transit: American communities finding smarter, cleaner, faster transportation solutions. The Environmental Defense Fund (www.edf.org). www.edf.org/documents/9522_Reinventing_Transit_FINAL.pdf

El-Geneidy A, Levinson D (2006) Access to destinations: development of accessibility measures. Center for Transportation Studies, University of Minnesota (www.cts.umn.edu). www.cts.umn.edu/access-study/publications

Ennis M (2010) Vanpools in the Puget Sound region: The case for expanding Vanpool Programs to move the most people for the least cost. Washington Policy Center for Transportation (www.washingtonpolicy.org). www.washingtonpolicy.org/Centers/transportation/policybrief/CompleteVanpoolPB.pdf

Evans JE, Pratt RH (2005) Vanpools and buspools; traveler response to transportation system changes. Chapter 5, TCRP Report 95, Transportation Research Board. www.trb.org

Ewing R, Bartholomew K, Winkelman S, Walters J, Chen D (2007) Growing cooler: the evidence on urban development and climate change. Urban Land Institute and Smart Growth America. www.smartgrowthamerica.org/gcindex.html

FHWA (2009) Economics: pricing, demand, and economic efficiency: a primer. Office of Transportation Management, Federal Highway Administration (www.ops.fhwa.dot.gov). www.ops.fhwa.dot.gov/publications/fhwahop08041/fhwahop08041.pdf

Hendrickson H, Cicas G, Matthews HS (2006) Transportation sector and supply chain performance and sustainability. Transportation Research Record 1983 (www.trb.org), pp 151–157

ICF and CUTR (2005) Analyzing the effectiveness of commuter benefits programs. TCTP Report 107, TRB (www.trb.org). http://gulliver.trb.org/publications/tcrp/tcrp_rpt_107.pdf

ITE (2010) Designing walkable urban thoroughfares: a context-sensitive approach. An ITE recommended practice. Institute of Transportation Engineers (www.ite.org) and Congress for New Urbanism (www.cnu.org). www.ite.org/css

Kasipillai J, Chan P (2008) Travel Demand Management: Lessons for Malaysia. J Public Transport, 11(3):41–55. www.nctr.usf.edu

Leather J (2009) Rethinking transport and climate change. Asian Development Bank (www.adb.org). www.transport2012.org/bridging/ressources/files/1/96,Rethinking_Transport_and_Climate_Chan.pdf

Litman T (1997) Distance-based vehicle insurance as a TDM strategy. Transp Q 51(3):119–138. www.vtpi.org/dbvi.pdf

Litman T (1999) Exploring the paradigm shifts needed to reconcile transportation and sustainability objectives. Transportation Research Record 1670. Transportation Research Board (www.trb.org), pp 8–12. www.vtpi.org/reinvent.pdf

Litman T (2003) Measuring transportation: traffic, mobility and accessibility. ITE J (www.ite.org) 73(10):28–32. www.vtpi.org/measure.pdf

Litman T (2005) Efficient vehicles versus efficient transportation: comparing transportation energy conservation strategies. Transport Pol 12(2):121–129; at www.vtpi.org/cafe.pdf

Litman T (2006a) Transportation market distortions. Berkeley Planning J. Issue theme: Sustainable Transport in the United States: From Rhetoric to Reality? (www-dcrp.ced.berkeley.edu/bpj) 19:19–36. www.vtpi.org/distortions_BPJ.pdf

Litman T (2006b) Parking management best practices. Planners Press: Chicago (www.planning.org). www.vtpi.org/PMBP_Flyer.pdf

Litman T (2007) Win-win transportation solutions: cooperation for economic, social and environmental benefits. Victoria Transport Policy Institute (www.vtpi.org). www.vtpi.org/winwin.pdf

Litman T (2008) Guide to calculating mobility management benefits. Victoria Transport Policy Institute (www.vtpi.org). www.vtpi.org/tdmben.pdf

Litman T (2009) Are vehicle travel reduction targets justified? Evaluating mobility management policy objectives such as targets to reduce VMT and increase use of alternative modes. Victoria Transport Policy Institute (www.vtpi.org). www.vtpi.org/vmt_red.pdf

Litman T (2010) Socially optimal transport prices and markets. Victoria Transport Policy Institute (www.vtpi.org). www.vtpi.org/sotpm.pdf

Meakin R (2004) Urban transport institutions. Sustainable Urban Transport Project (www.sutp.org) and GTZ (www.gtz.de). www.sutp.org/dn.php?file=1A-UDP-EN.pdf

Metschies G (2005) International fuel prices 2005, with comparative tables for 172 countries. German Agency for Technical Cooperation (www.internationalfuelprices.com)

NTHP (2010) Helping Johnny walk to school: policy recommendations for removing barriers to community-centered schools. National Trust for Historic Preservation (www.preservationnation.org). www.preservationnation.org/issues/historic-schools/helping-johnny-walk-to-school

Pozdena RJ (2009) Driving the economy: automotive travel, economic growth, and the risks of global warming regulations. Cascade Policy Institute (www.cascadepolicy.org). www.cascadepolicy.org/pdf/VMT%20102109.pdf

Pucher J, Dill J, Handy S (2010) Infrastructure, programs and policies to increase bicycling: an international review. Preventive Med 48(2), February; prepared for the Active Living By Design Program (www.activelivingresearch.org/resourcesearch/journalspecialissues). http://policy.rutgers.edu/faculty/pucher/Pucher_Dill_Handy10.pdf

Roth G, Villoria OG Jr (2001) Finances of commercialized urban road network subject to congestion pricing. Transportation Res 1747:29–35 TRB (www.trb.org)

Sakamoto K (2010) Financing sustainable urban transport. GTZ Sourcebook Module, Sustainable Urban Transport Project (www.sutp.org) Asia and the German Technical Cooperation (www.gtz.de/en). www.sutp.org/dn.php?file=1f-FSUT-EN.pdf

Schwaab JA, Thielmann S (2001) Economic instruments for sustainable road transport: an overview for policy makers in developing countries. GTZ (www.gtz.de) and the United Nations Economic and Social Commission for Asia and the Pacific (www.unescap.org). www.gtz.de/dokumente/Economic_Instruments_for_Sustainable_Road_Transport.pdf

Shoup D (2005) The high cost of free parking. Planners Press, Chicago (www.planning.org)

Sloman L, Cairns S, Newson C, Anable J, Pridmore A, Goodwin P (2010) The effects of smarter choice programmes in the sustainable travel towns: summary report. Report to the Department for Transport (www.dft.gov.uk); at www.dft.gov.uk/pgr/sustainable/smarterchoices/smarterchoiceprogrammes/pdf/summaryreport.pdf

Turnbull KF, Levinson HS, Pratt (2006) HOV facilities – traveler response to transportation system changes. TCRB Report 95, Transportation Research Board (www.trb.org); available at http://onlinepubs.trb.org/onlinepubs/tcrp/tcrp_rpt_95c2.pdf

USDOT (2010) Transportation's role in reducing U.S. greenhouse gas emissions: volume 1. Report to Congress, U.S. Department of Transportation (www.dot.gov), at http://ntl.bts.gov/lib/32000/32700/32779/DOT_Climate_Change_Report_-_April_2010_-_Volume_1_and_2.pdf

USEPA (2006) Parking spaces/community places: finding the balance through smart growth solutions. Development, Community, and Environment Division (DCED); U.S. Environmental Protection Agency (www.epa.gov/smartgrowth/parking.htm)

VTPI (2009) Urban transport performance spreadsheet. Victoria Transport Policy Institute (www.vtpi.org). www.vtpi.org/Transit2009.xls

VTPI (2010) Online TDM Encyclopedia. Victoria Transport Policy Institute (www.vtpi.org/tdm)

Wilbur Smith (2008) Traffic & Transportation Policies and Strategies in Urban Areas in India. Ministry of Urban Development (www.urbanindia.nic.in). www.urbanindia.nic.in/programme/ut/final_Report.pdf.

Wright L (2007) Bus rapid transit. Module in the Sustainable Transport: A Sourcebook for Policymakers in Developing Cities, published by the Sustainable Urban Transport Project – Asia (www.sutp-asia.org), Deutsche Gesellschaft fur Technische Zusammenarbeit (www.gtz.de), and the Institute of Transportation and Development Policy (www.itdp.org); at www.itdp.org/index.php/microsite/brt_planning_guide

WSDOT (2009) Least cost planning guidance. Washington State Department of Transportation (www.wadot.wa.gov). www.wadot.wa.gov/NR/rdonlyres/FDBC2704-7998-49D9-9F70-B16F5D1A0B2E/0/LeastCostPlanningexampledefinitionsfordiscussion.pdf

Chapter 15
Automobiles and Climate Policy in the Rest of the OECD

Michael P. Walsh

Abstract Over the course of the past decade, there has been a fundamental change in the approach to regulating fuel economy from road vehicles, mainly induced by concerns of human-induced climate change. The number of countries adopting some form of regulation has grown dramatically. Moreover, the form of the standard is starting to shift away from a mass based approach toward a footprint based approach, which will open up additional opportunities to take advantage of lightweighting as a key element of a control strategy. This chapter describes the history and most recent developments (up to the beginning of 2011) on greenhouse gas emission and/or fuel economy standards from non-EU OECD countries around the world, namely in Canada, Japan, South Korea and the United States. Other OECD countries such as Australia and Mexico are also considering the implementation of similar standards. While command and control standards are expected to remain the backbone of control efforts, economic incentives or disincentives including fuel taxes are expected to play an even more important role in the future than they do today.

15.1 Introduction

The 33 member countries of OECD are: Australia, Austria, Belgium, Canada, Chile, Czech Republic, Denmark, Finland, France, Germany, Greece, Hungary, Iceland, Ireland, Israel, Italy, Japan, Korea, Luxembourg, Mexico, the Netherlands, New Zealand, Norway, Poland, Portugal, Slovak Republic, Slovenia, Spain, Sweden, Switzerland, Turkey, United Kingdom, United States. Twenty two of these countries are either members of the European Union or follow the vehicle emissions roadmap of the EU. Of the remaining – Australia, Canada, Chile, Israel, Japan, Korea, Mexico, New Zealand, Turkey and the United States – 4 have proposed, established, or are in the process of revising light-duty vehicle fuel economy or greenhouse gas (GHG) emission standards.

M.P. Walsh (✉)
International Council on Clean Transportation, Washington, DC, USA
e-mail: mpwalsh@igc.org

A number of different test procedures, formulas, baselines, and approaches to regulating fuel economy and GHG emissions have evolved over the last several decades. The policy objectives of these regulations vary depending on the priorities of the regulating body, but most standards are applied to new vehicles in order to reduce either fuel consumption or GHG emissions. There are important differences between these two approaches. Fuel economy standards seek to reduce the amount of fuel used by the vehicle per distance driven and also effectively reduce the amount of carbon dioxide (CO_2) emitted per kilometer driven. GHG emission standards generally target at least CO_2 but may also include other climate forcing emissions from the vehicle, such as refrigerants from the air conditioning system or nitrous oxide (N_2O) from the catalytic converter. GHG emissions standards may even extend beyond the vehicle to encompass the GHG emissions generated from the production of fuels.

Certification of GHG emission and fuel economy performance for new vehicles is based on test procedures intended to reflect real world driving conditions and behavior in each country. The European Union, Japan, and the U.S. have each established their own test procedures. China and Australia use the European Union's test procedures, while California, Canada, South Korea, and Taiwan, China follow the U.S. Corporate Average Fuel Economy (CAFE) test procedures. The following sections based on previous analysis by The International Council on Clean Transportation (ICCT) outline the history and most recent developments on GHG emission and fuel economy standards from non EU OECD countries around the world. (ICCT 2010).

15.2 Policies in Individual Countries

15.2.1 Japan

The Japanese government first established fuel economy standards for gasoline and diesel powered light-duty passenger and commercial vehicles in 1999 under its 'Top Runner' energy efficiency program. Fuel economy targets are based on weight class, with automakers allowed to accumulate credits in one weight class for use in another, subject to certain limitations. Penalties apply if the targets are not met, but they are minimal. The effectiveness of the standards is enhanced by highly progressive taxes levied on the gross vehicle weight and engine displacement of automobiles when purchased and registered. These financial incentives promote the purchase of lighter vehicles with smaller engines. For example, the Japan Automobile Manufacturers Association has estimated that the owner of a subcompact car (750 kg curb weight) will pay $4000 less in taxes relative to a heavier passenger car (1,100 kg curb weight) over the lifetime of the vehicle (JAMA 2007).

In December 2006, Japan revised its fuel economy targets upward, and expanded the number of weight bins from nine to sixteen (Fig. 15.1). This revision took place before the full implementation of the previous standards because the majority of

15 Automobiles and Climate Policy in the Rest of the OECD

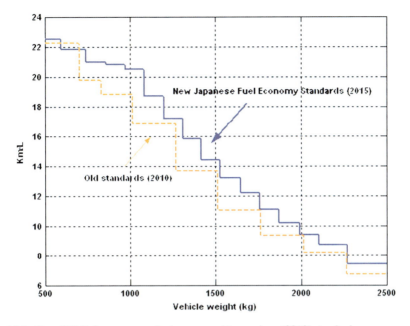

Fig. 15.1 New (2015) Japanese standards compared to previous (2010) standards

vehicles sold in Japan in 2002 already met or exceeded the 2010 standards. This new standard is projected to improve the fleet average fuel economy of new passenger vehicles from 13.6 kilometers per liter (km/L) in 2004 to 16.8 km/L in 2015, an increase of 24%. Based on ICCT's analysis, the new target reaches an equivalent average CO_2 emission of 125 g/km if it were to be measured on the New European Driving Cycle (NEDC) test cycle, which is the test cycle used in the EU (see Fig. 15.4 in next section).

In 2010 Japan will introduce a new test cycle, the JC08, to measure progress toward meeting the revised 2015 targets. Relative to the previous 10–15 test cycle, the JC08 test cycle is longer, has higher average and maximum speeds and requires more aggressive acceleration. These differences are illustrated in Fig. 15.2.

According to the Japanese government, the JC08 cycle's higher average speed,[1] quicker acceleration, and new cold start increased the stringency of the test by 9%. The government determined the relative stringency by measuring fuel economy of 2004 model year vehicles under each test cycle. The fleet average fuel economy for MY2004 vehicles was 15.0 km/L under the 10–15 test cycle (MLIT 2006) and 13.6 km/L under the JC08 test cycle (ANRE/MLIT 2006). The more rigorous JC08 test cycle serves to further increase the stringency of the 2015 standards beyond the difference seen in Fig. 15.1.

---

[1] Because Japanese vehicles are calibrated to slower driving conditions, increases in test speeds are claimed to increase fuel consumption, in contrast to results for vehicles sold in the U.S. market.

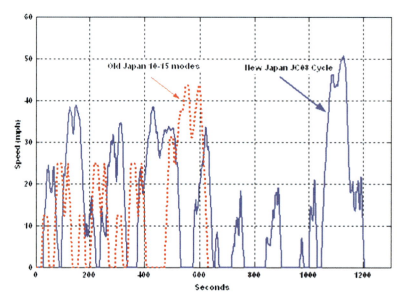

Fig. 15.2 New versus old Japanese vehicle emission test cycles

15.2.2 United States

Following the end of the Second World War, the car population in the United States exploded, rising from under 50 million to approximately 100 million by 1970. Fuelled by cheap gasoline prices and a strong economy, cars over this period got progressively bigger and bigger with greater horsepower and fuel consumption. All this changed in the early 1970s.

On October 17th, 1973, the Organization of Petroleum Exporting Countries (OPEC) slapped an embargo on oil exports. Although the embargo lasted only 5 months, it resulted in long lines of cars at filling stations and following a further blip late in the decade, the Iranian oil cut off, the addition of odd even day gas rationing. In response, the United States adopted a mandatory fuel efficiency program called Corporate Average Fuel Economy (CAFE). The Energy Policy and Conservation Act, passed in 1975 (to come into effect in model year 1978), amended the Motor Vehicle Information and Cost Saving Act to require new passenger cars to get at least 27.5 miles per U.S. gallon (8.55 L/100 km) by model year 1985, as measured by the existing U.S. Environmental Protection Agency (EPA 2009) test procedures; light-duty trucks, including jeeps, minivans, and SUVs, had to meet a more lenient corporate fuel economy standard. Fuel efficiency standards adopted through 2009 are summarized in Table 15.1.

NHTSA began setting CAFE standards for light trucks based on vehicle size as defined by their 'footprint' (the bottom area between the vehicle's four wheels). The new standard is based on a complex formula matching fuel economy targets

15   Automobiles and Climate Policy in the Rest of the OECD

Table 15.1 U.S. new-car fuel efficiency standards (cafe) (miles per U.S. Gallon)

Model year	Passenger car	Light truck		
		Combined	2WD	4WD
1978	18		–	–
1979	19		17.2	15.8
1980	20		16	14
1981	22		16.7	15
1982	24	17.5	18	16
1983	26	19	19.5	17.5
1984	27	20	20.3	18.5
1985	27.5	19.5	19.7	18.9
1986	26	20	20.5	19.5
1987	26	20.5	21	19.5
1988	26	20.5	21	19.5
1989	26.5	20.5	21.5	19
1990	27.5	20	20.5	19
1991	27.5	20.2	20.7	19.1
1992	27.5	20.2		
1993	27.5	20.4		
1994	27.5	20.5		
1995	27.5	20.6		
1996	27.5	20.7		
1997	27.5	20.7		
1998	27.5	20.7		
1999	27.5	20.7		
2000	27.5	20.7		
2001	27.5	20.7		
2002	27.5	20.7		
2003	27.5	20.7		
2004	27.5	20.7		
2005	27.5	21.0		
2006	27.5	21.6		
2007	27.5	22.2		
2008	27.5	22.5[a]		
2009	27.5	23.1[a]		

[a]Manufacturers may choose reformed or unreformed standard beginning Model Year 2008; 22.5 is the unreformed standard in model year 2008; 23.1 is the unreformed standard in model year 2009

with vehicle sizes. For the first 3 years, manufacturers can choose between truck-fleet average targets of 22.7 mpg in 2008, 23.4 mpg in 2009, and 23.7 mpg in 2010, or size-based (so called reformed) targets. Beginning in 2011, manufacturers will be required to meet the size-based standards that are expected to result in a fleetwide average of 24.0 mpg (CFR 2006).

Throughout the 1980s and 1990s, fuel prices dropped, however. The CAFE pressures to improve fuel efficiency were not tightened, and U.S. new-car fuel efficiency slipped steadily. Figure 15.3 and Table 15.2 depict time trends in car, light truck, and car-plus-light truck fuel economy. Since 1975, the fuel economy of the combined car and light truck fleet has moved through several phases:

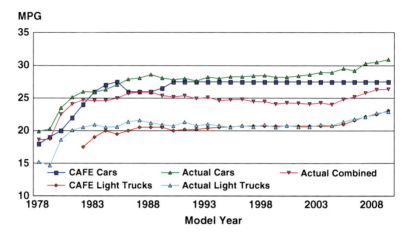

Fig. 15.3 Actual fuel economy performance versus CAFE standards in the US
Note: Actual values reflect the sales weighted averages for each model year based on the actual production and sales for that model year

1. A rapid increase from 1975 through 1981;
2. A slow increase until reaching its peak in 1987;
3. A gradual decline until 2004; and
4. An increase beginning in 2005.

Significantly, U.S. efficiency improvements began with the industrialized world's least-efficient car fleet. Only after the dramatic improvements observed to date are typical U.S. cars generally as efficient as those in the same weight class in other countries. But because vehicles tend to be bigger and heavier in the United States – where light trucks and SUVs account for approximately half of new light-duty vehicles sales – the fuel consumption and CO_2 emissions per mile driven tend to be the highest in the world (on early US efforts and difficulties in reducing greenhouse emissions from motor vehicles, see Walsh and MacKenzie 1990).

As it has in the past with 'conventional' air pollutants, California took the lead in the United States in addressing greenhouse gas emissions from vehicles. In 2002, the California legislature adopted the so called Pavley bill (AB 1493), directing the California Air Resources Board (CARB) to achieve the maximum feasible and cost-effective reduction of greenhouse gases from California's motor vehicles.[2] In response, CARB adopted near-term standards to be phased in from 2009 through 2012, and midterm standards to be phased in from 2013 through 2016. (CARB

---

[2] AB 1493, also known as the California Vehicle Global Warming Law, was signed into law by Governor Gray Davis on July 22, 2002.

15 Automobiles and Climate Policy in the Rest of the OECD

Table 15.2 Measured CAFE performance

Model year	Cars	Light trucks	Combined
1975	15.8	13.7	15.3
1976	17.5	14.4	16.7
1977	18.3	15.6	17.7
1978	19.9	15.2	18.6
1979	20.3	14.7	18.7
1980	23.5	18.6	22.5
1981	25.1	20.1	24.1
1982	26.0	20.5	24.7
1983	25.9	20.9	24.6
1984	26.3	20.5	24.6
1985	27.0	20.6	25.0
1986	27.9	21.4	25.7
1987	28.1	21.6	25.9
1988	28.6	21.2	25.9
1989	28.1	20.9	25.4
1990	27.8	20.7	25.2
1991	28.0	21.3	25.4
1992	27.6	20.8	24.9
1993	28.2	21.0	25.1
1994	28.0	20.8	24.6
1995	28.3	20.5	24.7
1996	28.3	20.8	24.8
1997	28.4	20.6	24.5
1998	28.5	20.9	24.5
1999	28.2	20.5	24.1
2000	28.2	20.8	24.3
2001	28.4	20.6	24.2
2002	28.6	20.6	24.1
2003	28.9	20.9	24.3
2004	28.9	20.8	24
2005	29.5	21.4	24.8
2006	29.2	21.8	25.2
2007	30.3	22.1	25.8
2008	30.5	22.7	26.3
2009	30.9	22.9	26.4

2004a, b) California also sought a waiver from EPA to impose CO_2 emissions standards on cars, which was initially denied by the George W. Bush administration but later granted by the Obama administration in 2009.

The California standards cover the whole suite of GHG emissions related to vehicle operation and use. These include:

- CO_2, methane (CH_4) and N_2O emissions resulting directly from vehicle operation;
- CO_2 emissions resulting from energy consumption in operating the air conditioning (A/C) system;

- Hydrofluorocarbon (HFC) emissions from the A/C system due to either leakage, losses during recharging, or release from scrappage of the vehicle at the end of life[3]; and
- Upstream emissions associated with the production of the fuel used by the vehicle.

Following their passage, the California standards were adopted by numerous other states.

In the last 3 years, American regulators have taken significant steps to improve fuel economy and reduce greenhouse gas emissions from motor vehicles. In April 2007, the U.S. Supreme Court ruled (*Massachusetts v. EPA*), in a 5–4 decision, that GHG emissions are air pollutants potentially subject to federal regulation under the Clean Air Act. In response, the Bush Administration signed an executive order directing the U.S. EPA, in collaboration with the Departments of Transportation and Energy, to develop regulations that could reduce projected[4] oil use by 20% within a decade (Executive Order 2007). The Administration suggested that the 'Twenty in Ten' goal be achieved by: (1) increasing the use renewable and alternative fuels, which will displace 15% of projected annual gasoline use; and (2) by further tightening the CAFE standards for cars and light trucks, which will bring about a further 5% reduction in projected gasoline use. The federal Energy Independence and Security Act of 2007 raised the U.S. fuel economy standard for passenger vehicles and light trucks to 35 mpg by the year 2020.

On May 19, 2009, President Barack Obama announced a policy that called for a standard of 35.5 mpg by 2016, essentially requiring light-duty vehicles nationally to meet the same requirements as California. President Obama called for a joint rulemaking by EPA and the National Highway Transportation Safety Administration, which was issued in September 2009 (*Federal Register* 2009).

Responding to consumer complaints, EPA has readjusted the fuel economy test procedures to more accurately report real world consumer experience. While this does not affect the CAFE standard or compliance by automakers, it does give consumers a more accurate reflection of expected fuel use. EPA's new testing method – which applies to model year 2008 and later vehicles – includes the city and highway tests used for previous models along with additional tests to represent faster speeds and acceleration, air conditioning use, colder outside temperatures, and wind and road surface resistance.

In a May 21, 2010 memorandum, President Obama directed EPA and DOT to issue a Notice of Intent (NOI) that would lay out a coordinated plan, to propose regulations to extend the national program and to coordinate with CARB in developing

---

[3] The industry-standard mobile air conditioner refrigerant HFC-134a has a Global Warming Potential of 1300; alternative refrigerants such as HFC 152a and CO_2 have GWPs of 120 and 1, respectively (CARB 2004a, b).

[4] Choosing to set reductions goals from a baseline of *projected* emissions rather than a firm baseline, such as the year in which the policy was adopted or a point in the past, can limit the total expected emission reductions substantially.

a technical assessment to inform the NOI and subsequent rulemaking process. NHTSA and EPA, recently announced they will begin the process of developing tougher greenhouse gas and fuel economy standards for passenger cars and trucks built in model years 2017 through 2025, building on the first phase of the national program covering cars from model years 2012–2016.Continuing the national program will help make it possible for manufacturers to build a single national fleet of cars and light trucks that satisfies all federal and California standards.

Consistent with the presidential memorandum, the NOI includes an initial assessment for a potential national program for the 2025 model year and outlines next steps for additional work the agencies will undertake. Next steps include issuing a supplemental NOI that would include an updated analysis of possible future standards by November 30, 2010. As part of that process, the agencies will conduct additional study and meet with stakeholders to better determine what level of standards might be appropriate. The agencies aim to propose actual standards within a year.

The results of the interim technical assessment are summarized in the NOI and presented in a separate document, which NHTSA, EPA and CARB also jointly released. The assessment also considers the costs and effectiveness of applicable technologies, compliance flexibilities available to manufacturers, potential impacts on auto industry jobs, and the infrastructure needed to support advanced technology vehicles. This assessment was developed through extensive dialogue with automobile manufacturers and suppliers, non-governmental organizations, state and local governments, and labor unions.

The U.S. EPA and the U.S. Department of Transportation (DOT) have also announced the first national standards to reduce greenhouse gas (GHG) emissions and improve fuel efficiency of heavy-duty trucks and buses. They are proposing new standards for three categories of heavy trucks: combination tractors, heavy-duty pickups and vans, and vocational vehicles. The categories were established to address specific challenges for manufacturers in each area. For combination tractors, the agencies are proposing engine and vehicle standards that begin in the 2014 model year and achieve up to a 20% reduction in CO_2 emissions and fuel consumption by 2018 model year. For heavy-duty pickup trucks and vans, the agencies are proposing separate gasoline and diesel truck standards, which phase in starting in the 2014 model year and achieve up to a 10% reduction for gasoline vehicles and 15% reduction for diesel vehicles by 2018 model year (12% and 17% respectively if accounting for air conditioning leakage). Lastly, for vocational vehicles, the agencies are proposing engine and vehicle standards starting in the 2014 model year which would achieve up to a 10% reduction in fuel consumption and CO_2 emissions by 2018 model year.

15.2.3 Canada

Canada's Company Average Fuel Consumption (CAFC) program was introduced in 1976 to track the fuel consumption of the new light duty vehicle fleet. CAFC is similar to the U.S. CAFE program. Also, the CAFC program was voluntary since Canadian automakers made a commitment to meet the targets set out by the program in the early 1980s. The fuel consumption goals set out by the program have historically been equivalent to CAFE standards. Since Canadian consumers tend to buy more fuel-efficient vehicles than U.S. consumers, the auto industry, as a whole, has consistently met or exceeded CAFC targets.

In 2000, the Government of Canada signalled its intention to seek significant improvements in fuel efficiency under a voluntary agreement with automakers. Negotiations culminated in 2005 with the signing of a voluntary Memorandum of Understanding (MOU) between the government and automakers. Under the MOU, the automakers committed to reducing on-road GHG emissions from vehicles by 5.3 megatonnes CO_2 equivalent ($CO_{2\ eq}$) per year in 2010 (MOU 2005). The 5.3 Mt target is measured from a 'reference case' level of emissions based on a 25% reduction target in fuel consumption that is designed to reflect the actions of automakers that would have occurred in the absence of action on climate change. Under the MOU, automakers can receive credits for reductions in: CO_2 achieved by reducing vehicle fuel consumption; exhaust N_2O and CH_4 emissions; HFC emissions from air-conditioning systems; and reductions in the difference between lab-tested and actual in-use fuel consumption. Since the MOU covers all GHGs emitted by both the new and in-use vehicle fleet, the need to improve new vehicle fuel efficiency will depend on what other GHG reductions will be achieved by industry and counted toward the target.

In October 2006, the Canadian government announced a number of additional measures to reduce air pollutants and GHG emissions. Among these measures was a commitment to formally regulate motor vehicle fuel consumption beginning with the 2011 model year, signalling the end of the voluntary CAFC program.

In the 2007 budget, the Canadian Government also introduced a program called the Vehicle Efficiency Incentive (VEI), which came into effect March 2007. The program includes a rebate and tax component, both of which are based on vehicle fuel efficiency. The performance-based rebate program, run by Transport Canada, offers $1000 to $2000 for the purchase or long-term lease (12 months or more) of an eligible vehicle. Transport Canada maintains a list of the eligible vehicles, which currently includes new cars achieving 151 g CO_2/km or less and 36 mpg or better, new light trucks getting 198 g CO_2/km or less and 28 mpg or better, and new flexible-fuel vehicles with combined fuel consumption E85 ratings of 302 g CO_2/km or less and 18 mpg or better (Transport Canada 2007). The new excise tax, called a 'Green Levy', is administered by the Canada Revenue Agency on inefficient vehicles. The sliding tax of up to $4000 applies only to passenger cars with a weighted average fuel consumption of 302 g CO_2/km or greater and 18 mpg or less (Canada Revenue Agency 2007).

In addition to actions taken by the federal government, some Canadian provinces have also announced their own plans to further reduce GHG emissions from motor vehicles by aligning their programs with California's GHG emission standard.

On October 1st, 2010 Canada's environment minister unveiled final regulations to impose progressively more stringent greenhouse gas emissions standards for new passenger automobiles and light trucks for the 2011–2016 model years. The regulations largely mirror U.S. standards. The regulations are expected to reduce vehicles' greenhouse gas emissions in the 2016 model year by 25% from 2008 levels. They apply to companies that manufacture or import for sale in Canada new passenger automobiles and light trucks for the 2011 and subsequent model years.

They establish fleet average emissions standards aligned with the U.S. national fuel economy program, with special provisions for the 2011 model year; create an emissions credit trading system; and include mandatory annual reporting of fleet average emissions performance.

For the 2011 model year, the regulations require companies to establish a unique fleet average standard based on the size and number of vehicles sold. They require the inclusion only of vehicles manufactured after the date the regulations come into force, but companies may elect to include all vehicles in the model year. Companies unable to meet the standard may purchase credits from the federal government at a rate of C$20 ($19.60) per metric ton of CO_2-equivalent emissions. For the 2012 and later model years, companies must comply with unique fleet average emissions standards for each model year that become more stringent over time. The regulations also establish, from 2012 on, separate limits for tailpipe emissions of nitrous oxide and methane.

The emissions trading system established in the regulations will provide credits to companies that outperform the annual fleet average standard for a given model year. The credits can be used to offset failure to meet the standard in a given model year. In general, the regulations give emissions credits a 5-year life span, permit them to be traded, and require that deficits be offset within three model years.

Special provisions recognize non-conventional technologies to reduce greenhouse gas emissions, such as reducing air conditioning refrigerant leakage or improving the efficiency of air conditioning systems. A special incentive is provided for advanced technology vehicles, including electric, plug-in hybrid, and fuel cell vehicles.

The regulations include early adopter credits for companies that exceed specified standards for the 2008–2010 model years, special rules for companies selling small volumes of vehicles, and treatment of dual-fuel vehicles that is consistent with U.S. regulatory approaches.

On October 16th, Environment Canada published formal notice of its intent to develop more stringent greenhouse gas emissions standards for new passenger automobiles and light trucks for the 2017 and later model years. Nine days later, on October 25th, the agency issued a consultation document outlining proposed elements of future regulations designed to reduce greenhouse gas emissions from on-road heavy-duty vehicles, starting with the 2014 model year. Environment Canada said the consultation is 'intended to seek early stakeholders' views on potential elements in advance of developing the proposed regulations,' which it hopes to publish in mid-2011. Final regulations for heavy-duty vehicles are expected to come out in December 2011.

The regulations for both light-duty and heavy-duty vehicles will be developed in collaboration with the United States. The standards for passenger cars and light trucks also will be harmonized with California's standards for model years 2017–2025, Environment Canada said in a notice published in the Canada Gazette, Part I.

The department said development of the new regulations would take into account technological, environmental, and economic factors, including projected composition of the future Canadian new vehicle fleet; cost, emissions reduction potential, and availability of conventional and emerging technologies; the need for flexibility to minimize the compliance burden on industry; and the need for regulatory mechanisms to continue encouraging development and deployment of technologies including electric vehicles, plug-in hybrid vehicles, and fuel cell vehicles.

The proposed regulatory framework for heavy-duty vehicles would apply to any entity that manufactures heavy-duty vehicles or engines in Canada or imports them into the country. The proposed rules would cover a range of on-road heavy-duty vehicles, including full-size pickup trucks; combination tractors used to haul commercial trailers; vocational vehicles such as freight, delivery, service, cement, garbage and dump trucks; and buses.

Environment Canada said it is working with the U.S. Environmental Protection Agency to develop a 'common approach' to regulate greenhouse gas emissions from heavy-duty vehicles, including an emissions testing protocol.

15.2.4 South Korea

South Korea established mandatory fuel economy standards in 2004 to replace a voluntary system. Starting in 2006 for domestic vehicles and 2009 for imports, standards are set at 34.4 CAFE-normalized mpg for vehicles with engine displacement under 1,500 cubic centimeters (cc) and 26.6 mpg for those over 1,500 cc. Credits can be earned to offset shortfalls.

On 6 July 2009, the Green Growth Council (GGC) set up under the President announced a new industry average target, target year (phase-in) and change in the fuel economy test cycle, while the details of the regulation remain to be decided in cooperation with Ministry of Environment and Ministry of Knowledge Economy.

The regulation will apply to all new passenger cars (with a maximum of 10 seats, including the driver). Manufacturers will have the option of complying with either a fuel economy target of 17 km/L or a CO_2 emissions standard of 140 g/km. These standards will be phased in during the years 2012 to 2015 at a rate of 30% the first year, 60% the second, 80% the third and finally 100% in 2015. The penalty for noncompliance has yet to be decided. The current fuel economy test cycle will be revised to adopt the US combined mode test cycle which incorporates both city and highway mode tests. The current Korean test cycle only includes a city mode test, leading the same vehicle model to exhibit 15–18% worse fuel economy than when it is tested in the US combined mode.

Small volume manufacturers will likely be given additional flexibility. Very efficient or clean vehicles emitting 50 g CO_2/km or less would be provided with incentives although the level of incentives still needs to be determined. Manufacturers exceeding their average fuel economy standard or CO_2 emissions target would earn credits which may be applied to any of the three consecutive years immediately before or after the year for which the credits are earned. Emission trading between manufacturers would also be allowed.

Demand side policies are also under consideration by the government including:

- Providing an incentive or disincentive to consumers based on the fuel economy of the cars purchased
- Adoption of a CO_2 based vehicle taxation scheme (currently the tax is based on engine displacement)
- Government support for R&D for green cars (hybrid cars, PHEV, fuel cells, clean diesel cars) and for development and commercialization

Other OECD countries such as Australia are considering voluntary standards and it is expected that Mexico will shortly propose requirements similar to those adopted by the US and Canada.

15.3 Comparing Vehicle Standards Around the World

To compare the fuel economy and greenhouse gas emissions standards adopted by key countries, updated through early 2010, the International Council on Clean Transportation (ICCT) developed a normalized comparison metric (ICCT 2010). As indicated in Figs. 15.4 and 15.5, the ICCT analysis shows further progress by the U.S. but it still indicates more stringent policies in effect in Europe and Japan – and even in China.

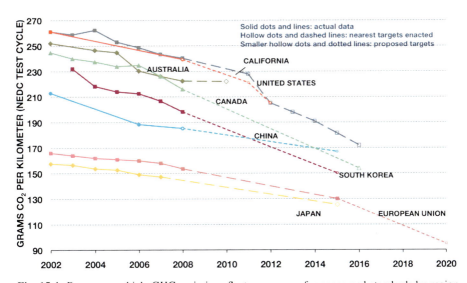

Fig. 15.4 Passenger vehicle GHG emissions fleet average performance and standards by region
Source: ICCT (2010)

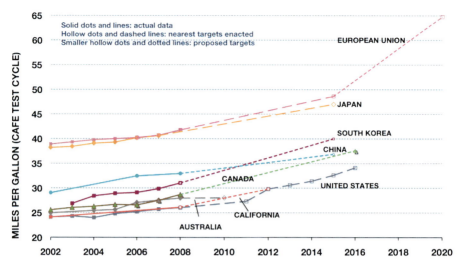

Fig. 15.5 Passenger vehicle fuel economy fleet average performance and standards by region
Source: ICCT (2010)

15.4 Conclusion

Over the course of the past decade, there has been a fundamental change in the approach to regulating fuel economy from road vehicles. While concerns over energy security and balance of payments still exist, a new dimension related to climate change has taken on much greater urgency and there is a clear shift toward regulation of at least CO_2 and in some cases other so called greenhouse gases (GHGs). The number of countries adopting at least some form of regulation has grown dramatically and is no longer limited to just the OECD countries. The country with the most rapidly growing vehicle population in the world, China, is in the late stages of developing its third generation of controls and India, another very fast growing country is expected to issue standards soon. Mexico regulations are expected soon as well. Furthermore, the form of the standard is starting to shift away from a mass based approach toward a footprint based approach, which will open up additional opportunities to take advantage of lightweighting as a key element of a control strategy. Finally, while regulation of light duty vehicles has been the primary focus to date, Japan has already adopted standards for heavy trucks and the US, Canada and the EU are expected to mandate similar controls in the next several years.

Finally, while command and control standards are expected to remain the backbone of control efforts, economic incentives or disincentives including fuel taxes are expected to play an even more important role in the future than they do today.

References

Agency for Natural Resources and Energy (ANRE) and the Ministry of Land, Infrastructure and Transport (MLIT) (2006) Concerning revisions of evaluation standards for automobile manufacturers with regard to energy efficiency: Joint Interim Report of the automobile evaluation standards subcommittee of the advisory committee for natural resources and energy and the automobile fuel efficiency standards subcommittee of the council for transport policy. December. (Japanese)

California Air Resources Board (CARB) (2004a) Staff proposal regarding the maximum feasible and cost-effective reduction of greenhouse gas emissions from motor vehicles. http://arb.ca.gov/cc/factsheets/cc_isor.pdf. Last accessed Mar 2011

California Air Resources Board (CARB) (2004b) Climate change emission control regulations. http://arb.ca.gov/cc/factsheets/cc_newfs.pdf. Last accessed Mar 2011

Canada Revenue Agency (2007) Web access: http://www.cra-arc.gc.ca/agency/budget/2007/excise-e.html. Last accessed Mar 2011

Code of Federal Register (CFR) (2006) Average fuel economy standards for light trucks, model years 2008–2011. National Highway Traffic Safety Administration, 49 CFR Parts 523, 533, 537, Docket No. 2006-24306, NIN 2127-AJ61

ICCT (2010) The international council on clean transportation, 'Passenger Vehicle Greenhouse Gas and Fuel Economy Standards: A Global Update', April 2010. http://www.theicct.org/passenger-vehicles/global-pv-standards-update/. Last accessed Mar 2011

Memorandum of Understanding between the Government of Canada and the Canadian Automobile Industry Respecting Automobile Greenhouse Gas Emissions (MOU) (2005) http://www.cvma.ca/supportfiles/mou-en-20050405.pdf. Last accessed Mar 2011

MLIT (2006) 2006 Survey of automobile fuel economy. http://www.mlit.go.jp/jidosha/nenpi/nenpilist/nenpilist0603.pdf (in Japanese). Last accessed Mar 2011

President Bush's Executive Order 'Twenty in Ten' (2007) http://www.whitehouse.gov/stateoftheunion/2007/initiatives/energy.html. Last accessed Mar 2011

The Japanese Automobile Manufacturers Association (JAMA) (2007) Taxes and automobiles (Japanese). http://www.jama.or.jp/tax/tax_system/tax_system_3t1.html (in Japanese). Last accessed Mar 2011

Transport Canada (2007) http://ecoaction.gc.ca/ecotransport/ecoauto-eng.cfm. Last accessed Mar 2011

U.S. Environmental Protection Agency (EPA) (2009) Light-duty automotive technology and fuel economy trends: 1975 through 2009

Walsh MP, MacKenzie J (1990) Driving forces: motor vehicle trends and their implications for global warming, energy strategies, and transportation planning. World Resources Institute Report, Washington, DC, December

Chapter 16
Transport and Climate Policy in the Developing World – The Region that Matters Most

Cornie Huizenga and James Leather

Abstract Sustained economic growth in emerging economies in Asia and other parts of the developing world is resulting in a rapid increase of the number of motorized vehicles, although overall motorization levels are still well below those of Japan, Europe and the United States. Motorized vehicles in developing countries, both for passenger and freight transport, will contribute the lion share of the projected global increase in greenhouse gases from transport in the years to come. Rapid motorization is also associated with worsening air pollution, congestion and increased road accidents. Transport policy in developing countries generally is still focused on 'predict and provide' which stimulates the expansion of transport infrastructure to cater for increased numbers of vehicles. This is increasingly resulting in an unsustainable growth trajectory for the transport sector. There are however countries and cities which are taking measures, sometimes aided by external assistance, which if replicated widely and scaled up to sector wide policies could have a meaningful impact on lowering future greenhouse gas emissions from transport. Future policies on transport and climate change in developing countries will have to be comprehensive, coordinated and integrated. Policies will need to combine restraining the growth in vehicles and the demand for transport by providing alternatives to individualized motorized transport through for example better public transport. The emphasis in transport policy needs to be on avoiding future emission through smart land use and transport planning, preventing a shift away from more sustainable modes of transport to private motorized vehicles and improvements in vehicle engine and fuel technologies.

16.1 Introduction

The image of transport in developing countries is becoming more and more diverse. The traditional pictures of chaotic and crowded roads with a mix of pedestrians, cyclists, animals, buses, trucks, cars and motor cycles competing for limited space

C. Huizenga (✉)
Partnership for Sustainable, Low Carbon Transport and Asian Development Bank, 200051 Shanghai, China
e-mail: cornie.huizenga@slocatpartnership.org

are starting to be replaced with pictures of endless gridlock of cars or in some cases gleaming state of the art subway systems. Both pictures will continue to reflect the transport reality on the ground in developing countries in the next decades.

China has been the largest producer of motor cycles for some years and in 2009 surpassed the United States of America as well as the largest market for cars. It is the only country, apart from Singapore, that has limited the number of new vehicles that can be sold in two of its main cities: Beijing and Shanghai. Over the last 14 years China has increased the production of electric bikes and scooters from 48,000 in 1996 to well over 20 million per year in 2009. China is also building the largest high-speed railway network in the world. Of course, China is the exception in the developing world and not (yet) the rule. It shows however the potential for motorization, both sustainable and less sustainable, that exists in the developing world.

To minimize the harmful impact of climate change more ambitious greenhouse gas (GHG) emissions reductions will be required between now and 2050. Emission reductions of over 50% will be required to avoid increases in the global temperature beyond 2°C. These can only be achieved if all sectors, including the transport sector contribute. With many OECD countries at, or close to, saturation levels for motorization, most of the global future motorization will take place in developing countries and a large part of transport related solutions to climate change will need to be in the developing world.

What are the patterns of motorization in the developing world, and what are the drivers? What are the solutions to climate change from transport in developing countries? What examples are out there that can serve as a basis for future policies on transport and climate change? How will these policies differ from those in the developed world? How can the rest of the world help the developing world to develop and implement effective policies on transport and climate change: what is the role for the United Nations Framework Convention on Climate Change (UNFCCC) and special climate instruments, what is the role of development assistance and what role is there for private sector in developed world? These are some of the questions this chapter tries to address.

16.2 Motorization in the Developing World

16.2.1 Drivers of Motorization

As personal wealth increases, people can afford private motorized vehicles. Economic growth improves the purchasing power of households and increasing numbers are in the position to buy a motor cycle or a car. History shows that countries have taken a different growth path when motorizing; Europe has maintained strong public transport while in the United States there is a strong dependence on private vehicles with a limited role for public transport. Policy decisions over the next 5–10 years taken by countries and cities in the developing world will determine which model they will follow and thereby the carbon footprint of the transport sector.

People exercise the possibility to obtain their own vehicle because of comfort or status but also because of an absence of viable mobility alternatives which are reliable, safe, fast and clean. In many parts of the developing world priority is given to the development of a road infrastructure before the development of high quality public transport. Also the upgrading of the non-motorized transport (NMT) infrastructure generally lags behind economic development and associated road development schemes. This support for private vehicle infrastructure reinforces the rapid growth in vehicles.

An important secondary driver of motorization in the developing world is urbanization. Statistics on urbanization in countries like China and India are mind-boggling. It is expected that just these two countries will add over half a billion persons to the urban population in less than 20 years (McKinsey Global Institute 2008, 2010). Other parts of Asia and Africa will also show an increase in the size of the urban population while some regions, e.g. Latin America, already have high levels of urbanization. In many cases the provision of public services including public transport can't keep up with the rapid pace of urbanization which encourages those who can afford it to resort to buying motor cycles or cars. Failing urban planning is also contributing to urban sprawl. Daily commutes become longer and average speeds decline. There are also examples, albeit on a much smaller scale, where urbanization has enabled the development of comprehensive public transport schemes and the creation of integrated logistics hubs. Unfortunately these are more limited in number.

Motorization in developing countries is also shaped by the prices of vehicles and the cost of driving. The vehicle industry is increasingly becoming a part of the economic growth strategy of developing countries. In some parts the initial emphasis was first on the manufacturing of motorcycles before moving to manufacturing cars, for example in Asia. In other parts, for example in Latin America, countries straight away took on producing cars. Generally, manufacturers of motor cycles, cars and trucks enjoy a status of protected industry which enables them to grow fast. The rapid increase in capacity in vehicle production in developing countries has the danger of a race to the bottom in terms of pricing. The trend toward cheaper cars is not only accelerating the rate of motorization in developing countries but is also having its impact felt in parts of the developed world, e.g. in Eastern Europe and parts of Western Europe. Lower prices generally will lead to more vehicles being sold. It also means that people will shift more quickly from a motorized two wheeler to a four wheeled vehicle.

The costs of owning and operating a motorized two or four wheeled vehicle in developing countries is generally lower than in developed countries and forms less of an impediment in purchasing a motor cycle or a car. Motorcycles in many developing countries only pay an one-time registration fee, parking for both two and four wheeled vehicles is generally cheap and many of the vehicles are either not or under-insured. Vehicle financing is increasingly available at favorable terms. Most important the wide spread of fuel subsidies encourages the purchase of motor vehicles and their wide-spread use. The IMF estimates that globally over $500 billion is spent on fuel subsidies, most of which is in developing countries (Coady et al. 2010).

16.2.2 Patterns of Motorization and Development of Transport Systems

The default option for many people in the developing world was (and in many cases continues to be) walking or cycling and public transport in the form of para-transit. Reliable numbers of how many people walk and cycle and for what distances are hard to come by in most of the developing countries and cities. However, it is widely believed that the share of walking and cycling in overall transport is declining. This is partly the result of alternatives in the form of public transport or cheap motor cycles and cars. For many the decision to walk or cycle less is not a voluntary decision but influenced by the growing distances to places of employment or education as well as the decreased attractiveness and safety for pedestrians and cyclists. Many people are however still too poor to be able to afford public transport, let alone being able to afford buying a motor cycle or a car (Roychowdhury 2010).

In a growing number of cities efforts are now being made to halt the decline in NMT, or to reverse it by improving the infrastructure for walking including the creation of pedestrian areas in downtown areas and by expanding cycling infrastructure and the development of public bicycle schemes modelled on the Paris 'Velib' public bike scheme. The impact of these efforts on NMT modal share has been limited so far.

For many people in developing countries their first and main contact with public transport is through para-transit which comes in many forms and shapes. Examples include the motorized three wheeled rickshaws in South Asia and tuk-tuks in Thailand; the bicycle rickshaws and pedi-cabs in Bangladesh and India; the matatus in East Africa, collective mini-vans in Latin America, jeepneys in the Philippines, and kijiangs in Indonesia. Para-transit, in common with NMT, is often not well documented in transport statistics. In many cases it operates in the informal sector, is fragmented in terms of ownership and under-capitalized, which makes it difficult to introduce more efficient operations and/or cleaner technologies. Countries and cities in the developing world are struggling to define the role para-transit can (or should) play in the modernization of transport systems. The large numbers of people employed in para-transit can make it an untouchable sector preventing rationalization and modernization. In many cases this leads to a situation that para-transit and formal public transport systems exist in parallel.

The trend for formal public transport systems has been that while the overall number of trips by bus and rail (light rail or metro) has increased the overall share of formal public transport is declining. Generally, the supply of formal public transport cannot keep up with demand. While the number of cities with a metro or light rail system is increasing, the high costs of well up to $100 million per kilometer have prevented many cities realizing their ambition to develop a metro as the backbone of an integrated public transport system. Overall the number of cities in developing countries with a subway or light rail network is still well below one hundred. For example, until recently less than 30 cities in India had a formal public transport system of which only 4 cities had a rail based public transport system.

Of much more importance is the passenger kilometers travelled by buses as this better represents its overall use. Bus systems come in many forms and shapes in developing countries and in some cases are receiving a new lease on life through Bus Rapid Transit (BRT). In recent years the number of cities planning or constructing a BRT, due to its generally lower cost per kilometer, has surpassed the number of cities planning or implementing rail based public transport systems.

The introduction of budget airlines as well as expansion by other carriers combined with larger purchasing power has resulted in large increases in the numbers of people travelling by air in many developing countries. This has often been at the expense of rail travel, which has received limited attention of policy makers, and underinvestment has eroded the ability of railways to compete with air travel or road transport for passengers. A clear exception is China which by 2012 will have a network of 12,000 km of high speed railway and on some parts of the network which are operational already, airlines have stopped competing for passengers.

Similar to passenger travel, freight transport in developing countries is often characterized by a wide range of modes. Freight is being carried by foot, on bicycles, by animals as well as by motor cycles, pick-ups, light and heavy duty trucks, boat, rail and air. The importance of transporting freight by road is increasing at the expense of other traditional means such as inland waterways and rail. At the same time rapid increases in air freight can also be observed. Compared to the modernization of passenger transport, the modernization of freight transport has received less attention and what efforts do exist are mostly geared toward enabling and facilitating road based freight transport.

With economic growth and trade, overall traffic activity (both in terms of passenger kilometers and ton kilometers) will increase greatly (see Fig. 16.1). Motor cycles will, especially in large parts from Asia, initially continue to be an important contributor to the passenger kilometers. At the same time the number of traditional motor cycles is increasing in Africa and Latin America, where they traditionally were very few in numbers.

Light duty vehicles will increase rapidly in numbers (see Fig. 16.2); 14 and 16 fold increases have been estimated up to 2035 for India and China respectively. Notwithstanding these rapid increases vehicle ownership ratios will be still well below that of the developed world by 2035.

16.3 Greenhouse Gas Emissions and Other Impacts from Transport in the Developing World

Information on GHG emissions from land transport in developing countries is patchy and very much out of date. Formal numbers are included in the National Communications, submitted by individual countries on a periodic basis as part

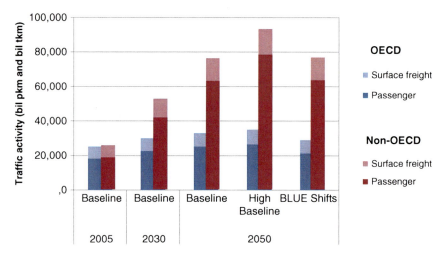

Fig. 16.1 Development of traffic activity (passenger and freight), 2005–2050 for OECD and Non-OECD countries
Source: IEA/OECD (2009)

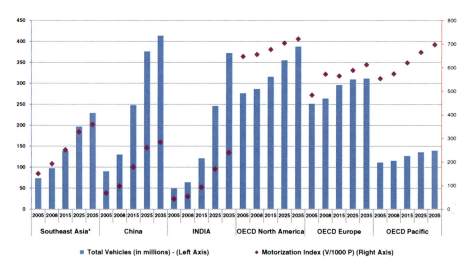

Fig. 16.2 Total vehicles and motorization index 2005–2035
Source: Schipper et al. (2009)

of their obligations under the 1992 Framework Convention on Climate Change UNFCCC. Developing countries, also known as non-Annex I countries, have no formal emission reduction obligations under the Kyoto protocol which was approved in 1997 and came into force in 2005. In their first National Communication non-Annex I countries were requested to report on their 1994 level GHG emissions; in the second National Communication, countries are generally reporting 2000 or

2002 GHG emission levels; and the most up to date emission data for transport in the fourth national communication of Mexico are for 2006. As of March 2011, of the 153 non-Annex I Countries 140 had submitted their initial national communications, 40 their second national communications, 2 their third national communication and only one party their fourth national communications.[1]

Transport is covered as a sub-sector of the energy sector and transport GHG emissions are generally reported based on aggregated fuel sales. Information is broken down by transport sub-sector with road transport being one of the sub-sectors. There is no forward projection required of GHG in the National Communications. Considering the rapid motorization in developing countries, it is clear that with the current guidelines for emission reporting National Communications are not a good source of information on transport GHG emissions and that because of the aggregated level of reporting without making use of activity data they are not a good basis for policy making.

In addition to the national communications GHG emissions are also reported in other country specific reports which are being drawn up by a range of stakeholders, sometimes from within the country, in other cases external stakeholders that have an interest in transport and climate change. Figure 16.3 summarizes ten studies on vehicle numbers and GHG emissions in India. It shows that because of methodological differences reported emissions for identical reporting years vary greatly (Gotha 2010).

The Fourth Assessment Report of the Intergovernmental Panel on Climate Change (IPCC) states that in 2004 the global transport sector accounted for 6 billion tons of CO$_2$ equivalent (GtCO$_{2eq}$) or 13% of total energy related GHG emissions (Kahn Ribeiro et al. 2007). In a 'business as usual' (BAU) scenario these are projected to increase by over 80% by 2050, with the bulk of the increase taking place in developing countries (IEA 2009). This means that at some point between 2020 and 2025 the developing world (non-OECD) countries will overtake the developed

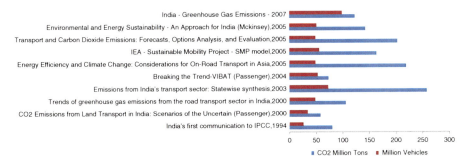

Fig. 16.3 Greenhouse gas emission estimates for India from different reports
Source: Gotha (2010)

---

[1] Submitted National Communications from Non-Annex 1 Countries: http://unfccc.int/national_reports/non-annex_i_natcom/submitted_natcom/items/653.php [last accessed March 2011].

world as the main source of transport GHG emissions. This must be a worrisome development considering the limited capacity for suitable policy development and implementation in the developing world.

In addition to GHG emissions, transport in developing countries has other negative impacts which affect social and economic development. As the number of motor vehicles continues to grow congestion is becoming more of a problem, especially in cities. Another problem is the increase in the number of road accidents which globally in 2008 were responsible for 1.3 million deaths per year, 90% of which were in low-income and middle-income countries, and which had only 48% of the world's vehicles (WHO 2009). Air pollution from transport is a problem in its own right and is resulting in thousands of premature deaths each year. The World Health Organization's global burden of disease study estimates that outdoor air pollution is responsible for about 0.8 million premature deaths. This burden occurs predominantly in developing countries; 65% in Asia alone (Cohen et al. 2005). The overlap between air pollution and climate change has become more pronounced following the growing body of evidence on the contribution of black carbon (a short-lived pollutant that is part of particulate emissions of amongst others diesel vehicles) to climate change (UNEP 2009).

In many cases it is the other impacts like congestion, road safety and air pollution which are the local drivers on transport policy and investments in developing countries, rather than concerns about the contribution of transport to climate change. So far, there are no integrated assessment frameworks which integrate traditional transport evaluation on mobility with local and global environmental impacts, or with the broader social impacts. This is hampering effective action to address local and global sustainability, either on a separate or integrated basis.

16.4 Policies for Sustainable, Low Carbon Transport in the Developing World

16.4.1 Current Transport Policy in the Developing World

Transport policy in developing countries can best be characterized by the 'Predict and Provide' approach to policy development and implementation. Forecasting studies indicate a rapid increase in motorized vehicles, resulting in policy makers tending to respond by creating additional road infrastructure. The underlying rationale has usually been that an increase in transport activity is required to stimulate economic and social development. When the newly constructed roads soon started to fill up this was seen as a justification of the policy choices made.

The 'Predict and Provide' approach was often financed by other sectors of the economy based on the assumption that the economy as a whole would benefit from development of road infrastructure. As part of this emphasis on infrastructure provision, the vehicle industry is often designated as a strategic industry which – in countries with an emerging auto manufacturing sector – is often the recipient of special incentives. The creation of a domestic market is then seen as important to protect investments in the local vehicle industry, and the sale and usage of vehicles is incentivized through fuel subsidies, low taxes on vehicle registration and low parking fees.

An exception to this model has been the motor cycle industry in China which is the largest in the world in terms of output.[2] Interestingly, motor cycles have been banned in the central parts of 170 of Chinese cities.[3] Similarly, the city of Shanghai has already for several years a vehicle quota system, modelled on the Singapore example, in which a set number of vehicle licenses are auctioned off. Currently the quota of new licenses is about 7000–8000 per month. This, notwithstanding that Shanghai is one of the centers of automobile manufacturing in China. More recently the Beijing city government has also announced an annual quota of 240,000 new vehicles. Unlike the case of Shanghai, in Beijing the licenses will allocated through a lottery.

Another departure from the 'Predict and Provide' approach are the efforts to limit the use of vehicles through the introduction of congestion charging as in the case of Singapore. Recently Beijing and Chongqing, two major cities in China, have indicated that they will also introduce congestion charging-like approaches.

The almost exclusive focus on provisions for private motor vehicles, especially cars, has resulted in a situation where large parts of the transport sector including NMT, para-transit and motor cycles have largely fallen outside the scope of transport policy makers and transport policies. Yet, these parts of the transport sector are important when it comes to improving sustainability of transport and providing access and mobility for all, especially the poorest in society.

Neither climate change nor local sustainability has been a strong driver of development policy in most of developing countries and for transport this is even more the case. Transport has had difficulties to integrate environmental considerations in policy making. Traditionally efforts on improving sustainability have been aimed at improving fuel quality and vehicle emission standards and, although progress has been made in some countries, overall emission and fuel quality standards in developing countries are still well behind most of the developing world. Especially the enforcement of in-use vehicle standards leaves much to be desired.

There are several reasons for the limited focus on sustainability of the transport sector. There is no tradition to incorporate negative externalities in the appraisal of transport policies, programmes and projects. Because of this, transport decision makers in developing countries are not confronted with the implications of their decisions in terms of impacts on environment, congestion and social factors. The transport sector in developing countries is characterized by a plethora of institutional arrangements and responsibilities whereby local considerations often win out

---

[2] China produced 27.5 million bikes in 2008, nearly half the world's production – see http://www.motorcycle.com/events/2009-chinese-motorcycle-show-part-1-88775.html [last accessed March 2011].

[3] http://factsanddetails.com/china.php?itemid=316&catid=13&subcatid=86 [last accessed March 2011].

over national policies. This while the experience of the developed countries shows that the mainstreaming of sustainability criteria often requires national leadership combined with local capacity. Environmental departments in developing countries are weak at the national level and especially at the local level; the structures which regulate their involvement in policy making and enforcement are frail, and the data based on which they could make their case for action on sustainable, low carbon transport are mostly absent.

Transport policy in developed countries has benefitted from well developed capacity among local stakeholders including private sector, academia and civil society. In addition an important driver in the case of Europe has been the European Union and to a lesser extent the OECD for its member countries. Such integrated policy making bodies with attached knowledge management functions are mostly absent so far in the developing countries. Over the last years we have seen an increase in activities to promote knowledge management, capacity development and policy advocacy through for example the United Nations Centre for Regional Development – Environmentally Sustainable Transport (EST) Forum, the Partnership for Sustainable, Low Carbon Transport (SLoCaT), the Better Air Quality conferences of the Clean Air Initiative for Asian Cities, and the Bridging the Gap initiative.[4] It is clear however that these efforts have had so far a limited impact on policies for transport and climate change in developing countries.

16.4.2 Sustainable, Low Carbon Transport Policies for the Developing World

The emphasis of transport policy in the developing world needs to be on strengthening the overall sustainability of transport systems. 'Low-carbon, sustainable transport reduces short and long term negative impacts on the local and global environments, has economically viable infrastructure and operation, and provides safe and secure access for both persons and goods.' (Dalkmann and Huizenga 2010).

To implement this definition a paradigm shift is required away from the 'Predict and Provide' approach toward a policy which actively seeks to control the growth in the number of private motor vehicles and their use and match it by providing an alternative growth trajectory for the transport sector in developing countries. The Avoid-Shift-Improve (A-S-I) approach embodies these two policy strands of controlled motorization and providing alternatives to meet the demand for access and mobility in support of still much needed economic and social development.

The A-S-I approach originates from the State of North Rhine-Westphalia in Germany but its applicability to the developing world is now being acknowledged by a growing number of international organizations. There is increasing agreement

---

[4] See www.uncrd.or.jp/env/est; www.slocat.net; www.cleanairinitiative.org; and www.transport2012.org [last accessed March 2011].

that GHG emissions from the transport sector can be reduced with no impact on economic progress through an integrated and comprehensive approach which *avoids* the need for travel through sound land-use and telecommuting, which *shifts* travel to the most efficient modes, e.g. public and non-motorized transport for passenger transport and rail and in-land waterway for freight transport; and which *improves* vehicle and fuel technologies as well as transport facility management and operations to reduce emissions from individual vehicles.

In applying the A-S-I approach to transport in developing countries notice should however be taken of differences between transport in the developed and developing world. For example, compared to the developed world the impact of fuel economy standards, part 'Improve', will be relatively more limited in the developing countries because of the low baseline in the number of vehicles and the rapid growth. The large increase in the size of urban population in several of the developing countries, on the other hand, increases the potential impact of measures under the 'Avoid' component because of the possibilities in designing new cities or city expansion in a transit-oriented development which would reduce the need for travel. Lastly, while in the developed world the emphasis under the 'Shift' component will be on shifting car users to NMT or public transport, in developing countries the emphasis will need to be more on encouraging NMT and public transport users to remain with their mode and not shift to individual, motorized transport.

An important characteristic of sustainable, low carbon transport is the integration of environmental, social and economic dimensions. Transport related policies and investment programmes in developing countries are taken for different motives and require the support of a wide range of stakeholders. The effectiveness of transport policies in the developing world in altering the GHG trajectory from transport will depend on the extent to which local and global stakeholders will be able to agree on making co-benefits a central policy principle. For this to work it is important that all stakeholders feel that they will benefit from co-benefits oriented policy making. This can best be accomplished by applying a comprehensive cost-benefit analysis of policies and programmes, which incorporates both local and global environmental benefits (e.g. air pollution and GHG emission reduction) as well as social (e.g. road safety and accessibility) and economic (e.g. congestion and cost of travel) benefits. The co-benefits dimension of sustainable, low carbon transport policies and projects can be safeguarded by avoiding incentive schemes which put a disproportionate emphasis on one dimension of the overall sustainability of such policies and projects.

A growing number of countries and cities in the developing world are already implementing policies, programmes and projects in line with the A-S-I approach. See Table 16.1 for a selective overview.

The challenge facing the developing world is to replicate and scale up these promising examples in a manner and a speed which will prevent countries and cities from locking-in on a high carbon development model for their transport sectors.

To meet this challenge four key elements need to be in place, and these need to be aligned with each other to be assured of genuine success.

Table 16.1 Examples of application avoid-shift-improve approach

Example	Location
Avoid	
Vehicle quota and license auctioning	Singapore, Shanghai, and Beijing
Eco cities, urban redevelopment with mixed land-use and transit oriented development	Several cities in China
Shift	
Bus Rapid Transit Systems	Bogota, Mexico City, Ahmedabad, Cape Town, Guangzhou and over 70 other cities where BRT is in operation, being planned or under construction
Metro	Delhi, Shanghai, and a range of other cities
Public bike schemes and cycling infrastructure	Bogota, Hanghzou and over 50 other cities
Development of water based freight transport and other green freight transport concepts	
Improve	
Biofuels	Brazil and several other countries
Electric vehicles	20 million e-bikes per year in China and pilots for 4 wheeled electric vehicles and charging infrastructure under development in several countries.
Fuel economy standards	China and other countries standards under development

1. Knowledge, including data on the composition of the transport sector, activity data, fuel use and emissions but also awareness on sustainable transport concepts and best practices;
2. Capacity for the formulation, implementation and monitoring of sustainable, low carbon transport. Effective institutions are required at the national and local level within government, civil society, academia and private sector which work together where and when required;
3. Policies that create an enabling legal and regulatory environment for sustainable, low carbon transport and which are accompanied by policy instruments which can catalyze the implementation of the A-S-I approach;
4. Financing for the implementation of A-S-I approach as well as pricing mechanisms based on the user pays principle and which rewards sustainable, low carbon transport and punishes unsustainable behavior.

The Jawaharlal Nehru National Urban Renewal Mission (JNNURM) in India is an example of the kind of integrated approach that is required to accomplish broad-based change. The JNNURM makes financing available to selected large cities to develop amongst others their transport services. An important pre-condition is that the investment plans are in line with the National Urban Transport Policy which is strongly oriented toward sustainable transport. Financing is also conditional upon the implementation of certain institutional reforms at the city level aimed at creating institutional and financial sustainability for the activities funder through the JNNURM. The Ministry of Urban Development has now started a benchmarking of service delivery in urban transport by cities under JNNURM. This will help to guide future prioritization of JNNURM funding to the transport sector.

16.4.3 GHG Scenarios for the Developing World

According to the Fourth Assessment Report of the IPCC global GHG emissions, in order to avoid dangerous levels of climate change, will have to peak within the next decade and be reduced by more than 50% in 2050 compared to 1990 levels. In terms of emission reduction objectives this translates into 25–40% compared to 1990 levels for developed countries by 2020, while the contribution by developing countries needs to be 15–30% below business as usual by 2020 (den Elzen and Höhne 2008). Given a baseline projection of 4.3 GtCO$_{2eq}$ for the transport sector, this would translate into 0.6–1.3 GtCO$_{2eq}$ per year reduction in 2020. For comparison: the European transport emissions in 2006 were approximately 1 GtCO$_{2eq}$ (Huizenga and Bakker 2010).

The reduction of 15–30% below BAU by 2020 for the developing world is not a politically agreed upon target and neither has it been broken down in objectives for individual sectors. It is too early therefore to arrive at firm conclusions on whether a reduction of 15–30% below BAU by 2020 is feasible for the developing world and whether countries are on track to achieve emission reductions in this magnitude.

It is possible, however, to arrive at an initial view on likely scenarios for GHG emissions from transport in the developing world based on a combination of policy statements by developing countries and analytical scenario studies conducted over the last years:

- In its overall potential assessment the IPCC in the Fourth Assessment Report concludes '(t)he mitigation potential by 2030 for the transport sector at the global level is estimated to be about 1600–2550 Mt CO$_2$ for abatement costs up to 100 US$/tCO$_2$. This is only a partial assessment, based on biofuel use throughout the transport sector and efficiency improvements in light-duty vehicles and aircraft and does not cover the potential for heavy-duty vehicles, rail transport, shipping, and modal split change and public transport' (Kahn Ribeiro et al. 2007).
- The IEA/OECD (2009) in a forecasting study concludes that: 'overall, with the efficiency, low-GHG fuels and advanced vehicles, and modal shift taken together, CO$_2$ emissions in transport can be cut globally by 40% in 2050 compared to 2005, and by 70% compared to the baseline in 2050. This represents a 10 Gt reduction from the 14 Gt that would otherwise be emitted by the transport system in 2050 in the Baseline and a 14 Gt reduction compared to the 18 Gt in the High Baseline'.
- South Korea is committed to a reduction of 33–37% below BAU by 2020 for transport, which is equivalent to a 20–24% reduction by 2020 compared to 2005 GHG emissions (Park 2010).

- China has announced a 40–45% reduction of CO_2 emissions per unit of GDP below 2005 levels by 2020 (Duscha et al. 2010) and a reduction in energy consumption by 2020 of respectively 16%, 20% and 5% for commercial trucks, commercial ships and commercial buses on per units basis compared to 2005 (Dai 2010).
- In the MEDEC study (Johnson et al. 2009), overall baseline emissions in Mexico by 2030 are estimated to be 1137 Mt CO_{2eq}. A package of nine transport interventions could reduce emissions in the Mexican transport sector by 131 Mt CO_{2eq} by 2030 (around 11.5% of overall emissions in 2030).
- In a study for East Asian countries, the World Bank (2010) estimates a potential of 35% reduction compared to the baseline for urban transport. This can be achieved by a combination of urban planning (7%), improved public transport (8%), Transport Demand Management (TDM) (7%) and fuel standards in line with the EU targets (15%).
- According to another World Bank study, road transport GHG emissions in India can be reduced by 19% against the dynamic BAU baseline by 2032 by improving public transport and light-duty vehicle technology (World Bank 2009).

It is clear that there is not yet an approach for assessing the GHG emission reduction potential in the transport sector which uses similar benchmark years and which can be used to assess land transport sector in its entirety or specific parts such as urban transport. It is unlikely that such a common approach will come about for the transport sector, or any sector for that matter, until there is a political agreement on a post 2012 climate agreement including the role of the developing countries.

Looking at the diverse assessments conducted so far it appears that a 15% reduction below BAU in 2020 (representing the less ambitious objective of the 15–30% range) is more likely to be achieved in land transport in developing countries than the 30%.

There are too many uncertainties to make a meaningful statement on 2050 emissions. It is clear however that incremental change will not result in the kind of large scale emission reductions in absolute terms required by 2050; countries and cities in the developing world would have to adopt a leapfrog approach to make this happen. Examples of such leapfrogging could include: generations of would-be car users switching to vehicles powered by electricity from clean renewable sources; new eco-cities built around clean public transport; and large scale adoption of e-commerce and telecommuting.

16.5 External Assistance Policies for Sustainable, Low Carbon Transport in Developing World

The development of sustainable, low carbon transport in the developing world is primarily the responsibility of national and local stakeholders with the government playing a lead role. What makes the situation in developing world special is the role of international assistance. This includes private investments, development assistance and special climate instruments.

16.5.1 Private Investments

International private investments in transport infrastructure and transport services in developing countries are not well documented and no authoritative data is available. Public private partnerships have focused more on ports and air ports; international private investments in land transport have been modest and were mostly directed at development of toll-roads, logistics hubs, and in incidental cases urban transport systems. So far, private investments are not leveraging policy changes in favor of sustainable, low carbon transport nor are they being used to fund specific sustainable, low carbon projects. For this to change international private sector investors would have to be assured of adequate returns on their investment. For this to happen policy changes would be required that internalize the full costs and benefits of transport.

16.5.2 Development Assistance

The impact of development assistance can be assessed in terms of financial support provided or in terms of bringing in new ideas. Both are important in the case of sustainable, low carbon transport in the developing world. Broadly speaking, financial assistance is being provided by the multilateral development banks e.g. the African Development Bank (AfDB), the Asian Development Bank (ADB), the Inter-American Development Bank (IDB) and the World Bank. In addition, examples of important bilateral providers of financial assistance include Agence Française de Développement, Japan International Cooperation Agency, and the KfW Entwicklungsbank from Germany. It is estimated that the total financial assistance for transport in developing countries amounts to US$ 20–25 billion per year.[5]

Utilization of external financial assistance has been so far predominantly for the construction and rehabilitation of roads. Recently some of the development banks such as ADB and IDB have indicated that the transport sector is an important sector when it comes to mitigation of climate change in developing countries. Internationally efforts are now well underway to conduct carbon foot-printing of external assistance to the transport sector (CTF 2009, ITDP 2010). Some of the development banks, e.g. ADB, have indicated their intention to broaden assistance into more transport sub-sectors for the provision of transport infrastructure and strengthening of transport services (ADB 2010). This is expected to result in more funding for public transport and NMT in cities as well as for freight and logistics.

---

[5] Authors' estimate based on published statistics of multi- and bilateral development banks.

The shift in approach of the financial institutions has been influenced by the efforts of non-governmental organizations like CAI-Asia; EMBARQ: the World Resources Institute for Sustainable Transport; and the Institute for Transportation and Development Policy (ITDP). Organizations like these have been successful in raising the awareness of both stakeholders in developing countries and in the development community on the need and possibility to strengthen the sustainability of transport systems in developing countries. Their direct involvement in flagship pilot BRT projects in cities like Ahmedabad, Guangzhou and Mexico City has helped to create models which are now being replicated and considered for scaling up across the developing world.

The efforts of the development community on sustainable, low carbon transport are facilitated as well by the growing interest shown by developing countries in sustainable transport. An important role in the development of a regional policy consensus on transport in Asia is being played by the EST Forum which was convened for the first time in 2005 by the United Nations Centre for Regional Development (UNCRD). Since then the EST Forum has met five times and adopted a series of policy statements and declarations, the most recent of which is the Bangkok Declaration on Environmentally Sustainable Transport 2010–2020, which contains 20 common goals related to EST. Several of these goals are directly related to the reduction of GHG from transport. A similar intergovernmental forum is now also being established in Latin America and a separate forum for Africa is also being considered.

16.5.3 Climate Instruments

The transport sector in the developing world has been able to access a range of special climate instruments which were set up following the entering into force of the United Framework Convention on Climate Change in 1994. These include the Clean Development Mechanism (CDM) and the Global Environment Facility (GEF) which are instruments of the UNFCCC, and the Clean Technology Fund (CTF) which was set up as an interim financing arrangement in 2009 to fill an immediate financing gap pending an agreement on the future (post 2012) climate regime. A key difference between these three instruments is that CDM supports projects in developing countries to offset emissions in the developed countries, while GEF and CTF fund GHG emission reduction projects and programmes in developing countries without such an off-setting objective. These instruments have not been able to generate substantial reductions in GHG emissions, or provide the financial contribution required to support transport activities in developing countries (see Table 16.2).

Why were these instruments used so seldom to develop projects in the transport sector? Unlike the power or industrial sector, the transport sector consists of a large amount of emission sources which as individual entities emit relatively small amounts of GHG emissions, although collectively the total emissions are large. Because of the large number of individual sources in the transport sector it has

Table 16.2 Overview of transport projects in existing climate instruments

	Year of 1st project	No. of projects	Funding [$ million]	Reported/expected emission reductions [MtCO$_{2eq}$/year]
CDM	2006	30 (3)[a]	672 (CERs) (63)[b]	3.1 (0.3)
GEF1-4	2006	37	201 (grants)	3.2[c]
CTF	2009	7	600 (loans)	10[d]

[a] In pipeline: registered, requesting registration and at validation, total CERs realized will most likely be lower than the number indicated, brackets values for registered projects
[b] CERs: Certified Emission Reductions; expected total undiscounted revenues at 10 $/CER, 3×7 years crediting, excluding transaction cost; brackets values for registered projects
[c] Direct impact, annual emission reductions calculated based on assumed 10 years lifetime
[d] Annual emission reductions calculated based on assumed life time of 10–20 years depending on type of investment
Source: Bakker and Huizenga (2010)

been difficult to come up with reliable methodologies to forecast and validate GHG emission reductions linked to the implementation of specific projects. The transport sector has also faced problems in proving the additionality of activities supported by climate financing. Since climate financing is in most cases only a very small part of the overall funding for transport projects it is difficult to argue that projects would not happen if not for the additional climate financing. These methodological challenges have been a stumbling block especially for CDM, which of the three climate instruments has the most stringent methodological requirements.

An important new climate instrument are Nationally Appropriate Mitigation Actions (NAMAs) which – following agreement at the 16th Conference of Parties to the UNFCCC (COP 16) in Cancun, Mexico – will come into force as part of a new post 2012 global climate agreement. NAMAs will be a turn-around in the sense that while under the Kyoto Protocol developing countries had no formal obligation apart from acting as host country for CDM projects, in the future developing countries will take on a new form of participation in global climate governance. Developing countries will implement NAMAs in the context of sustainable development, supported and enabled by technology, financing and capacity building, in a measurable, reportable and verifiable manner (UNFCCC 2010). Funding for NAMAs will come from a new climate fund which was first proposed at the climate talks in the 15th Conference of Parties to the UNFCCC (COP 15) in Copenhagen in December 2009 and was confirmed in COP 16 in Cancun in December 2010. The Copenhagen Accord and the Cancun Agreements state that such a climate fund by 2020 would have to amount to US$ 100 billion per year. As of September 2010, 46 countries had proposed NAMAs to the UNFCCC, 28 of which had a transport component (Binsted et al. 2010). Provided that it is possible to overcome the barriers which have hampered the participation of the transport sector in existing climate instruments (CDM, GEF and CTF), NAMAs could become an important source of funding for sustainable, low carbon transport in developing countries.

16.5.4 External Assistance Challenges

The future effectiveness of external assistance will be influenced by the manner in which the following challenges are addressed:

1. Developing countries and external organizations both are changing their approach to transport. At present each side has its own learning curve and while the two sides interact the learning curves are not really synchronized.
2. External agencies have contributed to the development of a range of examples on sustainable, low carbon transport. The effectiveness of their further contribution to a wide scale implementation of sustainable, low carbon transport would be guided by policy changes in developing countries including changes in the pricing of transport. External agencies so far have limited interaction with governments on policies, including on pricing, which could restrain the growth in private motorization and support the implementation of sustainable, low carbon transport.
3. The financing needs for sustainable, low carbon transport are significant and without a substantial private sector involvement the influence of external agencies will be limited. There is a need for a dialogue between private sector and development community on how to mutually support each other in the promotion and financing of sustainable, low carbon transport in developing countries.
4. The contribution of development assistance to transport in developing countries has been far larger than assistance provided by climate financing. The realization of a US$ 100 billion climate fund has the potential to increase funding for sustainable transport although this fund will be used for a range of sectors. How can development assistance and climate financing complement each other? This is a question to which there is currently no answer. Both types of assistance by themselves will not be able to transform transport systems or the growth trajectory of GHG in developing countries. If used in tandem the combined leverage of both types of assistance can be greater.

16.6 Conclusions

The developing world will be the most important source of GHG emissions from transport in the very near future, and substantial changes will be required in its growth path in order for the transport sector in the developing world to be part of the solution to climate change rather than to be part of the problem. Early action will be required to avoid a lock-in effect to a high carbon growth path resulting from continued rapid growth of individual motorized vehicles enabled by an aggressive expansion of road infrastructure and aided by a pricing system of transport which continues to promote rather than constrain further growth of private motorization.

The likelihood of effective early action is threatened by weak current policies and limited capacity in many of the countries to formulate and implement comprehensive transport policies. Many countries and cities are struggling with the challenge to expand and modernize their transport sector in support of economic and social development. So far the 'Predict and Provide' approach that produced an in-tandem growth of vehicle numbers and infrastructure has prevailed in transport policies of developing countries and also in development assistance. This has generally been at the expense of those people who can't afford a motor cycle or a car. At the same time current policies are resulting in rapidly growing GHG emissions and are responsible for growing congestion and air pollution as well as an increase in road accidents.

To make transport in developing countries more sustainable, a paradigm shift is required away from the 'Predict and Provide' approach toward a policy which actively seeks to control the growth in the number of private motor vehicles and their use and match it by providing an alternative growth trajectory for the transport sector in developing countries. The A-S-I approach has the potential to help control motorization and provide alternatives to meet the demand for mobility in support of the still much needed economic and social development.

For sustainable, low carbon transport in developing countries to become a reality, priority needs to be given to the replication and scaling up of examples reflecting the A-S-I approach. Policies on sustainable, low carbon transport in developing countries need to acknowledge that the growth trajectory of the transport sector in developing countries is different from developed countries and that this will result in different application of policy instruments. External assistance – whether it is from development assistance or through special climate instruments – should be aimed at facilitating and accelerating domestic action rather than to be project based.

Climate change is a powerful driver for more sustainable transport in developing countries. If it is used wisely it can facilitate and accelerate positive change, if used in isolation it carries the danger of stunting the development of comprehensive transport policies and thereby allowing the development of a transport sector which will ultimately have a larger carbon footprint.

References

ADB (2010) Sustainable transport initiative – operational plan. http://www.adb.org/documents/policies/sustainable-transport-initiative/Sustainable-Transport-Initiative.pdf. Accessed 9 Jan 2011

Bakker S, Huizenga C (2010) Making climate instruments work for sustainable transport in developing countries. Nat Res Forum 34:314–326

Binsted A, Davies A, Dalkmann H (2010) Copenhagen Accord NAMA submissions implications for the transport sector. Bridging the Gap Initiative. http://www.transport2012.org/bridging/ressources/files/1/913,828,NAMA_submissions_Summary_030810.pdf. Accessed 9 Jan 2011

Coady D, Gillingham R, Ossowski R, Piotrowski J, Tareq S, Tyson J (2010) Petroleum product subsidies: costly, inequitable, and rising. IMF Staff Position Note, SPN 10/05

Cohen AJ, Ross Anderson H, Ostro B, Pandey KD, Krzyzanowski M, Künzli N, Gutschmidt K, Pope A, Romieu I, Samet JM, Smith K (2005) The global burden of disease due to outdoor air pollution. J Toxicol Environ Health Part A 2005 Jul 9–23;68(13–14):1301–1307

CTF (2009) Clean technology fund result measurement system. CTF/TFC.3/8. http://www.climateinvestmentfunds.org/cif/sites/climateinvestmentfunds.org/files/CTFresultsmeasurement.pdf. Accessed 8 Jan 2011

Dai D (2010) Moving towards sustainable transport development in China. Presented at ADB transport forum 2010 http://www.cleanairinitiative.org/portal/ADBTransportForum2010?page=2. Accessed 9 Jan 2011

Dalkmann H, Huizenga C (2010) Advancing sustainable low-carbon transport through the GEF. Prepared on behalf of the Scientific and Technical Advisory Panel of the Global Environment Facility. http://www.thegef.org/gef/sites/thegef.org/files/documents/C.39.Inf_.17 STAP-Advancing Sus. Low-carbon Transport.pdf. Accessed 9 Jan 2011

den Elzen M, Höhne N (2008) Reductions of greenhouse gas emissions in Annex I and non-Annex I countries for meeting concentration stabilisation targets. Climatic Change 91(3–4):249–274

Duscha V, Graichen J, Healy S, Schleich J, Schumacher K (2010) Post-2012 climate regime. How industrial and developing nations can help to reduce emissions – assessing emission trends, reduction potentials, incentive systems and negotiation options. On behalf of the German Federal Environment Agency. http://www.umweltdaten.de/publikationen/fpdf-l/3954.pdf. Accessed 9 Jan 2011

Gotha S (2010) My critique on India's GHG emissions 2007 – transport estimates. http://cai-asia.blogspot.com/2010/05/india-transport-emissions-2007.html. Accessed 8 Jan 2010

Huizenga C, Bakker S (2010) Climate instruments for the transport sector. Consultants report prepared for Asian Development Bank and Inter-American Development Bank

IEA (2009) World energy outlook 2009. ISBN: 978 92 64 06130 9, Paris

IEA/OECD (2009) Transport, energy and CO_2. Moving toward sustainability. ISBN 978-92-64-07316-6, Paris

ITDP (2010) Manual for calculating greenhouse gas benefits of global environmental facility transportation projects. Global Environment Facility-Scientific and Technical Advisory Panel. GEF/C.39/Inf.16

Johnson T, Alatorre C, Romo Z, Liu F (2009) Low-carbon development for Mexico. The World Bank, Washington, DC

Kahn Ribeiro S, Kobayashi S, Beuthe M, Gasca J, Greene D, Lee DS, Muromachi Y, Newton PJ, Plotkin S, Sperling D, Wit R, Zhou PJ (2007) Transport and its infrastructure. In: Metz B, Davidson OR, Bosch PR, Dave R, Meyer LA (eds) Climate change 2007: mitigation. contribution of working group III to the Fourth assessment report of the intergovernmental panel on climate change. Cambridge University Press, Cambridge, UK and New York, NY, USA

McKinsey Global Institute (2008) Preparing for China's urban billion. http://www.mckinsey.com/mgi/reports/pdfs/China_Urban_Billion/China_urban_billion_full_report.pdf. Accessed 9 Jan 2011

McKinsey Global Institute (2010) India's urban awakening: building inclusive cities, sustaining economic growth. http://www.mckinsey.com/mgi/reports/freepass_pdfs/india_urbanization/MGI_india_urbanization_fullreport.pdf. Accessed 9 Jan 2011

Park JY (2010) Low carbon growth path for the transport sector in Korea. Presented at ADB transport forum 2010 http://www.cleanairinitiative.org/portal/ADBTransportForum2010?page=2. Accessed 9 Jan 2011

Roychowdhury A (2010) Slow murder: the insidious link between vehicular pollution – public health – climate- and urban poor in India. http://www.scribd.com/doc/43063065/Slow-murder-%E2%80%93-How-can-fuel-and-emission-systems-be-made-more-pro-poor. Accessed 7 Jan 2010

Schipper L, Fabian H, Leather J (2009) Transport and carbon dioxide emissions: forecasts, options analysis, and evaluation. ADB Sustainable Development Working Paper Series, No. 9. http://www.adb.org/documents/papers/adb-working-paper-series/ADB-WP09-Transport-CO2-Emissions.pdf. Accessed at 28 Mar 2011

UNEP (2009) Black carbon e-bulletin 1(2), 2009. http://www.unep.org/climateneutral/Portals/0/Image/BC%20e-Bulletin%20Sep%2009.pdf. Accessed 8 Jan 2011

UNFCCC (2010) Outcome of the work of the ad hoc working group on long-term cooperative action under the convention. Decision -/CP.15. Advance unedited version. http://unfccc.int/files/meetings/cop_16/application/pdf/cop16_lca.pdf. Accessed 9 Jan 2011

WHO (2009) Global status report on road safety. http://www.who.int/violence_injury_prevention/road_safety_status/2009/en/. Accessed 8 Jan 2011

World Bank (2009) India: options for low-carbon development. http://siteresources.worldbank.org/INTINDIA/Resources/LCD-Synopsis-Dec2009.pdf. Accessed 9 Jan 2011

World Bank (2010) Winds of change. East Asia's sustainable energy future. World Bank, Washington, DC, April 2010

Chapter 17
Epilogue – The Future of the Automobile: CO_2 May Not Be the Great Decider

Lee Schipper[†]

Abstract This volume has illustrated the strong link between automobiles and CO_2 emissions associated with climate change. In thinking about the future of the automobile it is tempting to blame the car for its contribution to climate change. Yet it is us the drivers who have chosen to create a world of large cars and in most Western nations established a very automobile-dependent lifestyle. The automobile has given many of its owners and users greater choices on where and how to live. But it is clear that those choices increasingly impinge on all drivers, and, more important, on all others trying to move in increasingly crowded cities or between urban areas on crowded motorways. The situation in developing countries is dire at a tenth or less of the motorization rate industrialized countries. People are frozen in most large cities. It is thus hard to foresee expansion in car ownership to high levels forecast by some international organizations and analysts. Does this mean the future of the automobile is grim? Yes, if individuals, their elected officials and stakeholders in fuel and vehicle companies continue as if there are not profound problems confronting the choices automobiles give their users. In any case, CO_2 is not the deciding factor over the future of the automobile, rather more fundamental issues such as the difficulty of fitting in so many individual vehicles to so little space. Technology can help somewhat, but the larger issues are what people decide to do with technology.

17.1 Introduction

This volume has presented an exhaustive survey of how carbon emissions from travel, principally in light duty vehicles (LDVs), can be reduced in Europe. For some of us the discussion is a necessary 'deja-vu all over again'. Lew Fulton, Celine Marie and I wrote a book for the International Energy Agency in 2000, 'The Road From Kyoto' (IEA 2000) that with less sophistication outlined much of the same strategy as IEA member countries (five of which were in Europe) saw

---

L. Schipper (Deceased)
Precourt Energy Efficiency Centre, Stanford University, Stanford, CA, USA

Global Metropolitan Studies, University of California, Berkeley, CA, USA

the problem a little over 10 years ago. Since then policy instruments have gotten tighter, but we're still not there.

Among the points we made in that volume these stand out as more important than ever:

- Timing and stock turnover. It takes almost 20 years to replace all the vehicles on the road. Without a continual tightening of emissions limits, it may take two or three turnovers before emissions are back down to even 1990 levels, not to mention below those levels.
- Although the focus of this book is on LDVs, we should not lose sight of the broader issues of transport. I made the point illustrated by the 'ASIF' identity (Schipper et al. 2000) that lower emissions means not only technological progress with the '*I*ntensity' and '*F*uel' terms in ASIF, but shifting to less travel **A**, particularly the share **S** in cars (the most energy and CO_2 intensive mode) and a larger share **S** of travel in collective and non-motorized modes. This requires understanding the drivers of mobility demand itself, which Bleijenberg has addressed in Chapter 2 of this volume. Nevertheless, improving by a large amount the efficiency of LDVs is an important part of the overall reduction in CO_2 emissions
- More fundamentally, we found at the time that few European nations were watching carefully how both transport volume (the *A* in *ASIF*) and its modal shares *S* were evolving, as well as how the fuel intensity of each vehicle and the modal intensity (or fuel use per passenger-kilometer) of each mode was changing. At the time there was little careful effort do to this in a systematic way in all but a handful of European nations. Now, the five European countries we reviewed (France, Germany, Netherlands, the United Kingdom, and Sweden) all watch closely only the test fuel intensity and emissions of new LDVs, but have undertaken much more effort to monitor how these are used on the road, and how they really perform. After all, that is what determines how much carbon is emitted, not simply the test value of emissions times a 'typical' yearly driving distance.

In this light, the present volume adds a great deal to what we know about ways to reduce emissions in the future, while also illustrating the uncertainties we need to understand better. In a way technology per se is the smallest uncertainty: experts understand what technological possibilities exist to reduce the carbon content of fuels and reduced the fuel used per kilometer to move a car. Larger uncertainties revolve around how much more or less we will drive in cars, what the costs of the technological improvements will be, how fast they will be implemented, what policies, at what strength will be implemented, and what the overall results will be, i.e., what we will monitor. For the majority of the world, these uncertainties are even larger. The future of the automobile may be uncertain, but I doubt that either CO_2 or fuel contributes as much to the uncertainty as the problems the transport system itself faces, at least in the developing world.

17.2 Automobiles and Climate Change

The first uncertainty is how much more Europeans will want to move around? Some of the articles in this book (Bleijenberg in Chapter 2, Chlond in Chapter 12, Matthews and Nellthorp in Chapter 13, and Litman in Chapter 14) address ways in which changes in transport policies and user fees would reduce travel in cars. But with few exceptions (Stockholm and London) little has been done to raise the cost of using a car per se. For better or worse, higher fuel prices (and other factors noted in Millard-Ball and Schipper (2010)) have slowed or reversed per capita car use in Europe, North America, Japan and Australia through 2009.

The current plateau in per capita travel (Millard-Ball and Schipper 2010) offers a breather. I think we'll see very sluggish growth in car use for many reasons, including congestion and higher fuel prices, but also because of saturation of travel, although some of that travel may be appearing as international air travel and thus not captured in the national statistics we reviewed.

The drivers of mobility, as Bleijenberg reviews in Chapter 2, can continue to be important in Asia and Latin America as urbanization increases, space grows scarce and congestion dominates urban traffic, while in Africa poverty hinders the thoughtful expansion of urban and intercity road networks. And in all regions of the developing world, informal transport, whether three, four or more wheels ('colectivos' in Hispanic Latin America, 'clandistinos' in Brazil, 'trotros' in Ghana, watatu in East Africa, three-wheelers and 'six-seaters' in South Asia etc.) still provide the largest share of motorized urban transport in most regions. The explosion in electric bikes in China, on top of gasoline fuelled ones, is a testimony to the difficulties of moving around in cities on anything larger. Yet this same boom shows how much consumers seem to value individual motorized mobility.

On the other hand, travel time is relatively constant across regions, incomes, and culture (Schäfer et al. 2009). The recent plateau or even peak in per capita car use suggests changes in the works although centuries of shifts to faster modes (air, high speed rail) may take up some of the slack. In particular, international air travel is not counted in the Millard-Ball and Schipper (2010) work. Schäfer et al. (2009) foresee such travel as possibly adding 5000 passenger-kilometers per capita globally by 2050, more than the present per capita level of travel for all developing countries today. While air and high speed rail modes are currently less energy and CO_2-intensive than car travel, they are so fast that travellers move 10–20 times faster in a given time. The overall impact on energy use is upward. Since most high speed rail (HSR) travel is in a limited range of 100–1000 km, its overall impact is small compared to that of air travel (Kosinski et al. 2011). Even if HSR were energized by carbon-free electricity, the demand for jet fuel as we know it today (kerosene) would increase carbon emissions from travel (and air freight) by a very significant amount (Schäfer et al. 2009).

Will congestion in airports and higher fuel prices dampen the upward march of air travel? Will the aging of populations in IEA countries and even in China slow or reverse the upward march of distance covered? Will today's global majority still without motorized transportation motorize in mostly collective transport (as in Latin

American cities today), in one kind of two wheeler or another (as increasingly in Asian cities), or rush toward cars, as seems to be the rage in China? Aging is something we all have to face, but airport, road, and port congestion can be addressed in part by massive investment in more capacity. If these investments are somehow charged to travellers, then the increases in travel might be moderated considerably, as congestion pricing in Stockholm and London has shown. Winston and Shirley (1998), in a classic paper, argued that simply throwing money at public transport when drivers are undercharged for congestion is a waste of funds. I would argue air pollution needs to be factored in, again with a considerably externality cost (Small and Kazimi 1995, Parry et al. 2007). The result would be considerably lower LDV use in and around metro regions than is the case today in Europe.

Reliable but conventional forecasts hold increased demand for automobile travel as the main driver of future emissions. But will automobile travel increase? And with rapid increases in fuel prices, is it not possible that the upward spiral of car power and weight will slow or even reverse (as it has recently among new cars in Europe) long enough for technology and continued down-sizing to reduce fuel use per km faster than km increase? The IEA projections illustrated by Lew Fulton in Chapter 3 are a good start at understanding what continued motorization means for global energy and CO_2 emissions. Yet even without considering the recent plateau of car use in developed countries, we have to ask about the new 'Great Wall of China' – solid queues of cars stuck for hours with similar experiences from India that mimic what Bangkok, Manila, Jakarta, and many Latin American cities have experienced for decades. Will this immobility lower the value of a car so much as to both suppress ownership and use and possibly shrink what we think of as a car so that more can fit into the available road and parking spaces?

Thus the first uncertainty remains just that: how many more cars can the world hold, how far can they go, and how much will all of that cost?

17.3 Policies at the EU Level

There should be no doubt that transport and energy or CO_2 policies can shape car use and emissions. The uncertainty is clearly over how strong those policies should be, and how strong they will be after the political process takes its toll on policies.

EU has achieved a great deal at both the Community-wide and individual country level (IEA 2000), yet the overall results in terms of fuel economy improvements and emissions reductions from passenger LDVs have been modest until recently. The sales-weighted average test CO_2 emissions of new cars in eight EU countries in 1995 fell at the rate of 0.9% per year below what it was in 1980, according to national and (after 1995) EU data.[1] From 1995 to 2009, the rate of decline accelerated to

---

[1] The EU-8 are Belgium, France, West Germany, Italy, Netherlands, Spain, Sweden and the UK. The EU-15 also include Austria, Denmark, Finland, Greece, Ireland, Luxembourg, and Portugal. Germany after 1994 is united Germany.

1.7% per year, probably because of the Voluntary Agreement (VA) after 1995, bolstered by higher real fuel prices starting in 2003, when the decline accelerated from 1.5% per year (1995–2003) to 2% per year between 2003 and 2009. Much of this can be attributed to the VA on CO_2 emissions that took force in 1998. Even it if did not reach its initial objective of an EU-wide test emissions level of 140 grams per kilometer (g/km), emissions/km started to fall in the late 1990s. That decline accelerated after prices for crude oil, and in some cases fuel taxes (particularly on diesel) started to rise in the early 2000.

Fuel use or total emissions depend both on the (slow) turnover of the stock as well as on increases in car ownership and changes in car driving distances. The emissions/100 km from actual use (all fuels) is shown in Fig. 17.1. On-road emissions are given from each country's official or authoritative derivations, as collected and analyzed in Schipper (2010a). The figures are calculated by calculating emissions/km from fuel economy and yearly driving distances of gasoline, diesel, and liquefied petroleum gas (LPG) cars using estimates of yearly distance by fuel type and the CO_2 content of each fuel. The result is then divided by the CO_2 content of a liter of gasoline to give a gasoline equivalent l/100 km figure. Note that this procedure reduces the apparent advantage of the rise in diesel cars. The improvements (declines) in fuel or emissions intensity are not dramatic, but they are real.

During the time period illustrated, per capita car ownership and use continued to rise slowly. But by 2000 the downward pressure on emissions intensity, and the slowdown and plateau in car use began to have an impact on per capita emissions as can be seen from Fig. 17.2.

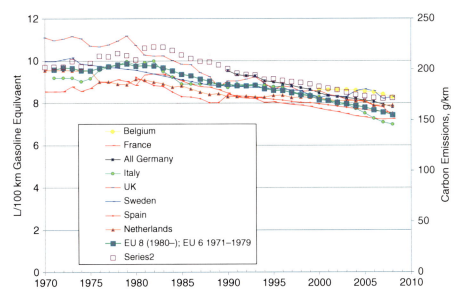

Fig. 17.1 On-road emissions per km from automobiles for eight European countries
Source: Schipper (2010a), IEA (2000) and national data sources

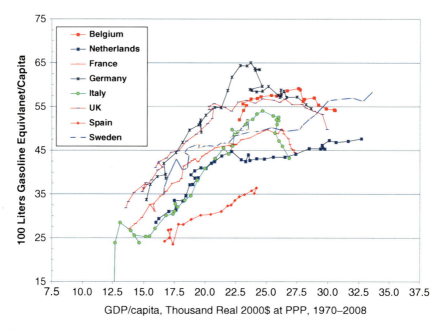

Fig. 17.2 Fuel use per capita from automobiles vs. GDP per capita (the latter expressed in constant purchasing power-adjusted US dollars of year 2000)

Figure 17.2 is given in per capita terms, since population growth as such is not considered as 'anti'-climate. By portraying emissions against GDP per capita, we eliminate to first order the important effects of income driving up car ownership and emissions. Still, flat per capita behavior does not mean flat or falling absolute emissions. The results then appear disappointing. Stock turnover is slow, the reduction in new vs. actual fleet emissions (when 'new' is raised roughly 20% to reflect real traffic (Smokers et al. 2006)) is not dramatic, hence the overall path of emissions is stagnant or slowly upward, but not downward in any dramatic sense. In other words, the 140 gCO_2/km target, which works out to somewhat under 170 gCO_2/km on the road, was not bold enough to offset modest growth in per capita car ownership and population growth in the original EU-12, not to mention growth in the new EU members, some of which had less than 100 cars per 1000 people in the late 1980s before the fall of communism.

Before one is disappointed by lack of a dramatic decline in emissions per capita (not to mention total emissions), it is worth recalling that the real cost of fuel for 1 km has not changed much despite wide swings, and through 2008 was still below its real peak (for most countries, including Sweden) in the 1980–1982 period as shown in Fig. 17.3. Again, these figures (Schipper 2010a) take into account the real mix of diesel, gasoline and LPG used by passenger LDVs in each country and the individual prices of each fuel. The rising importance of diesel, until recently significantly less costly than gasoline, and the residual importance of low-cost LPG in a few countries (Italy, Netherlands) helped keep the cost of fuel for a kilometer

17 Epilogue – The Future of the Automobile: CO_2 May Not Be the Great Decider

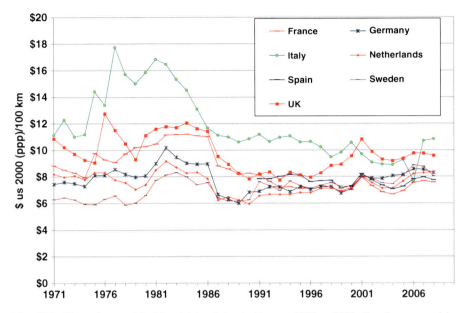

Fig. 17.3 The real cost of fuel for driving 1 km in Europe, 1970 to 2008. Cost is expressed in constant purchasing power-adjusted US dollars of year 2000, per 100 km

down, just as the improvements in on-road fuel economy also helped.[2] The values for 2009 were not available at the time of this writing, but since new vehicles were less fuel intensive than the stocks, the on road fuel consumption per 100 km is lower than 2008, while the real price of fuel fell considerably from 2008 peaks.

With such little price pressure (until recently at least), perhaps the policy pressure from EU and member states was too weak? Relying only on a voluntary agreement among manufacturers selling in the EU as the principal policy instrument seems to me in retrospective to have been a weak move, even if the sales-weighted average emissions in 2009 came within 5% of the 140 gCO_2/km voluntary target. The problem was two fold: car power and weight increased significantly to eat up much of the benefit of truly more efficient engines and aerodynamics, while the much touted shift to diesel was more than offset by a faster increase in power and weight for diesels than for gasoline cars. By 2009 the sales weighted emissions of new diesels in the EU24 was only 4% below that of gasoline cars (European Commission 2010, Schipper and Fulton 2009, Schipper and Hedges 2011).

Where did we go wrong? The answer seems to be that the VA was not really backed up by strong price signals, at least not until the past few years. But an additional answer comes from two key studies (Ryan et al. (2009) for EU, Morrow

---

[2] The calculations start with the real price of each fuel (year 2000 real local currency converted to 2000 US Dollars at purchasing power parity), weighted by the amounts purchased and then expressed as dollars per litre gasoline equivalent. The result is multiplied by the fuel consumed on road per 100 km to give the cost per100 km.

et al. (2010) for the US). Both found that carbon reduction policies based only on standards or incentives are much weaker and slower to act than those that also include strong pricing signals for carbon. Equally important, pricing signals to capture important transport externalities could also be brought to bear (European Commission 1996, ECMT 1998, IEA 2000) because these tend to be more costly, in terms of Eurocents/km, than additional carbon taxes piled on top of the existing taxation of fuels in Europe. Even the high carbon tax in Sweden still had only a modest effect on bringing down Sweden's high new vehicle test emissions, which finally fell from the unenviable position of the highest in Europe. Finally, as mentioned in Chapters 6 and 7 of this book, there is clear evidence that company car schemes undermined the drive for both fuel economy and lower car use.

How much price pressure could be brought to bear? A carbon charge of roughly $85/tonne CO_2 (the level recommend by the Stern (2006) report) corresponds to only 15 Eurocents/l (at $1.30 to the Euro), or slightly above 1 Eurocent/km at an on-road fuel intensity of 7–8 l per 100 km (see Fig. 17.1). Worse, company car policies undermined some or all of this and other variable costs, boosting the affordability (compared to no company car tax concessions) of larger than otherwise cars by up to 50% of new car buyers, and then leaving these cars to filter through the stock after 2–3 years as used cars. Thus it is no surprise that the real improvement in on-road CO_2/km is relatively timid in Europe. Imposition of road user charges, and charging for car insurance in part by the actual number of kilometers driven could raise the variable cost of using cars by a larger amount, but to date there are no countries committed to this direction (other than the Netherlands). In other words, the policy instruments reviewed are available and potent, but leaders have backed away from using them!

Thus policy uncertainty remains – will EU leaders insist on both stronger emissions targets and the pricing and other policies that reinforce these targets, in part by halting the previously incessant upward march of power and weight? Interestingly, power and weight stopped increasing in 2006/7 new models, and turned slightly downward through 2009. Emissions intensity fell even more, so the near-achievement of the VA target by 2009 (147 gCO_2/km, or within 5%) came both because of slight downsizing and some continued reductions in the emissions per unit of power or weight of new car.

17.4 National Policies

A number of national policies contributed to lower CO_2 emissions intensities of new vehicles. 'Green car' policies as described by Beser-Hugosson and Algers in Chapter 11 or Kågeson in Chapter 6, seem to be popular. Yet as recent data from the Swedish traffic authority (Trafikverket 2011), Kågeson's work, and my own studies of the diesel market suggest, 'green' cars are not always bought by 'green' people.

In the case of Sweden, buyers of 'green', i.e. low-carbon cars tend to be wealthy and more often supported by company car schemes, as data in Chapter 11 show.

Were the cars they bought lower carbon than those bought by non-company car users? Were the cars they bought lower CO_2 than what they would have bought without the green car incentives? And how much of the advantage of a green car is eaten away by the higher driving distances of company car owners, even after correcting for the small natural rebound effects of greater efficiency? More important, should policies focus primarily on the properties of new cars, or on how they are used?

What does seem to be having an impact is the 'bonus/malus' scheme, also called feebates (discussed here by Braathen and Kågeson in Chapters 8 and 6 respectively, but see also Bunch and Greene (2010) and Tessier and Meunier (2010)). In one form or another, this scheme raises or lowers the taxes on new cars according to their rated CO_2 consumption around a balance or neutral point. That balance point could be the Voluntary Agreement level of 140 gCO_2/km or the 2016 EU target of 130 gCO_2/km to be achieved from vehicle technology. How steeply the 'malus' rises or falls from the target level is a matter of policy. Over time the balance point (between taxes or refunds) can fall to lower and lower CO_2 emissions.

Coupled with strong price incentives, this approach has the advantage of affecting all new car purchases, not just those that happen to fit the various 'green car' incentives. The importance of this approach has been made clear by a series of papers on 'notches' as cited by Kågeson in Chapter 6. The problem with many policies and even some technologies is that they have 'notches', i.e. discrete boundaries. The most obvious one: why does a hybrid vehicle qualify for a tax break or other incentive, but not a conventional vehicle with the same test carbon emissions? Why were new vehicles in Europe taxed by weight or engine size or power in bands, rather than as a continuous function of the parameter in question? Not surprisingly, new European cars always clustered around the notches of taxation, i.e. 1799 cubic centimeters (cc) was a popular engine capacity if the tax on new cars rose at 1800 cc. And manufacturers of the largest and most fuel intensive cars sold in the US clustered their products with just enough fuel economy to fit snugly under the notch of fuel economy (the inverse of fuel economy really) where the 'gas guzzler tax' would be lower, as Sallee (2010) and Sallee and Slemrod (2010) showed.

Thus the question of how to frame national or EU-wide incentives remains one of how to affect all vehicles, not just those that meet some discrete boundaries, and how to prevent company car benefits from soaking up some or most of the benefits of emissions saving technology as larger cars and greater driving?

A final note about policy impacts is important. As illustrated in Fig. 17.4, the question of policy impact should be asked compared to a counterfactual with no policies, not simply by comparing present emissions from cars with say the emissions from 10 years ago. This approach is important both because of small but important rebound effects (Schipper and Grubb 2000) as well as the presence of other factors that are related to transport policies that may raise or lower emissions. As noted elsewhere (Millard-Ball and Schipper 2010), likely causes of the present stagnation in car use in Europe could be attributed to factors besides policies – higher fuel prices, an aging population, congestion, switches to air travel etc. Sorting these impacts out is important in order to be able to attribute causes

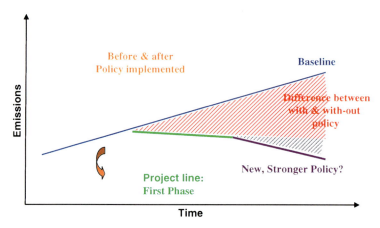

Fig. 17.4 Symbolic diagram of evaluation of policy or technology impact. Adapted from Schipper et al. (2009a)

to affects. This is particularly important for transport related policies such as congestion pricing in London and Stockholm. Good transport models are needed to be able to discern impacts of policies both on specific emissions (g/km) and on total kilometers driven. Relying simply on total sales of one kind of fuel or another is hardly sufficient, particularly with the prominence of diesel fuel and in some cases biofuels, compressed natural gas (CNG) and LPG.

There is one danger in focusing too strongly on national policies when the overriding driving force is the CO_2 reduction agreement between EU and car manufacturers. Efforts in any one EU state that bring emissions well below the EU target can cause 'leakage' of emissions to other States, as Goulder et al. (2010) show for California compared with the rest of the US. That is, if one or more larger countries succeed in reducing emissions well below the EU average (France in 2009, for example), manufacturers could sell larger or less efficient cars in other states and still achieve the overall standard. Once again we see where a well-meant policy not backed up by clear price signals only achieves part of its objective.

17.5 The Broader International Context

Part of the answer to the previous question about restraint in emissions beyond that focused on carbon per se lies in the review of transport policies in the Litman paper (Chapter 14). As noted above, most studies find that the externalities in transportation cost more, per kilometer of car use, than those from CO_2 even at a high value of CO_2 damages. But with some exceptions (Stockholm and London with congestion charging, Scandinavian countries with variable taxation on fuel by quality), few authorities have imposed variable cost charges on car users and other travellers. Thus a potential policy resource that has a potentially large impact on CO_2 – though

17 Epilogue – The Future of the Automobile: CO_2 May Not Be the Great Decider

not focusing on carbon alone – has not been employed as continually urged by so many authorities.

This book has focused on the European situation with a concentration on Western Europe. Not surprisingly, the fall of the Berlin Wall and loosening of the reins of former regimes brought car ownership and use up rapidly in Eastern Europe, fed first by a flood of used cars from west to east as both western and previously eastern European manufacturers modernized existing factories and built new ones. The result was a relatively quick catch up from east to west (Pucher and Buehler 2005), as also illustrated by Ščasný for the case of the Czech Republic in Chapter 10. Indeed, the authoritative German transport data sources ('Verkehr in Zahlen'), which kept data on Eastern and Western Germany separated through 1994, shows that by the late 1990s when the regions are considered as one there is hardly any gap, with a simple extrapolation of West Germany car ownership and use alone. East met West on the Autobahn.

This observation has fundamental implications for considering the developing world, as Huizenga (and this writer) have pointed out (Schipper et al. 2009b, Ng et al. 2010). Relative to GDP, Latin America (including Mexico) still had far more cars than other developing regions, and was projected to increase its car ownership relative to GDP faster than other regions (WBCSD 2004). But with GDP per capita growing much more rapidly in China and now India and much of the rest of Asia (specifically the ASEAN region) car ownership is shooting upward. Whether car use is following is not certain, and the two (ownership and use) should not be equated.

Figure 17.5 shows how the ownership of automobiles, household light trucks and Sport Utility Vehicles (in vehicles per 1000 people) has grown with GDP per

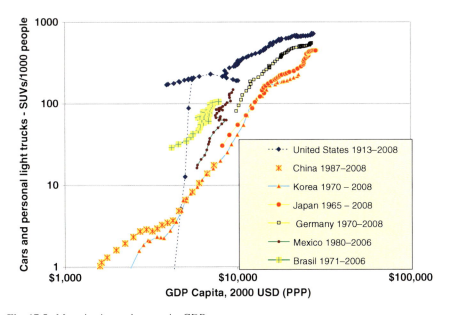

Fig. 17.5 Motorization and per capita GDP

capita. For each country, the data run over the years shown in the legend. The US data include the Depression years, when GDP per capita fell and car ownership fell slightly, then the lurch to the right when GDP rose and car ownership fell (World War II), and finally the slow saturation at levels of over 720 vehicles per 1000 people.

Notable in the figure is the high position of Mexico and Brazil, which are chasing the US level of cars for a given GDP. Germany (for which united Germany after 1998 matches almost perfectly the extrapolation of pre-unification West Germany) represents a high (but not the highest) level of ownership for Europe, while Japan and Korea represent well the high-income denser countries of Asia.

In my previous scenario studies of China (Schipper et al. 2001, Schipper and Ng 2005, Ng et al. 2010) I projected car ownership per capita in China by taking a projected GDP per capita for China and assigning China the same number of cars per capita that Korea had at that GDP per capita level. The data in Fig. 17.5 show this relationship has held remarkably well for 20 years, suggesting the approach was a valid way of making projections. But understating the rate of GDP growth in China and possibly the car per GDP ratio meant that China in 2009 passed the business-as-usual scenario originally developed, labelled politely as 'The Road Ahead' in Schipper and Ng (2005), but originally called 'car collapse' in Schipper et al. (2001) because of my sense that the number of cars in China could not grow that rapidly (Fig. 17.6).

I was wrong. China in 2009 surpassed our highest projections for 2010. Yet the emergence in 2010 and 2011 of a new 'Great Wall in China' composed of cars stuck in urban traffic and cars and trucks stuck on intercity roads suggests perhaps 'car collapse' was an appropriate name for where China's motorization has been headed. Moreover, the lackluster sales of the Nano in India (which I labelled the 'No No' because of its potential to worsen an already bad urban traffic situation) may also be a harbinger of what is to come in India. Not really built for India's tough

Fig. 17.6 Traffic Jam in Beijing December 2005; there are more than twice as many cars in 2011

17 Epilogue – The Future of the Automobile: CO_2 May Not Be the Great Decider

Fig. 17.7 Pune, India, 2004. Loans for two wheelers, but no sidewalks in front of the bank

but relatively high-speed intercity roads, the Nano was foreseen as the ideal urban vehicle. Yet with city travellers increasingly stuck in traffic, two wheelers maintain their advantage. In Fig. 17.7, the bank in question (in Pune, the motor vehicle capital of India) offered quick loans for private individuals buying two wheelers, but there were no sidewalks in front of the bank!

In Hanoi, which I studied extensively (Schipper et al. 2008) two wheelers still dominate, but car sales are booming. Is there room? I think not. Projections for Hanoi done by an outside agency suggested that an alternative vision for the city would have far less car travel and somewhat less motorcycle travel than was projected. As Fig. 17.8 shows, Hanoi was already stuffed with two-wheeler traffic in 2006. Where would the projected cars go?

Fig. 17.8 Two wheelers in mixed traffic in Hanoi

This leads to the key point implied by Huizenga and Leather in Chapter 16. The rapid growth in car ownership in Asia, and continued growth in Latin America and elsewhere is projected by the IEA to lead to emissions growth that swamps efforts to reduce emissions in the car-intensive developed world. But this rapid growth implies even less sustainable transportation than we observe today in the developing countries. Emissions of local pollutants may continue to fall absolutely, but increased congestion and traffic deaths from the rapid growth in car travel means that a situation that is already bad for travellers can only get worse. I labelled Asia's situation 'hypermotorization' (Schipper 2010b) not because motorization is bad, but because the speed of its increase outstrips all efforts to build roads and maintain or create better collective transport, and above all squeezes opportunities for the large share of trips (if not total travel) on foot, pedals, and even under animal power in the developing world.

Moreover, the large role (and in many regions, dominance) of informal transport among motorized modes (Cervero and Golub 2007) means continued political and social difficulties in organizing collective transport to serve the masses in a cleaner, safer, faster and more affordable way than car travel (World Bank 2008). Focusing only on the carbon emissions of light duty vehicles without reducing their domination and the ever presence of small collective transport on two, three or four wheels will reduce CO_2 emissions from transport, yet leave it unsustainable (Fig. 17.9).

Fig. 17.9 'Air cooled transport'? Colectivo on the Pan American Highway outside of San Salvador, El Salvador, 2000

17 Epilogue – The Future of the Automobile: CO_2 May Not Be the Great Decider

Sadly few countries or urban regions have acted to oppose hypermotorization, particularly as the car industry has taken off (e.g. in China and now in India). In other words, the evolution of passenger transport is driven almost entirely by private investments in individual vehicles, supported somewhat more slowly by public investments in road and parking infrastructure. A few cities in the world (Singapore, Curitiba) have maintained relatively low shares of car use in total travel through a variety of demand side policies (Singapore's high car ownership and congestion charging) or rapid development of bus rapid transit (BRT) integrated with other bus services (Curitiba's supply-side approach.) Both cities have strong land-use controls. Advances in BRT in corridors previously dominated by cars (e.g. in Bogota, but also now in Mexico City and increasingly in other Latin American and Asian cities), and some success with metro (notably Shanghai, which also has tried to control the number of new vehicles) suggest changes from the business-as-usual projections (IEA 2009) are possible. Yet through 2009, cars are winning. Figure 17.10, taken from the front of a bus moving in its own counter-flow lane in Mexico City against cars, symbolizes this apparent victory.

Transport policies that charge for externalities and other policies can make an enormous difference. This is because the externalities in transport from other than fuel and CO_2, when expressed per vehicle-kilometer, are much larger (Parry et al. 2007). The will to impose taxes, however, is not strong except in a few smaller countries (notably Sweden, which has congestion pricing in its largest city, a $140/tonne carbon tax on road fuels, and differential taxes on fuels from the least polluting (lowest tax) to the most polluting (highest tax). This leads to a conundrum – policies that might lead to the greatest restraint in CO_2 emissions from transport are not those that aim at CO_2 directly, or even those aimed at fuel economy. Instead, they aim at the other problems of transport and reap lower-than-otherwise CO_2 emissions as a 'co-benefit'.

A single illustration from Mexico City may suffice. Figure 17.11 – from Schipper et al. (2009b), summarized in Schipper et al. (2011) – shows reduction in CO_2

Fig. 17.10 Mexico City. Cars in the counterflow bus lane darting out of the way as the bus plows forward

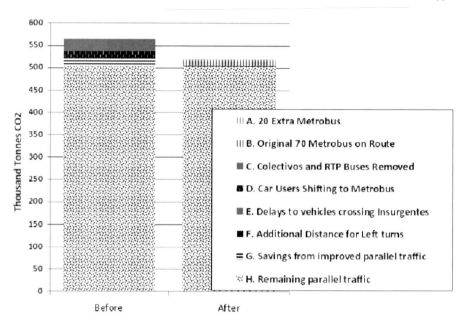

Fig. 17.11 CO_2 emissions from all traffic in the insurgentes corridor of Mexico City in 2005, before and after Metrobus was established
Source: Schipper et al. (2009b)

emissions along a major corridor in Mexico City (Insurgentes) resulting from the replacement of two lanes of vehicle traffic with a BRT corridor called Metrobus. The modest savings arose principally from three almost equal parts: larger, modern buses substituting for mini-buses ('colectivos'), car users shifting to Metrobus, and smoother parallel traffic. The latter effect was large and occurred because hundreds of errant mini-buses and some city buses ('RTP buses') were removed from traffic. There was a small increase in emissions as a strict no-left turn policy caused drivers to have to deviate to the right to cross Insurgentes, and because some cross traffic dwelt longer at red lights so that Metrobus could pass.

When the social benefits of time saved, lower air pollution, less wear on the roads, and the direct value of fuel savings were compared with the 50,000 metric tonnes per year of CO_2 saved from these shifts, the CO_2 savings were tiny (at $5/tonne the Mexico City earned by selling carbon credits) and only 20% of total project benefits if the CO_2 had been valued at the Stern (2006) value of $85/tonne. For the European situation, the clean air benefits would be smaller because air is much cleaner, but time savings would be much greater because of a much higher wage rate. And the calculation based on the figure did not include the value of fewer accidents and lives lost in traffic on this busy corridor.

The lesson, thus, is that some of the largest benefits of CO_2 reduction come as indirect benefits of other strategies to improve transportation. Indeed, replacing the new, conventional diesel Metrobus vehicles with parallel hybrids available at the time (and in use in Seattle, Washington) would have saved an additional 3000 tonnes of CO_2 per year at a marginal cost of at least $15 million to get hybrid buses. By contrast the initial project cost $80 million and led to the much greater savings at no cost of saving CO_2. Ironically, then, a focus only on CO_2 might be short sighted, leading to lower real savings at much higher cost than a broader focus on transport improvements. This indeed was one of the conclusions of our 2000 study (IEA 2000).

Reducing emissions per kilometer from every kind of vehicle is important, but must rest in the larger context of making transport work, both in the developing world and in developed countries. That was the message of the EU 1996 document (European Commission 1996) and many since. We must not lose focus on the larger transport picture because of the importance of making LDVs less carbon intensive. And given that domestic or international freight emissions have risen faster than those from passenger travel (Kamakate and Schipper 2009, Eom et al. 2011), one must also focus on this part of the transport system to achieve real, lasting emissions reductions.

17.6 Monitoring – A Big Task

The calculations for Mexico City were enabled by almost 20 years of data collection motivated largely by concerns over air quality. The Secretariea de Media Ambiente (Environmental Secretary) has a relatively accurate mobile source emissions inventory built in the ASIF approach – total vehicle-kilometers, share by vehicle, fuel use per kilometer, and pollution emissions (including CO_2) from each vehicle-fuel combination. Santiago and Sao Paolo now have similar inventories. Unfortunately, most of the rest of the world, developed and developing, has very poor data on passenger- and tonne-kilometers, coupled to the vehicles moving and fuels used. Less than a handful carry out regular surveys of these parameters, among those Australia (Apelbaum 2009) stands out as the most complete. Little of these data are available from developing countries, and almost none can be traced to well-defined data sources that provide reliable data year for year (Schipper et al. 2008, 2009a).

Indeed, even the EU voluntary agreement on fuel economy only monitors the sales-weighted fuel intensity and emissions of new vehicles by fuel, manufacturer group (i.e., Europe, Japan, Korea and other) and country. There are no systematic, regular estimations of how far cars are driven or how much fuel economy they attain on road. France is one of the few countries with regular vehicle use and fuel consumption surveys, while a number of other European countries only follow vehicle use but not fuel consumption. To be sure, at least a dozen EU member governments do publish key data for most modes. While these are useful for describing general trends, they are rarely accurate enough to spot the impacts of new policies, technologies and other trends on a year-by-year basis. Thus there is little opportunity for feedback from the impacts of policies and technologies on both vehicle use and fuel intensity to allow policy-makers to adjust their policies.

Consequently it came as a surprise to many to learn that in the race to satisfy the EU Voluntary Agreement on new vehicle emissions, the switch to new diesel cars per se had little impact on overall emissions per kilometer (Schipper and Hedges 2011). Instead it was the decline in emissions/km of both diesel and gasoline new cars that accounted for the drop to 146 gCO_2/km by 2009, just short of the goal of 140 gCO_2/km set for 2008. What was surprising was that emissions from new gasoline cars alone dropped more than those from diesel. What happened was that power and weight of both car types rose, with that of diesel rising more rapidly than that of gasoline. And diesel cars were driven far more than gasoline cars. Much of these could be explained by the fact that the average diesel car was newer and owned by a wealthier user than a gasoline car, or more likely to be a company car. Little of this was noted by the EU 'Monitoring' reports, because they – like the VA itself – were focused only on new vehicles.

This behavior was not unnatural, given that the fuel costs of driving 1 km did not rise significantly until after 2005 and remained below those of the early 1980s in most of EU. Rather, it suggests that while the VA was a useful element of an overall CO_2 policy it was insufficient to cause significant changes in both LDV emissions and use. The latter might have helped boost the use of other less carbon intensive modes even more than the small shifts that have occurred through 2008 (Millard-Ball and Schipper 2010).

Three other issues arise that need monitoring. The first is the gap between test and actual emissions (Smokers et al. 2006, Schipper and Tax 1994). The gap depends both on driving conditions and who is driving, and robs us of saved emissions. The second is the much touted rebound effect, that is, the impact of lower-than-otherwise driving costs on how much we drive. For car use, this is found to be small (Hymel et al. 2010), roughly 10% in the US but could be larger in Europe because more drivers live close to alternatives of mass transit. But another kind of rebound effect occurred in Europe (and the US) from 1980 onward. Technology that could have saved fuel and emissions at constant vehicle properties (e.g. weight or power) instead gave only slowly falling emissions per kilometer (in Europe) and no decline (until 2004) in the US while weight and power rose. Sprei (2010) notes that this ate up a considerable share of the total technological improvement in vehicles in Sweden, the country that already had the largest cars in Europe. Can we reclaim this from downsizing? Careful monitoring of both vehicle use and new car properties is important to see whether the hoped for reductions in emissions are realized. As implied in the introductory chapter, such monitoring should cover all modes through regular national travel surveys, not simply attention to cars.

The final issue is that of life cycle analysis of not only fuels (as explained by Dixson-Declève and Dings in Chapters 5 and 7 respectively) but also of vehicles and transport systems. Fortunately for the analyst, the energy and emissions from combustion or electric power generation dominate the picture of emissions over the life cycle of a car (Chester 2008). Still confusion reigns. The picture in Fig. 17.12 was taken in a parking lot in Oakland, Ca. in February 2011. The car on the left runs on Compressed Natural Gas. The driver must have thought that CNG combustion

17 Epilogue – The Future of the Automobile: CO_2 May Not Be the Great Decider

Fig. 17.12 Sorting out life cycle emissions: Which registration plate is closer to the truth?

emitted no greenhouse gases! The car on the right belongs to the author. A third car, shown below, is a two-seater electric car with the registration 'No COII'. Despite many attempts to intercept the owner (who lives near me), I have not been able to determine where she gets her zero-CO_2 electrons! Surely the life cycle analyses are complex (Chester 2008). But they must be carried out as part of a monitoring process to be sure that the CO_2 that does not flow through the tailpipes of cars is not leaking out elsewhere.

The adage 'you cannot master what you cannot meter' certainly applies here. European countries are close to being able to monitor the key *ASIF* components of all modes of travel, including even international trucking, rail, and air travel and freight. The same is true for the life cycle of fuels. But developing countries, where growth in car use is fastest, are far from even being able to count how many cars there are and how far they go. Thus there is a big monitoring job ahead.

17.7 What then is the Future of the Automobile?

This volume has illustrated the strong link between automobiles and CO_2 emissions associated with climate change. In thinking about the future of the automobile it is tempting to blame the car for its contribution to climate change. Yet it is us, the drivers (and the policies that support us, policies enacted by those we voted for), who have chosen to create a world of large cars and in most Western nations established a very automobile-dependent lifestyle. The automobile (or other individual motorized transportation modes) has given many (but not all) of its owners and users greater choices on where and how to live. But it is clear that those choices increasingly impinge on all drivers, and, more important, on all others trying to move in increasingly crowded cities or between urban areas on crowded motorways. The situation in developing countries is dire at a tenth or less of the motorization rate (in cars per 1000 people) of IEA countries. People are frozen in most large cities. It is thus hard to foresee expansion in car ownership to the level forecast by the IEA (2009) or Sperling and Gordon (2009).

Does this mean the future of the automobile is grim? Yes, if individuals, their elected officials and stakeholders in fuel and vehicle companies continue as if there are not profound problems confronting the choices automobiles give their users. Yet it has been hard to recognize that continuing to provide these choices is both prohibitively expensive to the public or private sector (to build so many roads that we could not possibly fill them up), and increasingly expensive to health (air pollution and accidents) and to oil security and climate. The WBCSD (2004) study recognized the CO_2 problem and leaned toward low-CO_2, and almost non-polluting automobiles. But the report skirted the real issue of whether there was room on the road for billions of cars.

My own conclusion is that CO_2 is not the deciding factor over the future of the automobile, rather more fundamental issues raised by Huizenga and Leather in Chapter 16 of this book, namely the difficulty of fitting in so many individual vehicles to so little space. Technology can help somewhat, but the larger issues are what people decide to do with technology. The pictures in Fig. 17.13 illustrate those choices well. Ironically, the Cadillac was photographed in Timmernabben, Sweden, while the Fiat was found in West Berkeley, California. Clearly, as the figure shows, there are many choices over the future of the automobile.

Fig. 17.13 Automobile choices

References

Apelbaum J (2009) A case study in data audit and modelling methodology – Australia. Energy Policy 37:3714–3732

Bunch DS, Greene DL (2010) Potential design, implementation, and benefits of a feebate program for new passenger vehicles in California: interim statement of research findings. Institute of Transportation Studies, University of California, Davis, Research Report UCD-ITS-RR-10-13

Cervero R, Golub A (2007) Informal transport: a global perspective. Transport Policy 14:445–457

Chester M (2008) Life cycle inventory of passenger transportation in the United States. PhD Thesis. http://www.sustainable-transportation.com/. Last accessed Mar 2011

Eom J, Schipper L, Thompson L (2011) We keep on Truckin': trends in freight energy use and carbon emissions in 10 IEA countries. Proceedings of the 2011 ECEEE Summer Study. European Council for an Energy Efficient Economy, Stockholm and Paris

European Commission (1996) Towards fair and efficient pricing. Office for Official Publications of the European Communities, Luxembourg

European Commission (2010) Report from the commission to the European parliament and the council: monitoring the CO_2 emissions from the new passenger cars in the EU: data for the year 2008. COM (2009) 615 final. Brussels

European Conference of Ministers of Transport (ECMT) (1998) Efficient transport in Europe: internalization of external costs. ECMT/OECD, Paris (now International Transport Federation)

Goulder L, Jacobsen M, van Benthem A (2010) Unintended consequences from nested state & federal regulations: the case of the Pavley greenhouse-gas-per-mile limits. Stanford University Department of Economics, Palo Alto

Hymel K, Small K, Van Dender K (2010) Induced demand and rebound effects in road transport. Transportation Res B 44:1220–1241

IEA (2000) The road from Kyoto. International Energy Agency, Paris, France

IEA (2009) Transport, energy and CO_2. Moving towards sustainability. International Energy Agency, Paris, France

Kamakate F, Schipper L (2009) Trends in truck freight energy use and carbon emissions in selected OECD countries from 1973 to 2005. Energy Policy 37:3743–3751

Kosinski A, Schipper L, Deakin E (2011) Analysis of high-speed rail's potential to reduce CO_2 emissions from transportation in the United States. 2011 Annual meeting of the transportation research board. TRB, Washington, DC. http://amonline.trb.org/13hpb5/13hpb5/1

Millard-Ball A, Schipper L (2010) Are we reaching peak travel? Trends in passenger transport in eight industrialized countries. Transport Reviews. First published on 18 November 2010 (iFirst). http://dx.doi.org/10.1080/01441647.2010.518291

Morrow WR, Sims-Gallagher K, Collantes G, Lee H (2010) Analysis of policies to reduce oil consumption and greenhouse gas emissions from the U.S. transportation sector. Energy Policy 38:1305–1320

Ng WS, Schipper L, Yang C (2010) Motorization in China. New directions for crowded cities. J Transportation Land Use 3:5–25. doi: 10.5198/jtlu.v3i3.151. http://jtlu.org

Parry IWH, Walls M, Harrington W (2007) Automobile externalities and policies. J Econ Literature 45(2):373–399

Pucher J, Buehler R (2005) Transport policies in central and Eastern Europe. In: Button K, Hensher D (eds) Handbook of transport strategy, policy and institutions, vol. 6. Elsevier Press, Handbooks in Transport, Amsterdam, pp 725–743

Ryan L, Ferreira S, Convery F (2009) The impact of fiscal and other measures on new passenger car sales and CO_2 emissions intensity: evidence from Europe. Energy Econ 31(3):365–374

Sallee JM (2010) The taxation of fuel economy. NBER Working Paper No. 16466, National Bureau of Economic Research, Cambridge, MA

Sallee JM, Slemrod J (2010) Car notches: strategic automaker responses to fuel economy policy. NBER Working Paper No. 16604, National Bureau of Economic Research, Cambridge, MA

Schäfer A, Heywood JB, Jacoby HD, Waitz IA (2009) Transportation in a climate-constrained world. MIT Press, Cambridge, MA

Schipper L (2010a) Automobile use, fuel economy and CO_2 emissions in industrialized countries: encouraging trends through 2008? Transport Policy 18:358–372

Schipper L (2010b) Car Crazy: the Perils of Hypermotorization in Asia. Global Asia 4(4). http://www.globalasia.org/l.php?c=e243. Last accessed Mar 2011

Schipper L, Fulton L (2009) Disappointed by diesel? The impact of the shift to diesels in Europe through 2006. Transportation Research Record 2139

Schipper L, Grubb M (2000) On the rebound? Feedback between energy intensities and energy uses in IEA countries. Energy Policy 28:367–388

Schipper L, Hedges E (2011) The impact of new passenger vehicle changes and the shift to diesel on the European Union's CO_2 emissions intensity. 2011 Transportation Research Board Meeting, Washington, DC

Schipper L, Ng W-S (2005) In: Baumert KA, Bradley R (eds) Growing in the greenhouse: protecting the climate by putting development first. World Resources Institute, Washington, DC

Schipper L, Tax W (1994) New car test and actual fuel economy: Yet another gap? Transport Policy 1(2):1–9

Schipper L, Deakin E, McAndrews C (2011) Carbon dioxide emissions from urban road transport in Latin America: CO_2 reduction as a co-benefit of transport strategies. In: Rothengatter W, Hayashi Y, Schade W (eds) Transport moving to climate intelligence. Springer, Dordrecht

Schipper L, Fabian H, Leather J (2009a) Transport and carbon dioxide emissions: forecasts, options analysis, and evaluation. ADB Sustainable Development Working Paper Series No. 9, Asian Development Bank, Manila, Philippines

Schipper L, Marie-Lilliu C, Gorham R (2000) Flexing the link between transport and greenhouse gases: a path for the world bank. International Energy Agency, Paris, France

Schipper L, McAndres C, Deakin E, Scholl L, Frick K (2009b) Considering climate change in Latin American and Caribbean Urban transportation: concepts, applications, and cases. Prepared for the World Bank. Global Metropolitan Studies, University of California, Berkeley

Schipper L, Tuan LA, Oern H, Cordeiro M, Ng WS, Liska R (2008) Measuring the invisible: quantifying emissions reductions from transport solutions – Hanoi case study. EMBARQ - World Resources Institute, Washington, DC, March. http://pdf.wri.org/measuringtheinvisible_hanoi-508c_eng.pdf. Last accessed Mar 2011

Schipper L, Price-Davies G, Marie-Lilliu C (2001) Rapid motorisation in the largest countries in Asia: implication for oil, carbon dioxide and transportation. International Energy Agency, Paris, France.

Small K, Kazimi C (1995) On the costs of air pollution from motor vehicles. J Transport Econ Policy 29(1):7–32

Smokers RTM, Vermeulen R, van Mieghem R, Gense R, Skinner I, Fergusson M, MacKay E, ten Brink P, Fontaras G, Samaras Z (2006) Review and analysis of the reduction potential and costs of technological and other measures to reduce CO_2-emissions from passenger cars. Final Report Contract nr. SI2.408212, 31 October 2006

Sperling D, Gordon D (2009) Two billion cars. Oxford University Press, New York

Sprei F (2010) Energy efficiency versus gains in consumer amenities. Examples from passenger cars and the Swedish building sector. PhD. Thesis. Chalmers University, Gothenburg, Sweden

Stern N (2006) Stern review: the economics of climate change: executive summary. http://www.hm-treasury.gov.uk/stern_review_report.htm

Tessier O, Meunier L (2010) Commisariat du plan. Le Point Sur 53, May 2010 and Boutin X, Haultfoeuille X, Givord P, (2010) The environmental effect of green taxation: the case of the French Bonus-Malus. Prepared for the 11th CEPR Conference on Applied Industrial Organization. Toulouse France

Trafikverket (2011) Ökade utsläpp från vägtrafiken trots rekordartad energieffektivisering av nya bilar. PM 2011-02-18. Boerlaenge: Trafikverket (Swedish Traffic Authority)

Winston C, Shirley C (1998) Alternate route. Towards efficient public transportation. Brookings Institution, Washington, DC

World Bank (2008) A strategic framework for urban transportation projects: operational guidance for world bank staff. Prepared by Slobodan Mitrič. Transport Papers. Washington, DC, TP-15, January 2008

World Business Council for Sustainable Development (WBCSD) (2004) Mobility 2030: meeting the challenges to sustainability: the sustainable mobility project. http://www.wbcsd.org/plugins/DocSearch/details.asp?type=DocDet&ObjectId=NjA5NA

Index

A

Abatement cost, 7, 11–12, 127, 139, 143, 154, 196, 383
Accessibility, 9–10, 20, 31–34, 37, 257, 287, 290, 292, 329, 348, 381
Accidents, 2–3, 9–10, 149, 197–198, 201–202, 212, 216, 287, 296, 299, 302, 317, 327–328, 330, 335, 341, 344, 349, 371, 378, 389, 408, 412
ACEA, 60, 75–77, 83, 91, 104, 135, 137, 203, 296
Africa, 46, 113–114, 182, 343–344, 373–375, 386, 395
Agriculture, 31, 101, 108, 122
Air conditioning, 77, 82–83, 147, 169, 174, 356, 361–365
Air pollutants, 2, 8, 101–102, 181, 197, 202, 215–216, 360, 362, 364
Air quality, 99, 101–102, 104–107, 109, 125, 380, 409
Air travel, 21, 27–28, 38–39, 44–46, 51, 58, 375, 395, 401, 411
Aromatics, 107–108, 157
Asia, 46, 329, 371, 373–378, 386, 395, 403–404, 406
Australia, 10, 308–310, 320, 355–356, 367, 395, 409
Automotive industry, 74, 82, 85, 91, 137, 140, 271, 285
Auto oil programme, 5, 100, 102, 104, 107, 121, 124–125
Aviation, 19, 27–30, 38–40, 44–45, 49, 98, 300
'Avoid-Shift-Improve' approach, 10, 380, 382

B

Biofuels, 2, 48, 51–52, 56–60, 79–80, 82, 84–85, 88, 98, 100–101, 103–104, 108–112, 114–123, 138, 148, 153, 156–157, 168–169, 175, 249, 254, 259, 266–267, 305, 331, 382–383, 402
Bonus-malus, 135, 138, 142–148
Brazil, 39, 40, 63, 90, 121, 267, 343, 382, 395, 404
Bus rapid transit, 375, 382, 407

C

California, 61, 157, 210, 296, 356, 360–364, 366, 402, 412
Canada, 10, 61, 65, 114, 181, 183–184, 188, 204, 208, 309–310, 312, 327, 355–356, 363–368
Car attribute, 263
Carbon footprint, 6, 58, 114, 123, 155–157, 175, 373, 389
Carbon leakage, 118–119, 129, 154, 196
Car ownership, 1–2, 8, 21, 82, 206–208, 210, 229–237, 244–245, 262, 265, 269, 271, 273, 278, 281–283, 285, 288, 291, 295–297, 393, 397–398, 403–404, 406–407, 412
Car performance, 20
Car sharing/ridesharing/carpooling, 206, 282–283, 289–291, 329, 335–338, 340
Car use, 4, 8, 19–41, 197, 202, 206, 223, 227, 233, 237, 248, 249, 261, 263, 270, 273–274, 276, 280, 282–284, 287–288, 292, 297–299, 301, 320, 381, 384, 395–397, 401–403, 407–408, 410–411
China, 1, 13, 39–40, 46, 49, 55, 61–62, 65, 67, 90, 139, 156, 325, 356, 367–368, 371–373, 375–376, 379, 382, 384, 395–396, 403–404, 407
Clean Development Mechanism (CDM), 111, 386–387
Clean Technology Fund (CTF), 385–387
Command-and-control, 6, 10–11, 212

Commuting, 24, 26, 39, 273, 278, 280–283, 286, 306, 381, 384
Company car, 5, 38, 132, 140, 156, 158, 160, 185–186, 192–193, 251, 259–260, 263, 265–266, 400–401, 410
Compressed Natural Gas (CNG), 52–54, 56–57, 59, 89, 100, 112, 115, 117–118, 136, 249, 254, 402, 410
Congestion, 2–3, 9–11, 19–20, 26, 29, 31–37, 40, 55, 63, 129, 149, 181, 197–198, 201, 212, 215–216, 251, 254–256, 259, 291, 296, 298–299, 301–302, 305, 307–309, 312–317, 327–335, 337, 339, 341–342, 345, 348–351, 371, 378–379, 381, 389, 395–396, 401–402, 406–407
Congestion charge, 36, 251, 254–256, 259, 266, 298, 301–302, 313, 316, 345
Consumer preferences, 12, 264
Corporate Average Fuel Economy standards, 38, 356, 358
Cost-benefit analysis, 131, 381
Cycling, 9, 274, 282, 284, 288, 328–329, 332, 335–338, 340–341, 344, 348, 351, 374

D

Damage cost, 6, 296, 300
De-carbonization, 98–101, 106, 113–114, 117, 123–124
Demand management, 99, 101, 105, 313, 327, 332, 339, 348, 384
Demand system, 237
Developing countries, 10, 13, 43, 45–47, 49, 210, 213, 328, 332, 334, 342, 371–381, 383–388, 395, 406, 409, 411–412
Development assistance, 10, 372, 385–386, 388–389
Development bank, 10, 348, 385
Dieselization, 60
Discrete choice, 263
Distributional impact, 3, 199, 207, 213–214
Double dividend, 214–215
Driving cost, 265, 329, 415
Driving cycle, 357

E

E85, 89, 109, 133–134, 136, 138, 247, 249–250, 252–254, 257–260, 266–267, 364
Econometric, 206, 209–211, 231–232, 235
Economic growth, 12–13, 59, 86, 265, 269, 372–373, 375
Efficient pricing, 330, 349
Electricity, 5, 20, 48, 50–54, 56–59, 64, 68, 84–88, 98–101, 111–112, 118, 129, 138–139, 143, 155–156, 223, 270, 314, 384, 395
Electric vehicle, 48, 58, 63–64, 66, 85–87, 111, 137, 143–144, 156, 250, 263, 366, 382
Electrification, 57, 85–86, 111, 123–124
Emission abatement, 11, 196
Emissions trading/cap-and-trade, 6, 81, 90, 127–128, 130, 132, 154–155, 203, 300, 365
Energy conservation, 206, 328, 330, 332, 334, 341–342, 349, 351
Energy efficiency, 6, 20, 46–48, 83–84, 91, 97–99, 110, 113–114, 123, 130, 141, 153, 155–156, 160, 169, 175, 205, 247, 356, 377, 393
Energy security, 78, 97–98, 106, 108, 119, 123–124, 206
Enforcement costs, 11
Equity, 7, 10, 197, 220, 224, 237, 319–320, 332, 334
Ethanol, 8, 50, 89, 98, 108–109, 133, 138, 249–254, 257, 261, 265
European Climate Change Programme, 86
Eurovignette, 315, 317–318
Excise tax, 203, 205, 213, 216, 222–223, 225, 239, 240, 244, 364
Externality/external cost, 12, 131, 201, 203, 214–216, 223–224, 296, 298–299, 302–303, 305–307, 315, 318, 334, 350, 396

F

Feebate, 135, 138, 142, 401
Financial incentives, 38, 40, 133, 149, 305, 328, 333, 336, 340, 356
Financing, 8, 10, 13, 86, 130–131, 135, 216, 312, 319, 382, 386–388
Fiscal policies, 3, 11
Flexifuel vehicle, 257–258
Food, 60, 110, 254, 347
Forecast, 11–12, 22, 27–28, 30, 87, 98, 100, 139, 206, 262, 264, 267, 378, 383, 387, 396, 412
Formal transport, 374–375
Fossil fuel, 2, 51–52, 59, 84, 88, 90, 97, 99–101, 112, 116–120, 123–125, 138, 148, 153, 157, 168, 175, 198, 252–254
Freight transport, 269, 338, 340, 375, 381–382
Fuel cell vehicle, 48, 63, 69, 156, 365–366
Fuel economy, 1–2, 4, 9–13, 38, 48, 50, 54–56, 59–64, 131–132, 139, 147–148, 153, 156, 174, 181, 198, 203, 206, 212, 356–360, 362–363, 365–368, 381–382, 396–397, 399–401, 407, 409
Fuel industry, 91, 118

Index 419

Fuel intensity, 1–2, 43, 54, 60–63, 207, 394, 400, 409
Fuel price, 2, 8, 38, 63, 77, 87, 128–129, 132, 142, 155, 159, 201–209, 213, 224–225, 227, 235, 238, 240, 242, 244–245, 253–254, 256–257, 262, 265, 277, 341, 349, 351, 359, 395–396, 401
Fuel quality, 82, 84, 88, 99–115, 119–120, 122, 125, 157–158, 175, 379
Fuel tax, 7, 9, 149, 153, 155, 158–162, 175, 198–199, 203–204, 207, 212, 216, 223, 231, 235–237, 239–241, 245, 250–251, 259, 265, 295, 297–298, 303, 305–306, 308, 310, 312, 320, 328

G

Gas flaring, 113
GDP growth, 47, 49, 206, 265, 404
Global Environment Facility, 386
Greenhouse gases, 3–4, 19–20, 35, 38–40, 43–69, 77, 79–80, 84–85, 88, 97, 108, 111–112, 127, 133, 149, 181, 247, 269, 292, 298, 301, 314, 317, 355, 360, 362–367, 375–378, 411

H

Heavy goods vehicle, 296, 298, 306, 309–311
High speed trains/high speed rail, 4, 24, 27–30, 35, 372, 375, 395
Household, 226–244
Household expenditure, 7, 225, 239, 243, 245
Hybrid vehicle, 59, 140, 163, 186, 252, 366, 401
Hydrogen, 48, 51–52, 56–57, 59, 63, 79, 85, 88, 89, 100–101, 116, 118, 124, 156

I

Income, 7, 23–24, 27, 44, 47, 52, 132, 158, 181, 199, 206–210, 213–214, 225–227, 229, 231–240, 242–245, 277, 329, 331, 334, 343, 378, 398, 404
India, 1, 13, 39–40, 46, 49, 55, 63, 90, 344–345, 347–348, 368, 373–377, 382, 384, 396, 403–405, 407
Indirect land use change, 88, 112, 120–123, 156–158, 168, 175
Industrial revolution, 21, 30–32
Informal transport, 395, 406
Infrastructure, 4, 8, 10, 13, 32–33, 36–37, 47, 58, 64, 67–69, 86, 136, 202, 223, 263, 271, 277–278, 282, 284–287, 295, 299, 302, 309, 312, 314–315, 317–319, 331, 342, 344, 346, 363, 373–374, 378, 380, 382, 385, 388, 407
Intergovernmental Panel on Climate Change, 132, 377
International Energy Agency, 4, 38, 87, 98, 202, 204, 393
Investment, 4, 10, 13, 35–36, 50, 58–59, 68, 100, 106, 112, 114, 131, 144, 277, 285, 291, 298, 313–314, 317–319, 331, 340–341, 347, 375, 378–379, 381–382, 385, 387, 396, 407

J

JAMA, 60, 75, 77, 356
Japan, 10, 54, 60–63, 65, 67, 101–102, 139, 156, 204, 206, 355–358, 367–368, 385, 395, 403–404, 409
Jinonice index, 235–236

K

KAMA, 60, 75–76
Kyoto protocol, 203, 376, 387

L

Land use, 9, 11, 13, 25, 48–50, 58, 84, 88, 101, 111–112, 115, 120–124, 140, 156–158, 168, 175, 206, 278–279, 328–329, 330–332, 335, 337–338, 340–342, 351, 371, 381–382
Latin America, 46, 373–375, 386, 395–396, 403, 406–407
Lead, 21, 24, 26, 33–34, 37–38, 49, 69, 88, 90, 101, 103, 107, 114, 119, 133, 154, 205, 210, 212, 214, 247, 260, 305, 332, 360, 373, 384, 406–407
Legislation, 5–6, 61, 75, 78, 80, 83, 87–88, 90, 97, 100–105, 108, 111–112, 115, 122, 124, 128, 156, 161–162, 164–165, 169, 203, 276, 277–279, 281, 284, 304–305, 313, 315
Life cycle, 67, 69, 84, 88, 100, 111, 113, 116–118, 265, 410–411
Lifestyle, 8, 90, 206, 271, 275, 278, 283, 412
Lifetime, 7, 66, 84, 91, 141, 143, 149, 186–196, 207, 213, 356, 387
Light rail, 277, 286–287, 290, 374–375
Light trucks, 46, 50, 58, 61, 130, 147, 184, 358–365, 403
Lightweight, 10, 164, 175, 355, 368
Liquefied Petroleum Gas (LPG), 52–54, 56–57, 59, 89, 98, 100, 112, 115, 117–118, 136, 317, 397–398, 402
Low carbon fuel standard (LCFS), 5, 82, 84, 110, 153, 157, 175

M

Market share, 21, 25–27, 29, 38, 48, 87, 156, 166, 184, 198, 253, 263–264, 318
Mass transit, 27, 34, 37, 46, 410
Methane, 113, 138, 144–145, 249, 361, 365
Mexico, 10, 355, 367–368, 377, 382, 384, 386–387, 403–404, 407–409
Mileage, 26, 28–29, 129, 131, 143, 146, 149, 206–208, 216, 223, 265–266, 331, 333–334, 339–340, 350
Mitigation potential, 383
Mobility, 1, 3–4, 9, 21–22, 33–39, 44–47, 69, 87–88, 146, 273–274, 285, 289, 291, 327–352
Mobility management, 9, 284, 327–352
Modal shift, 30, 48, 49, 57, 305, 383
Model, 45–46, 53, 115–119, 140–145, 262–266, 276, 286, 304–308, 360–361
Monitoring, 77–78, 81–83, 111, 113, 168, 382, 409–411
Motorcycles, 45, 306, 317, 373, 379, 405
Motorization, 4, 10, 13, 40, 372–377, 380, 388–389, 396, 403–404, 406–407
Multimodality, 269–292

N

Non-motorized transport, 13, 43, 58, 208, 344, 373, 381
Non-OECD, 45, 47, 49–50, 54–55, 63, 182, 376–377

O

OECD, 10, 19, 45–47, 49–50, 54–64, 118, 131, 133, 137, 139, 144, 149, 181–182, 185, 187, 196, 198, 206–207, 210, 355–368, 372–377, 380, 383
OECD Europe, 46, 55–59, 376
Oil price, 47, 49, 50, 85, 98, 124, 130, 155, 342
On-road fuel economy, 206, 399
Ozone, 105, 109, 202

P

Para-transit, 348, 374, 379
Parking, 26, 37, 251, 255, 260, 272, 279, 281, 283–285, 287–288, 327–343, 345, 348–349, 373, 379, 396, 407, 410
Passenger kilometres, 4, 27, 33, 39, 47, 50, 56, 305, 375, 394–395
Pay-As-You-Drive pricing, 334, 340
Peak travel, 302, 331
Pigovian tax, 12, 216
Planning reforms, 10, 328, 332
Plug-in hybrid vehicle, 43, 59, 69, 163, 366
Polycentric approach, 59, 163, 366
Population density, 208
Population growth, 23, 52, 56, 265, 398
Power generation, 1, 58, 73, 81, 410
Powertrain, 48
'Predict and Provide' approach, 10, 371, 378–380, 389
Price elasticity, 129, 155, 160, 203, 206–211, 215, 238
Pricing, 4, 11, 26, 35, 37, 39–40, 202, 216, 221, 225, 232, 237, 240, 242, 244–245, 283–284, 298, 305–307, 309, 311–317, 319, 328–331, 334–336, 338–340, 345, 349, 373, 382, 388, 396, 400, 402, 407
Progressive tax, 135, 356
Projection, 27–28, 35, 39, 47–59, 63, 210, 377, 383, 396, 404–405, 407
Propulsion system, 59, 63–64, 144
Public transport, 4, 8–9, 11, 25–27, 33–35, 52, 86, 90, 105, 123, 199, 206, 208, 213–214, 216, 222, 233, 240–241, 270–272, 274, 276–284, 286–292, 299, 317, 328–329, 332, 343, 347–348, 372–375, 381, 383–385, 396

R

Railway, 22, 32–33, 98, 244, 286–287, 290, 317, 348, 372, 375
Rebound effect, 146, 160, 212, 248, 261, 265–266, 401, 410
Registration tax, 6, 133, 135, 137–138, 141–142, 146, 149, 309
Regressive tax, 7, 213, 329
Regulations, 81–83, 85, 87, 127–150, 154–158, 162–171, 301, 336
Regulatory reforms, 339, 341
Renewable energy, 81, 84–85, 88, 110–112, 119, 120, 122, 130, 157, 168
Renewable fuel, 82, 84, 88, 99, 108, 115, 122, 175, 248–249, 265, 267
Research & development, 12
Revenue recycling, 239–240, 242
Road charge, 2, 9
Road pricing, 35, 202, 216, 284, 295, 298, 305, 309, 313–315, 319, 335, 340, 345

S

Sales, 8, 38, 45, 52–54, 58–59, 61, 64–65, 67–68, 134–135, 144, 158–160, 162–163, 166–167, 172, 175, 198, 204, 247–248, 250, 252–253, 255–256, 258, 260, 263, 266, 318, 360, 377, 396, 399, 402, 404–405, 409

Index

Scenario, 47–59, 87, 89, 156–158, 239–242, 306, 377, 383–384
Scrappage, 165–166, 261, 362
Social cost, 3–4, 6, 35–36, 149, 201, 216, 302, 305, 315
South Korea, 10, 156, 356, 366–367, 383
Spatial planning, 202, 278–279
Speed limit, 35, 130
Spill-over, 131
Stakeholder consultation, 86
Standards, 62–63, 103, 110, 122, 154–158, 164–165, 168, 300, 336, 367–368
Subsidy, 67–68, 134, 140, 143–144, 149–150, 158, 189–191, 251, 253, 256, 258–259, 330
Subway, 27, 32, 34, 372, 375
Suits index, 235–236, 245
Sulfur, 2, 101, 103, 107, 125, 157, 202
Sustainable transport, 10, 13, 26, 80, 125, 292, 345, 380, 382, 386, 388, 406

T

Tax exemption, 79, 156, 250, 253, 257, 266, 334
Tax harmonization, 149
Tax reforms, 224, 336
Technological change, 131, 133
Telecommunications, 48, 328–329
Tele-commuting, 381, 384
Toll, 149, 223, 298, 315, 318–319, 385, 396
Tradable permit, 214, 300–301
Traffic calming, 288, 328, 337, 339, 341, 344
Traffic safety, 287, 330, 338, 342, 344, 351
Transaction costs, 81, 130, 307
Transport funding, 331
Transport planning, 10, 248, 261, 279, 328–330, 341, 347, 371
Transport policy, 4, 26, 35–37, 40–41, 76, 80, 84, 86, 89, 204, 209, 304, 310, 313, 348, 371–372, 378–380, 382
Transport reforms, 330–332
Transport volume, 38, 90, 394
Travel speed, 4, 22, 25, 31, 33, 35

Travel time, 22–29, 33–34, 37, 39–40, 201, 273, 301–303, 308, 335, 395
Trucks, 10, 19, 36, 44, 46–47, 50–52, 58–59, 62–63, 130, 147, 153, 160–162, 175, 184, 212, 306, 318–319, 358, 360–366, 368, 371, 373, 375, 384, 403–404
Two wheeler, 49, 373, 396, 405

U

United Nations Centre for Regional Development, 380, 386
United Nations Framework Convention on Climate Change, 121, 372
United States, 10, 47, 54, 61, 65, 67, 77, 91, 101, 139, 143–144, 148, 181, 184, 186, 198, 204, 206–209, 211, 309, 355, 358–363, 372, 403
Urbanization, 19–20, 30–35, 37, 41, 47, 49, 279, 290, 292, 325, 346, 373, 395
Urban planning, 4, 8, 19, 25–26, 35, 373, 384
User charge, 36, 296, 303, 307, 315, 318, 320, 331, 400

V

Vehicle footprint, 164
Vehicle insurance, 331, 334
Vehicle kilometres, 318–319
Vehicle miles travelled, 214
Vehicle price, 89, 141, 197
Vehicle stock turnover, 210, 231
Vehicle tax, 7, 62–63, 135–136, 138, 140–141, 156, 184, 186, 205, 224, 264–265, 297
Voluntary agreement, 75–76, 78, 81–82, 162, 364, 397, 399, 401, 409–410

W

Walking, 9, 21, 32, 45, 328–329, 331–332, 336–338, 340–342, 344, 346–347, 351, 374
Welfare, 2, 10, 12, 26, 197–198, 212, 214–215, 225, 231, 239–243, 305–307
Well-to-wheel, 6, 20, 39, 58, 155, 157
World Energy Outlook, 47, 98
World Health Organization, 105